Fundamentals of Dynamical Systems and Bifurcation Theory

Fundamentals of Dynamical Systems and Bifurcation Theory

Milan Medveď

Mathematical Institute of the Slovak Academy of Sciences, Bratislava, Czechoslovakia

ADAM HILGER, BRISTOL, PHILADELPHIA AND NEW YORK

© M Medveď 1992
Translation © J Hajnovičová, D Halašová

All rights reserved. No part of this publication may be reproduced, stored in a retrieval system or transmitted in any form or by any means, electronic, mechanical, photocopying, recording or otherwise, without the preceding written permission of the copyright owner. Multiple copying is permitted in accordance with the terms of licences issued by the Copyright Licensing Agency under the terms of its agreement with the Committee of Vice-Chancellors and Principals.

British Library Cataloguing in Publication Data
Medveď, Milan
    Fundamentals of dynamical systems and bifurcation theory
    I. Title
    515
    ISBN 0-7503-0150-3

Library of Congress Cataloging-in-Publication Data
Medveď, Milan, RNDr.
    Fundamentals of dynamical systems and bifurcation theory/ Milan Medveď.
        p.   cm.
    Includes bibliographical references and index.
    ISBN 0-7503-0150-3
    1. Differentiable dynamical systems. 2. Bifurcation theory.
I. Title
QA614. 8. M44 1991
515'.35 – dc 20   91-20084
                                                                                                                 CIP

Published under the Adam Hilger imprint, the book imprint of the Institute of Physis, London, by IOP Publishing Ltd, Techno House, Redcliffe Way, Bristol BS1 6NX, England
335 East 45 Street, New York, NY 10017 - 3483, USA

US Editorial Office: IOP Publishing Inc., The Public Ledger Building, Suite 1035, Independence Square, Philadelphia, PA 19106, USA

Published in co-edition with Ister Science Press Ltd, Staromestská 6, 811 03 Bratislava, Czechoslovakia.

Distribution of this book is being handled by the following publishers:
    in Czechoslovakia and Poland
    by
    Ister Science Press Ltd
    P. O. Box 96, 811 03 Bratislava, Czechoslovakia
    in all remaining areas
    by
    IOP Publishing Ltd, Techno House, Redcliffe Way,
    Bristol BS1 6NX, England

Printed in Czechoslovakia

CONTENTS

INTRODUCTION vii

1 SELECTED READING ON THE FOUNDATIONS OF ALGEBRA, TOPOLOGY, MATHEMATICAL ANALYSIS AND THE THEORY OF DIFFERENTIAL EQUATIONS 1
  1.1 Some basic concepts and notation 1
  1.2 Relations on a set 2
  1.3 Linear, Euclidean and linear normed spaces 3
  1.4 Metric spaces 5
  1.5 Topological spaces 5
  1.6 Matrices 9
  1.7 Linear mappings 10
  1.8 Mathematical analysis 12
  1.9 Differential equations 18

2 FOUNDATIONS OF THE THEORY OF DIFFERENTIABLE MANIFOLDS AND DIFFERENTIABLE MAPPINGS 22
  2.1 C^r-manifolds 22
  2.2 C^r-mappings 26
  2.3 Tangent space to a C^r-manifold 27
  2.4 C^r-submanifolds 32
  2.5 C^r-manifolds in R^N 35
  2.6 Immersion and submersion theorems 36
  2.7 Regular and critical values of mappings 39
  2.8 Topology on the space of C^r-mappings 41
  2.9 Jets 45
  2.10 Transversality 46
  2.11 Stratification of algebraic and semi-algebraic manifolds 55
  2.12 Transversality to stratification 61

3 VECTOR FIELDS AND DYNAMICAL SYSTEMS 63
  3.1 Vector fields on differentiable manifolds 63

3.2 Limit properties of dynamical systems 75
3.3 Examples of vector fields 84
3.4 Generic properties of parameter-dependent matrices 86
3.5 Linear dynamical systems and some notions from the theory
    of non-linear dynamical systems 103
3.6 Grobman–Hartman Theorem 128
3.7 Normal forms of differential equations 143
3.8 Poincaré mapping 159

4 INVARIANT MANIFOLDS 172
4.1 Stable and unstable manifolds 172
4.2 Centre manifolds 183

5 GENERIC BIFURCATIONS OF VECTOR FIELDS AND DIFFEOMORPHISMS 203
5.1 Ljapunov-Schmidt Method 203
5.2 Generic bifurcations of 1-parameter systems of vector fields
    in neighbourhoods of singular points 214
5.3 Generic bifurcations of 1-parameter systems of diffeomorphisms 238
5.4 Generic bifurcations of 1-parameter systems of vector fields
    in neighbourhoods of periodic trajectories 258

6 COMPLEMENTARY NOTES ON THE CONTEMPORARY THEORY OF DYNAMICAL SYSTEMS 262
6.1 Generic bifurcations of multi-parameter systems of
    vector fields 262
6.2 Global theory of dynamical systems 270
6.3 Šil'nikov bifurcation 273
6.4 Global Hopf bifurcation 276
6.5 Attractors and chaotic sets 279

REFERENCES 282

SUBJECT INDEX 289

INTRODUCTION

Differential equations were first studied from the point of view of their qualitative properties by H. Poincaré [127] in 1880. His work gave an impulse to the formulation and development of an independent theory of dynamical systems. Soon after the introduction of differential calculus problems of a dynamical character were being solved by means of differential equations and even I. Newton understood the utility of the solutions of differential equations. Today it is difficult to imagine how various problems of physics, chemistry, biology and economy could be solved without them.

The work of J. Liouville on the insolubility of various differential equations in quadrature, i.e. on the impossibility of expressing in general their solutions in the from of a final combination of elementary and algebraic functions and their integrals, evoked a new qualitative approach to their study. Poincaré's outline of the qualitative theory of differential equations is the best starting point from this apparently hopeless situation. Since the beginning of this new trend in the theory of differential equations, the applications of which now reach not only natural and technical science but also social sciences, a very rich theory has developed.

The current theory of dynamical systems not only uses the methods of the theory of differential equations but it is also closely connected with functional analysis, differential and algebraic, geometry, differential and algebraic topology, the theory of probability and even with the theory of numbres.

The aim of this monograph is to explain the fundamentals of the theory of dynamical systems in such a way that after reading it one should be able to study actively the very extensive literature on dynamical systems and to apply the knowledge.

Although the reader is expected to have some basic knowledge of mathematical analysis, algebra and topology he can find all the necessary material together with information on references in this book. Some methods of differential topology are used in the theory of dynamical systems therefore one chapter of the book is devoted to this problem. Without knowing these methods it is impossible to understand the substance of the contemporary qualitative theory of dynamical systems — they are fundamental to it.

This book does not cover the whole problem of the theory of dynamical systems and it is difficult to imagine a monograph that could cover it completely, but the monograph, together with the cited references, does contain considerable information on the present state of the theory of dynamical systems. It is designed for graduates and students of mathematics, technical or economic scientists, as well as for scientific workers in the field of physics, chemistry, economics and biology and for those who solve problems of dynamical character.

The author is indebted to Professor Kurzweil, D. Sc., P. Brunovský, D. Sc. and Dr. Š. Schwabík for their critical comments on the manuscript. The author is also pleased to be able to take this opportunity to express his thanks to Dr. Poláčik and Dr. T. Žáčik who contributed considerably to removing inaccuracies from the first version of the manuscript. Last but not least, the author would like to thank Academician Š. Schwarz and Senior Lecturer L. Mišík, D. Sc. for creating good working conditions in the Mathematical Institute of the Slovak Academy of Sciences.

1 SELECTED READING ON THE FOUNDATIONS OF ALGEBRA, TOPOLOGY, MATHEMATICAL ANALYSIS AND THE THEORY OF DIFFERENTIAL EQUATIONS

1.1 Basic Concepts and Notation

Some notation wil be explained as it occurs in text, but the following will be used throughout the book.

Sets: R $(N; Z; C)$ — the set of the real (natural, integer, complex) numbers. $R^+(Z^+)$ — the set of positive real (integer) numbers. $\{a_n\}_{n=1}^{\infty}$ — the sequence $a_1, a_2, \ldots X \cup Y, X \cap Y, X \setminus Y, X \times Y$ — union, intersection, difference and Cartesian product of the sets X, Y, $R^n = R \times R \times \cdots \times R$, $C^n = C \times C \cdots \times C$, $N^n = N \times N \times \cdots \times N$, $Z^n = Z \times Z \times \cdots \times Z$ (n-times). $x \in X$ ($x \notin X$) — x is (is not) an element of the set X. $\{x \in X : V\}$ — the set consisting of $x \in X$ having the property V. $\overline{A}$, cA — closure and the complement of the set A. $X \subset Y$ — X is a subset of the set Y. $\langle a, b \rangle$, (a, b), $\langle a, b)$, $(a, b\rangle$ — is closed, open, closed from the left and open from the right or open from the left and closed from the right. $\emptyset$ — is an empty set.

Polynomials: $R[X_1, X_2, \ldots, X_n]$ — the set of polynomials in n variables, whose coefficients are real numbers, $\deg P$ — the degree of the polynomial P.

Matrices: $M(m, n)$ — the set of the $m \times n$ type matrices whose elements are real numbers. $M(n)$ — the set $M(n, m)$. $M_c(m, n)$ — the set of matrices of the type $m \times n$ whose elements are complex numbers. $M_c(n)$ — the set $M_c(n, n)$. $0_{m,n}$ (or just 0) — the zero matrix in $M(m, n)$. 0_n — the zero matrix $0_{m,n}$. I_n (or I, only) — the unit matrix in $M(n)$. $\text{GL}(n)$ — the set of regular matrices in $M(n)$. AB, $A+B$, $A-B$ — the product, sum, difference of the matrices A, B. λA — λ-multiple of the matrix A. A^k — the k-th power of the matrix A. $\text{Tr}\, A$ — the trace of the matrix A. $\text{rank}\, A$ — the rank of the matrix A. A^* — the transposed matrix to the matrix A. A^{-1} — the inverse matrix to the matrix A.

1

Chapter 1

Mappings: $f : X \to Y$, $x \mapsto f(x)$ (or $y = f(x)$) — the mapping of the set X to the set Y. $f(A)$ — the image of the set $A \subset X$ of the mapping $f : X \to Y$. Image f — the set $f(X)$. $f^{-1}(B)$ — the preimage of the set $B \subset Y$ of the mapping $f : X \to Y$. Ker f — the kernel of the mapping $f : X \to Y$, where X, Y are linear spaces, i.e. Ker $f = \{x \in X : f(x) = 0\}$ (0 — the zero element of Y). f/A — the restriction of the mapping $f : X \to Y$ to the set $A \subset X$. $f \circ g$ — the composition of the mappings f and g. f^{-1} — the inverse mapping to the mapping f. $\max_{x \in A} f(x)$, $\sup_{x \in A} f(x)$, $\min_{x \in A} f(x)$, $\inf_{x \in A} f(x)$ — the maximum, supremum, minimum, or infimum of the function $f : X \to R$ on the set $A \subset X$. graph (f) — the graph of the mapping f. id $_x$ (or just id) — the identical mapping (identity) on the set X. The mapping $f : X \to Y$ is called surjective if $f(x) = Y$ or injective if $f(x) = f(y) \implies x = y$ and bijective if it is at the same time both surjective and injective, where $p \implies q$ means that the statement q is implied by the statement p.

Other notations and concepts will be explained later in the text.

1.2 Relations on a Set

The relation on the set X is an arbitrary non-empty subset $R \subset X \times X$. If $(x, y) \in R$, then we write xRy. The relation R on a set X is called a relation of equivalence (or equivalence) on the set X, if

1. xRx for all $x \in X$;
2. $xRy \implies yRx$ for all $x, y \in X$;
3. $xRy, yRz \implies xRz$ for all x, y, z.

If R is an equivalence and xRy then we say that the elements x, y are equivalent to each other. The set $[x] = \{y \in X : yRx\}$ is called a class of equivalence represented by the element x. $X/R = \{[x] : x \in X\}$ — the quotient set X by R. $p : X \to X/R$, $p(x) = [x]$ — the natural projection of the set X to X/R.

The relation R on the set X is called a relation of partial ordering on X if

1. xRx for all $x \in X$;
2. xRy, yRx implies $x = y$ for all $x, y \in X$;
3. xRy, yRx implies xRz for all $x, y, z \in X$.

The relation of partial ordering R on X is called a relation of ordering (or *ordering*) on X if moreover it shows the property

4. $x, y \in X \implies$ either xRy or yRx.

In this case we write $x \leqq_R y$ instead of xRy or just $x \leqq y$. If $x \leqq_R y$ and $x \neq y$ then we write $x <_R y$ or just $x < y$.

2

1.1 Zorn Lemma. *Let $X \neq 0$ be a partially ordered set (i.e. on X there is a partial ordering R) and let for every ordered subset $A \subset X$ (i.e. R on A is the relation of ordering) there exist $a \in A$ such that $x \leq_R a$ for every $x \in A$. Then in X there exists the maximal element m, i.e. an element such that for no $x \in X$ it holds that $m <_R x$.*

1.3 Linear, Euclidean and Linear Normed Spaces

The following concepts are taken to be well-known and we shall not define them any further: linear (vector) space V over the field F ($F = R$ or $F = C$), linear subspace, linear dependence and independence in V, a basis in V, dimension of the space V (dim V), coordinates of the element $v \in V$ with respect to a given basis in V, the linear mapping from the linear space V to the linear space W, the scalar product on V, the orthonormal basis in V, on which a scalar product is defined.

If V is a linear space over R then the set $A \subset V$ is called convex, if for arbitrary $x, y \in V$, $\lambda \in R$, $0 \leq \lambda \leq 1$ is $x\lambda + (1-\lambda)y \in A$. If $B \subset V$, then the set co $B = \cap\{C : C$ is convex, $B \subset C\}$ is called a convex hall of the set B.

We say that the linear space V is a direct sum of its subspaces V_i, $i = 1, 2, \ldots, n$, if

1. $V_i \cap V_j = \{0\}$ for $i \neq j$;
2. for every vector $v \in V$ there exists just one vector $v_i \in V_i$ ($i = 1, 2, \ldots, n$) such that $v = v_1 + v_2 + \cdots + v_n$.

This direct sum is denoted as $V_1 \oplus V_2 \oplus \cdots \oplus V_n$. The vector space $V_i \oplus V_{i+1} \oplus \cdots \oplus V_n$ is called an algebraic complement to the space $V_1 \oplus V_2 \oplus \cdots \oplus V_{i-1}$ in V.

Let V, W be linear spaces. The linear mapping $h : V \to W$ is called isomorphism if it is bijective. If such an isomorphism exists, then we say that the spaces V, W are isomorphic.

1.2 Theorem. *Arbitrary n-dimensional linear space over R is isomorphic with R^n.*

The set of the matrices $M(n, 1)$ (i.e. the set of column vectors of dimension n) is isomorphic with R^n and throughout the book it will also be denoted as R^n.

The linear space V over R on which the scalar product s, is defined is called an Euclidean space and it is denoted either as (V, s) or V. We also say that on V an Euclidean structure is defined. The value $s(u, v)$ will also be denoted as (u, v).

1.3 Theorem. *Every Euclidean space of finite dimension has an orthonormal basis.*

Let (V, s) be an n-dimensional Euclidean space over C and $\mathscr{B} = \{a_1, a_2, \ldots, a_n\}$ be its basis. If $u, v \in V$ then

$$u = \sum_{i=1}^{n} u_i a_j, \quad v = \sum_{j=1}^{n} v_j a_j,$$

where $u_i, v_j \in R$ $(i, j = 1, 2, \ldots, n)$ and

$$s(u, v) = \sum_{i,j=1}^{n} u_i \bar{v}_j s(a_i, a_j).$$

The above representation of the scalar product s with respect to the basis $\mathscr{B}$ will be denoted as $(u, v)_{\mathscr{B}}$. If the basis $\mathscr{B}$ is orthonormal, then, obviously,

$$(u, v) = \sum_{i=1}^{n} u_i \bar{v}_i.$$

Let V be a linear space over F ($F = R$, or $F = C$). The function $p : V \to R$ (the value $p(x)$ is denoted by $\|x\|$) is called a norm over V if for arbitrary $x, y \in V$, $\lambda \in F$ it holds that

1. $\|x\| \geq 0$;
2. $\|\lambda x\| = |\lambda| \|x\|$;
3. $\|x + y\| \leq \|x\| + \|y\|$;
4. $\|x\| = 0$ if $x = 0$.

The value $\|x\|$ is called a norm of the vector x.

The linear space V over F on which the norm p is defined, is called a linear normed space (or a normed space) denoted either (V, p) or V.

Let (V, p), (W, q) be linear normed spaces. The mapping $f : V \to W$ is called an isometry, if it is an isomorphism and $q(f(x)) = p(x)$ for all $x \in V$.

If (V, s) is the Euclidean space, then the function $p : V \to R$, $p(x) = (s(x, x))^{1/2}$ is a norm on V, and thus, (V, p) is a normed space. In this case we say that norm p is derived from the scalar product s and we call it the Euclidean norm on V.

1.4 Theorem. *If (V, s) is an Euclidean space and p is an Euclidean norm derived from the scalar product s, then $|s(x, y)| \leq p(x)p(y)$ for all $x, y \in V$, whereby the equality holds iff there is linear dependence between the vectors x, y.*

We say that the norms p_1, p_2 on the linear space V are equivalent to each other if there exist such positive numbers c_1, c_2 that $c_1 p_1(x) \leq p_2(x) \leq c_2 p_1(x)$ for all $x \in V$.

1.5 Theorem. *Arbitrary two norms on a finite dimensional linear space are equivalent to each other.*

References: [81, 132].

1.4 Metric Spaces

The function $d: X \times X \to R$ is called a metric on the set X if for every $x, y, z \in X$ it holds that
1. $d(x, y) \geq 0$, where $d(x, y) = 0$ iff $x = y$;
2. $d(x, y) = d(y, x)$;
3. $d(x, y) \leq d(x, z) + d(z, y)$.

The set X on which the metrics d is defined, is called a metric space and it is denoted as (X, d) or X.

Let (X, d) be metric space. The set $B_r(x) = \{y \in X : d(y, x) < r\}$ is called an open sphere in X with the centre $x \in X$ and the radius $r > 0$. The set $A \subset X$ is called open in X if for every $x \in A$ there exists such a number $r > 0$ that $B_r(x) \subset X$. The point $x \in X$ is called the accumulation point of the set $A \subset X$ if for every $\varepsilon > 0$ there exists the point $y \in B_\varepsilon(x) \cap A$ different from x. The point $x \in A$ is called an isolated point of the set A if there exists such an $\varepsilon > 0$ that $A \cap B_\varepsilon(x) = \{x\}$. The sequence $\{x_k\}_{k=1}^{\infty}$ of the points from the metric space (X, d) is called a Cauchy sequence if for every $\varepsilon > 0$ there exists such a $K \in N$ that for all $m, n > K$ it is $d(x_m, x_n) < \varepsilon$.

The metric space (X, d) is called complete if every Cauchy sequence of the points from X converges to a certain point from X.

The linear normed space (X, p) is called a Banach space if the metric space (X, d) is complete, where $d(x, y) = p(x - y)$.

The Euclidean space (X, s) is called a Hilbert space if (X, p) is a Banach space where $p(x) = (s(x, x))^{1/2}$.

References: [32], [81], [132], [148].

1.5 Topological Spaces

A topological space is a couple (X, S), where X is a set and S is a system of its subsets of the properties as follows:
1. $\emptyset \in S$ and $X \in S$;
2. Arbitrary union and the finite intersection of the sets from the system S is again a set belonging to the system S.

The elements of the system S are open sets of the topological space

(X, S). The system S is called a topology on the set X. Often, instead of (X, S) we shall write just X.

Let (X, S) be a topological space. The set $A \subset X$ is called closed in X if its complement $cA = X \setminus A$ is an open set in X, i.e. $cA \in S$. The set $\cap \{B \subset X : A \subset B, B \text{ is closed}\}$ is called a closure of the set A and it is denoted by $\overline{A}$. The set $\cup \{B \subset X : B \subset A, B \text{ is open }\}$ is called an interior of the set A and it is denoted by int A. The set $\overline{A} \cap (c\overline{A})$ is called a boundary of the set A and it is denoted by ∂A.

Let (X, S) be a topological space and $A \subset X$. Then the system $S_A = \{U \cap A : U \in S\}$ fulfils the axioms of a topological space, and thus, (A, S_A) is a topological space. This topological space is called a topological subspace of the topological space X (or a topological subspace in X). The system S_A is called a relative topology on A induced by the topology S. The set $V \subset X$ is called a neighbourhood of the point $x \in X$, if there exists an open set $U \cap V$ containing the point x.

1.6 Theorem. *Let X be a set and let for every $x \in X$ exist such a system T_x of the subsets of the set X having the following properties:*
1. *$V \in T_x \implies x \in V$;*
2. *$X \in T_x$;*
3. *$U, V \in T_x \implies U \cap V \in T_x$;*
4. *$U \in T_x, U \subset W \implies W \in T_x$;*
5. *for every $U \in T_x$ there exists such $V \in T_x$ that $V \subset U$ and for every $y \in V$ it is $V \in T_y$.*

Then, there exists a unique topology S on the set X at which T_x is the set of all neighbourhoods of the point X for every $x \in X$ (the open set in X is defined as follows: $A \subset X$ is an open set in X iff $A \in T_a$ for every $a \in A$ or if $A = \emptyset$).

The basis for the topological space (X, S) is a system $\mathscr{B}$ of open sets in X such that arbitrary set of the system S is a union of elements of the system $\mathscr{B}$. We say that the topological space (X, S) fulfils the second axiom of countability if in X there exists a countable basis, i.e. such a basis that consists at most of a countable number of sets.

Let (X_i, S), $i = 1, 2, \ldots, n$ be topological spaces, where $S_i = \{U_{\alpha(i)}\}_{\alpha(i) \in S}$ (I_i is an index set) and let $X = X_1 \times X_2 \times \cdots \times X_n$, $I = I_1 \times I_2 \times \cdots \times I_n$. For every $\alpha = (\alpha(1), \alpha(2), \ldots, \alpha(n)) \in I$ we define the set $U = U_{\alpha(1)} \times U_{\alpha(2)} \times \cdots \times U_{\alpha(n)}$. The system $B = \{U_\alpha\}_{\alpha \in I}$ is a basis of the topology called the topology of the product or the product topology on X.

The topological space (X, S) is called a Hausdorff space if for arbitrary $x, y \in X$, $x \neq y$ there exist such neighbourhoods U, V of the points x, or y that $U \cap V = \emptyset$.

Order ID: 202-5883720-8769165

Thank you for buying from Mike Cooke on Amazon Marketplace.

Delivery address:	Order Date:	Tue, Aug 4, 2020
Tomy Duby	Delivery Service:	Standard
OAA Computing	Buyer Name:	Tomy
29 Hamilton Close	Seller Name:	Mike Cooke
Bicester		
Oxon		
OX26 2HX		
United Kingdom		

Quantity	Product Details
1	Fundamentals of Dynamical Systems and Bifurcation Theory [Hardcover] [1992] Medved, Milan SKU: 5L-F6WK-6SZL ASIN: 0750301503 Condition: Used - Acceptable Listing ID: 0206XD0FWLN Order Item ID: 04802412168755

Thanks for buying on Amazon Marketplace. To provide feedback for the seller please visit www.amazon.co.uk/feedback. To contact the seller, go to Your Orders in Your Account. Click the seller's name under the appropriate product. Then, in the "Further Information" section, click "Contact the Seller."

The Static View

ents change with respect to
ut also denying that events
is to move, it must start off
temporal position where it is
present so that it can move
t so that it can move further
nge with respect to its being
ove in the way the transient

pause to look more closely at
guage expressions confuse us
nd to show that the notion of
d into the transient view is
ion of 'the NOW' which
ansient view. I will then show
ne is mistaken, arguing that
ect to being past, present and
ble objects in space which are
ut simply related spatially to
ated temporally to each other,
other events, and later than
eally past or present or future.
sed statement to be true ('E is
e terms will be explained in the
it is that no tensed statement
ement (a statement which says
y to each other making no
make mention of McTaggart's
sputing with him his claim that
tensed (that is, if events do not
present and future).

If we wish to deny that time is dynamic
the tensed views of time, we can do th
view of time. On this view, events a
'earlier than' (or its logical opposite 'l
are so ordered is not cashed in term
events cease being future and become
terms of events moving in time. The s
there is no moving present, and ther
notions are simply mistaken. The riv
seemingly indispensable when we nee
our lives, our plans for the future and
been, is a fraud, having nothing useful
nature of time and events. Why we
misleads our thought about time wi
section.

Events are not intrinsically past, pre
not change in respect of being past, p
fact that we speak of them as if they
relation between events. If event E_1 o
express all there is to say about the t
cerning E_1 and E_2 by stating that E_1
does not consist in any further facts,
E_2 is even more future, or E_1 being
being past and E_1 being even more p
the fact that E_1 attains presentness be
of time denies that there is a present
not become present. Even though
present' and talk of experiencing eve

s

tially the same material as Chapter
ils to references are given in the

of Time (or, The Man Who Did

yth of Passage' pp. 103 ff. for more

1.7 Theorem. *Topological space (X, S) is a Hausdorff space iff the diagonal $\Delta = \{(x, x) \colon x \in X\}$ is a closed set in the topological space $X \times X$ with a product topology.*

Let (X, S) be a topological space, R be the equivalence relation on X, X/R is the quotient set X by R and $p \colon X \to X/R$ be a natural projection. The system $T = \{U \subset X/R : p^{-1}(U) \in S\}$ satifies the axioms of the topological space, and thus, $(X/R, T)$ is a topological space. Topology T is called a quotient topology on X/R.

A topological space (X, S) is called compact if for any of its open coverings (open covering — the system $\{U_\alpha\}_{\alpha \in I}$ such that U_α, $\alpha \in I$ is an open set in X and $X = \cup_{\alpha \in I} U_\alpha$) there exists its finite subcovering, i.e., there exist such $i_1, i_2, \ldots, i_n \in T$ that $X = \cup_{i=1}^n U_{i_k}$. The set $A \subset X$ is called compact in X if (A, S_A) is a compact topological space, where S_A is a relative topology on A induced by the topology S.

Let (X, d) be a metric space and $I = \{(x, r) : x \in X, r > 0\}$. The system $\mathscr{B} = \{B_r(x)\}_{(x,r) \in I}$ is the basis of the topology S on X, and thus, (X, S) is a topological space. The topology S is called a topology induced by the metrics d.

1.8 Theorem. *Let (X, d) be a metric space and (X, S) be a topological space with the topology induced by the metrics d. Then (X, S) is a compact topological space iff for every infinite sequence of the elements of the space X there exists its subsequence converging in the metric space (X, d) to an element z of X.*

1.9 Theorem. *Let (X, S) be a topological space. Then it holds that*
  1. if (X, S) is a Hausdorff space and $A \subset X$ is a compact set, then A is closed in X;
  2. if (X, S) is compact and $B \subset X$ is a closed set, then B is a compact set.

1.10 Theorem. *The topological space (X, S) is compact iff for every system $\{F_i\}_{i \in I}$ of closed sets whose arbitrary finite number of elements has a non-empty intersection, is $\cap_{i \in I} F_i \neq \emptyset$.*

1.11 Theorem. *If (X, S) is a Hausdorff topological space and $A_1, A_2 \subset X$ are compact mutually disjoint sets, then there exist such open sets U_1, $U_2 \subset X$ that $A_1 \subset U_1$, $A_2 \subset U_2$ and $U_1 \cap U_2 = \emptyset$.*

1.12 Theorem. *Every compact metric space is complete.*

1.13 Theorem. *Let (X, p) be a linear normed space and S be the topology on X induced by the metrics $d(x, y) = p(x - y)$. Then, if $A \subset X$ is a compact set in (X, S), then its convex hall coA is also a compact set in X.*

Let (X, S) be a topological space. The set $A \subset X$ is called *dense in the set* $B \subset X$ if $\overline{A} = B$. The set A is called *dense everywhere* if it is dense in X. We say that the set $A \subset X$ is *massive* in X, if it contains a set that can be expressed as a countable intersection of open everywhere-dense sets.

The topological space (X, S) is called a Baire space if every its massive subset is an everywhere-dense set.

1.14 Theorem. *Every complete metric space is a Baire space.*

The topological space (X, S) is called connected if there do not exist two non-empty and closed sets $A, B \subset X$ such that $A \cap B = \emptyset$ and $X = A \cup B$. The set $A \subset X$ is called connected if (A, S_a) is a connected topological space where S_A is a relative topology on A induced by the topology S. The set $A \subset X$ is called a connected component of the topological space (X, S) if it has the following properties:
1. A is a connected set
2. $A \subset B, B$ is a connected set $\implies A = B$.

Let (X, S), (Y, T) be topological spaces. The mapping $f : X \to Y$ is called continuous at the point $x \in X$ if for every neighbourhood V of the point $f(x)$ there exists a neighbourhood U of the point x such that $f(U) \subset V$. The mapping f is called continuous on the set $A \subset X$ if it is continuous at every point $x \in A$. If the mapping f is continuous over the entire space X, then we say that it is continuous. The mapping f is called a homeomorphism if it is bijective, continuous and the corresponding inverse mapping f^{-1} is also continuous. The mapping f is called a local homeomorphism at the point $x \in X$ if there exists such a neighbourhood U of the point X that the mapping f/U is a homeomorphism of the set U onto the set $f(U)$. The mapping f is called a local homeomorphic if it is a local homeomorphism at every point $x \in X$. The topological spaces (X, S), (Y, T) are called homeomorphic if there exists a homeomorphism of the space X onto the space Y. The sets $A \subset X$, $B \subset Y$ are called homeomorphic if the topological spaces (A, S_A), (B, S_B) are also homeomorphic, where S_A, S_B are relative topologies on A or on B.

1.15 Theorem. *If X, Y are Hausdorff topological spaces, whereby X is compact and $f : X \to Y$ is a continuous mapping, then $f(x)$ is a compact set in Y.*

1.16 Theorem. *Every continuous and injective mapping of a compact topological space into a Hausdorff topological space is a homeomorphism onto its image.*

1.17 Theorem. *If X is a compact topological space then the continuous function $f : X \to R$ is bounded.*

References: [4, 81, 132].

1.6 Matrices

We assume that all the notions concerning matrices as introduced in Section 1.2 are well-known, as well as the following ones: the eigenvalue, eigenvector of a matrix, characteristic equation of a matrix, characteristic polynomial of a matrix (denoted by $P_A(\lambda)$), diagonal and block diagonal matrix.

We say that the eigenvalue λ of the matrix $A \in M(n)$ has the multiplicity k if the values of the characteristic polynomial and of its derivatives up to the order $k-1$ at λ are zero, whereby the k-th derivative of the polynomial P_A at λ is different from zero.

If $A \in M(n)$, then the sequence of matrices $\{T_i\}_{i=1}^{\infty}$,

$$T_i = I_n + \sum_{j=1}^{i} \frac{1}{j!} A^j$$

is a Cauchy sequence in the metric space $(M(n), d)$ where

$$d(A, B) = p(A - B), \quad p(C) = \left(\sum_{i,j=1}^{n} |c_{ij}|^2 \right)^{1/2},$$

$C = (c_{ij})$. Since this metric space is complete, the above sequence of matrices converges to a matrix $T \in M(n)$. The matrix T is called the exponent of the matrix A being denoted as e^A, or $\exp A$.

The matrices $A, B \in M(n)$ are called similar to each other if there exists such a regular matrix $C \in M(n)$ that $B = CAC^{-1}$.

Let $A \in M(n)$, $A(\lambda) = A - \lambda I_n$ and $D_k(\lambda)(1 \leq k \leq n)$ is the greatest common divisor of all the minors of the matrix $A(\lambda)$ of order k. The polynomials

$$F_k(\lambda) = \frac{D_k(\lambda)}{D_{k-1}(\lambda)}, \quad k = 1, 2, \ldots, n,$$

where $D_0(\lambda) \equiv 1$, are called the invariant factors of matrix A. If $\lambda_1, \lambda_2, \ldots, \lambda_n$ are all eigenvalues differing from each other then $F_k(\lambda) = (\lambda - \lambda_1)^{l_{k_1}} (\lambda - \lambda_2)^{l_{k_2}} \ldots (\lambda - \lambda_m)^{l_{k_m}}$, where $l_{1i} \leq l_{2i} \leq \cdots \leq l_{ni}$,

$i = 1, 2, \ldots, n$. The polynomials $(\lambda - \lambda_s)^{l_{1s}}$, $(\lambda - \lambda_s)^{l_{2s}}, \ldots, (\lambda - \lambda_s)^{l_{ns}}$ ($s = 1, 2, \ldots, m$) that are not equal to 1 are called elementary divisors of the matrix A corresponding to the eigenvalue λ_s.

1.18 Theorem. *Let $\lambda_1, \lambda_2, \ldots, \lambda_m$ be all the eigenvalues of the matrix $A \in M(n)$ different from each other, whereby the eigenvalue λ_j ($j = 1, 2, \ldots, m$) has k_{ij} elementary divisors of multiplicity i ($i = 1, 2, \ldots, r_j$). Then the matrix A is similar to the block diagonal matrix B whose diagonal blocks are as follows: if the eigenvalue λ_j is a real number then there are k_{mj} blocks ($m = 1, 2, \ldots, r_j$) corresponding to it of the form $\lambda_j I_i + J_i$ where I_i is the unit matrix, $J_i \in M(i)$ being the matrix that has all along its main diagonal just the elements equal to 1 the rest of elements being zero. If the eigenvalue λ_j is of the complex form $\lambda_j = a_j + ib_j$, $b_j \neq 0$ then there are the corresponding k_{ij} blocks of the form $\text{diag}\{\Lambda_j, \Lambda_j, \ldots, \Lambda_j\} + N_{2i} \in M(2i, 2i)$, where $\Lambda_j = (\lambda_{pq}^j) \in M(2)$, $\lambda_{11}^j = \lambda_{22}^j = a_j$, $\lambda_{12}^j = -\lambda_{21}^j = b_j$, N_{2j} being a block matrix whose blocks over the main diagonal are equal to the unit matrix I_2, the rest of its elements being zero (the matrix B is called the Jordan canonical form of the matrix A).*

References: [39], [85].

1.7 Linear Mappings

Let V, W be linear normed spaces over the field F ($F = R$ or $F = C$) with the topology induced by the norms on V or W. The space of all the continuous linear mapping from V to W is denoted as $L(V, W)$ and $L(V) \stackrel{\text{def}}{=} L(V, V)$. The space $L(V, W)$ is linear over F (($T_1 + T_2)(x) \stackrel{\text{def}}{=} T_1(x) + T_2(x)$, $(\lambda T)(x) \stackrel{\text{def}}{=} \lambda T(x)$, $T, T_1, T_2 \in L(V, W)$, $\lambda \in F$). The space of all continuous linear isomorphisms from V to W is denoted as $\text{GL}(V, W)$ and $\text{GL}(V) \stackrel{\text{def}}{=} \text{GL}(V, V)$.

1.19 Theorem. *If V, W are linear normed spaces, whereby W is a Banach space, then*
1. the function $p : L(V, W) \to R$, $p(T) = \sup_{\|x\| \leq 1} \|Tx\|$, is a norm on $L(V, W)$ (the value of $p(T)$ is denoted as $\|T\|$);
2. $(L(V, W), p)$ is a Banach space.

From the above theorem it follows that if $V_1, V_2, \ldots, V_n$ are linear normed spaces $L(V_1, L(V_2, \ldots, L(V_n, W)))$ and this space is a Banach space. For example at $n = 2$, the space $L(V_1, L(V_2, W))$ is concerned. If V is a Banach space, then we define $L^1(V) \stackrel{\text{def}}{=} L(V)$, $L^n(V) \stackrel{\text{def}}{=} L(V, L^{n-1}(V))$ for $n > 1$

Let $V_1, V_2, \ldots, V_n, W$ be linear spaces over the field F. The mapping $T : V_1 \times V_2 \times \cdots \times V_n \to W$ is called *n-linear* (*bilinear*, if $n = 2$) if it is linear with respect to each variable. If V_i, $i = 1, 2, \ldots, n$ are linear normed spaces and W is a Banach space, then the space of continuous n-linear mappings from $V_1 \times V_2 \times \cdots \times V_n$ to W is denoted as $L(V_1, V_2, \ldots, V_n, W)$.

1.20 Theorem. *If (V_i, p_i), $i = 1, 2, \ldots, n$ are linear normed spaces and (W, q) is a Banach space, then the following can be defined.*
  1. *The function*

$$p : L(V_1, V_2, \ldots, V_n, W) \to R, \quad p(T) = \sup q(T(x_1, x_2, \ldots, x_n)),$$

where the supremum is taken over the set $\{x \in V_1 \times V_2 \times \cdots \times V_n : p_i(x) \leq 1, i = 1, 2, \ldots, n\}$ is the norm on $L(V_1, V_2, \ldots, V_n, W)$.
  2. *$(L(V_1, V_2, \ldots, V_n, W), p)$ is a Banach space.*
  3. *There exists an isometry of the space $L(V_1, V_2, \ldots, V_n, W)$ onto the space $L(V_1, L(V_2), \ldots, L(V_n, W)))$.*

The mapping $T \in L(V_1, V_2, \ldots, V_n, W)$ is called symmetrical if $T(v_1, \ldots, v_i, \ldots, v_j, \ldots, v_n) = T(v_1, \ldots, v_j, \ldots, v_i \ldots, v_n)$ for all $v_i \in V_i$, $v_j \in V_j$, $i, j = 1, 2, \ldots, n$. The set of all symmetrical mappings from $L(V_1, V_2, \ldots, V_n, W)$ is denoted as $L_s(V_1, V_2, \ldots, V_n, W)$ and $L_s^n(V, W) \stackrel{\text{def}}{=} L_s(V, V, \ldots, V, W)$.

We define the n-th power of the mapping $T \in L(V) : T^0 \stackrel{\text{def}}{=} \text{id}_V$, $T^n \stackrel{\text{def}}{=} T \circ T^{n-1}$ for $n > 0$. In the same way as the exponent of a matrix was defined, the exponent of the mapping T can also be defined through a power series in the space $L(V)$ that will be denoted as e^T or $\exp T$.

1.21 Theorem. *Let V, W be linear spaces, $\dim V = n$, $\dim W = m$ and $T \in L(V, W)$. If $x = (x_1, x_2, \ldots, x_n)$ are coordinates of the vector $p \in V$ with respect to the basic $\mathscr{B} = \{v_1, v_2, \ldots, v_n\}$ in V and $y = (y_1, y_2, \ldots, y_n)$ are coordinates of the vector $q = T(p)$ with respect to the basis $\mathscr{B}' = \{w_1, w_2, \ldots, w_m\}$ in W, then $y^* = [T]x^*$, where $[T] \in M(n, m)$ is a matrix whose i-th row is $(a_{i1}, a_{i2}, \ldots, a_{im})$ where a_{ij} $(j = 1, 2, \ldots, n)$ is the i-th coordinate of the vector $T(v_j)$ with respect to the basis $\mathscr{B}'$.*

The matrix $[T]$ from the above theorem is called the matrix representation of the mapping T with respect to the given basis V and W, respectively.

The linear subspace L of a linear space V is called an invariant subspace of the linear mapping $T : V \to V$ if $T(L) \subset L$.

Let V be a linear space over the field F ($F = R$ or $F = C$) and $T : V \to V$ be a linear mapping. If $T(v) = \lambda v$ for $v \in V$, $v \neq 0$, $\lambda \in F$ then the number λ is called an eigenvalue of the mapping T and the vector v is called an eigenvector of the mapping T corresponding to the eigenvalue λ.

1.22 Theorem. *Let V be a linear space over F ($F = R$, or $F = C$). If λ is an eigenvalue of the linear mapping $T : V \to V$ and $v \in V$*

is its corresponding eigenvector, then for arbitrary $a \in F$ av is also an eigenvector of the mapping T and $L_i = \{x \in V, x = av, a \in F\}$ is an invariant subspace of the mapping T.

1.23 Theorem. Let V be a linear normed space over R and $\lambda = \alpha + i\beta$, $\beta \neq 0$ be an eigenvalue of the linear mapping $T : V \to V$. Then, there exist non-zero vectors $u, v \in V$ such that $T(u) = \alpha u - \beta v$, $T(v) = \alpha v + \beta u$ and the two-dimensional subspace $L_2 = \{x \in V : x = au + bv,\ a, b \in R\}$ is an invariant subspace of the mapping T. The vectors u, v can be chosen so that $\|u\|^2 + \|v\|^2 = 1$.

1.24 Theorem.
1. λ is an eigenvalue of the linear mapping $T : V \to V$ if it is an eigenvalue of its matrix representation.
2. If λ is an eigenvalue of the linear mapping T, then e^λ is an eigenvalue of the mapping e^T.

1.25 Theorem. If (H, s) is a Hilbert space and $f : H \to R$ is a continuous linear mapping then there exists a unique element $x_0 \in H$ such that $f(x) = s(x, x_0)$ for all $x \in H$.

1.26 Closed Graph Theorem. Let V, W be Banach spaces and $T : V \to W$ be a linear mapping whose graph $\mathrm{graph}(T) = \{(x, Tx) : x \in V\}$ is a closed set in $V \times W$. Then T is a continuous mapping.

References: [32, 115, 132].

1.8 Mathematical Analysis

Let V, W be Banach spaces and $U \subset V$ is an open set. The mapping $f : U \to W$ is called differentiable at point $x_0 \in V$ (in the sense of Fréchet) if there exists a linear mapping $T(x_0) \in L(V, W)$ such that for all $h \in V$ (such that $x_0 + h \in U$) it holds that: $f(x) - f(x_0) - T(x_0)(x - x_0) = w(x_0, x)$, $\lim_{x \to x_0} \|x - x_0\|^{-1} \|w(x_0, x)\| = 0$ (for simplicity, the notation for the norms V and W are identical). The mapping $T(x_0)$ is called the Fréchet derivative of the mapping f at x_0 being denoted as $d_{x_0} f$. The mapping f is called differentiable on the set $U' \subset U$ if it is differentiable at every point $x \in U'$. If $f : U \to W$ is a differentiable mapping on U, then we say it is differentiable (in the sense of Fréchet). If the mapping $f : U \to W$ is differentiable then the mapping $df : U \to L(V, W)$, $df(x) = d_x f$ is called the derivative of the mapping f.

Let $U \subset R^n$ be an open set, $f : U \to R^m$ is a differentiable mapping and $x \in U$. The matrix $[d_x f]$ that is a representation of the linear mapping

$d_x f : R^n \to R^m$ with respect to the bases of unit vectors in R^n or in R^m is called the Jacobi matrix of the mapping f at x, being denoted as

$$\frac{df(x)}{dx}, \quad \text{or} \quad \frac{df(y)}{dy}\bigg/_{y=x}.$$

Let V, W be Banach spaces, $U \subset V$ an open set and $f : U \to W$. If for $x \in U$, $h \in U$ there exists $\lim_{t\to 0} t^{-1}[f(x+h) - f(x)] = \delta f(x,h)$ then this limit is called the first variation of the mapping f at x in direction h. The mapping $\delta_x f : V \to W$, $\delta_x f(h) = \delta f(x,h)$ in general, need not be linear. If this mapping is linear and continuous then we say that f is a differentiable mapping at x in the sense of Gateaux. The mapping $\delta_x f$ is called the Gateaux derivative at x being denoted as $f'(x)$. If the mapping f is differenatiable in the sense of Gateaux at every point $x \in U$, then we say that it is differentiable in the sense of Gateaux. The mapping $f' : U \to L(V,W)$, $x \mapsto f'(x)$ is called the Gateaux derivative of the mapping f.

1.27 Theorem. *Let V, W be Banach spaces and $f : V \to W$ then the following hold true.*

1. If f is a differentiable mapping at x in the sense of Fréchet, then it is differentiable at x also in the sense of Gateaux and $d_x f = f'(x)$.

2. If f is a differentiable mapping in the sense of Gateaux in a neighbourhood U of the point $x \in V$ and the mapping $f' : U \to L(V,W)$, $y \mapsto f'(y)$ is continuous at x, then f is a differentiable mapping at the point x in the sense of Fréchet and $d_x f = f'(x)$.

If $f : R \to W$ then the value $f'(x)1$ is denoted either as $f'(x)$ or $\dot{f}(x)$.

Let $V_1, V_2, \ldots, V_n, W$ be Banach spaces, $U = U_1 \times U_2 \times \cdots \times U_n \subset V = V_1 \times V_2 \times \cdots \times V_n$ is an open set, $f : U \to W$ is a differentiable mapping and $x = (x_1, x_2, \ldots, x_n) \in U$. The partial derivative f with respect to the k-th variable x_k is defined analogously as for the functions of n real variables being denoted as $(\partial_k)_x f$ or $(\partial_{x_k})_x f$. If $V_k = R^n$, $W = R^m$, then the representation of the mapping $(\partial_k)_x f : R^n \to R^m$ with respect to the bases of the unit vectors in R^n or R^m is the matrix $[(\partial_k)_x f]$ being denoted also as

$$\frac{\partial f(x)}{\partial x_k} \quad \text{or} \quad \frac{\partial f(y)}{\partial y_k}\bigg/_{y=x}.$$

1.28 Theorem. *If $V_1, V_2, \ldots, V_n, W$ are Banach spaces, $V = V_1 \times V_2 \times \cdots \times V_n$, $f : V \to W$ is a differentiable mapping, $x \in V$ and $h = (h_1, h_2, \ldots, h_n) \in V$ then $d_x f(h) = (\partial_1)_x f(h_1) + (\partial_2)_x f(h_2) + \cdots + (\partial_n)_x f(h_n)$.*

If $f : R^n \to R$ is a differentiable function then the mapping $df : R^n \to L(R^n, R)$ is called the gradient of the function f being denoted as grad f.

The value (grad $f)(x)$ is called the gradient of the function f at the point x being denoted as grad $f(x)$. The Jacobiho matrix of the mapping grad f at the point x is of the form

$$\frac{\mathrm{d}f(x)}{\mathrm{d}x} = \left(\frac{(\partial f(x)}{\partial x_1}, \frac{\partial f(x)}{\partial x_2}, \ldots, \frac{\partial f(x)}{\partial x_n}\right).$$

Let V, W be Banach spaces, $U \subset V$ be an open set and $f : U \to W$ be a differentiable mapping whereby also the mapping $df : U \to L(V,W)$ is differentiable. Then the mapping $d_x(df) \in L(V, L(V,W))$ is called the second derivative of the mapping f at the point x, the latter is denoted as $d_x^2 f$. The mapping $d^2 f : U \to L^2(V,W) = L(V, L(V,W))$, $d^2 f(x) = d(df)(x)$ is called the second derivative of the mapping f. The r-th derivative is defined as follows: $d^r f \stackrel{\text{def}}{=} d(d^{r-1}f) : U \to L^r(V,W)$, where $L^r(V,W) := L(V, L^{r-1}(V,W))$, $r > 2$. If the mapping $d^r f$ exists and it is continuous, then we say that it is of class C^r or that it is C^r-differentiable. If all the mappings $d^r f$, $r = 1, 2, \ldots$ exist and are continuous then we say that f is of class C^∞, or that it is smooth. The set of all C^r-mappings from $U \subset V$ to W is denoted as $C^r(U,W)$ or $C^\infty(U,W)$ if $r = \infty$. The mapping $f : U \to R^n$, where $U \subset R^m$ is an open set, is called C^r -differentiable on the set A (A is an arbitrary subset in R^m), if for any $x \in A$ there exists its open neighbourhood $U' \subset U$ and also C^r-differentiable mapping $F : U \to R^n$ that $F/U' \cap A = f/U' \cap A$.

1.29 Theorem. *Let V, W be Banach spaces, $U \subset V$ and $C_B^r(U,W)$ be the space of all C^r-mappings on U that are bounded along with their derivatives up to the order r. Then $(C_B^r(U,W), p)$ is the Banach space where $p(f) = \sup_{x \in U}\{\|f(x)\|, \|d_x f\|_1, \ldots, \|d_x^r f\|_r\}$ (the value of $p(f)$ is denoted also by $|f|_r$, $\|\ \|_r$ is the norm in $L^n(V,W)$).*

Of course, higher order partial derivatives of the function f in the variables $x_i \in V_i$, $i = 1, 2, \ldots, n$ can also be defined for which we use the notation of

$$\frac{\partial^{|\alpha|} f(x)}{\partial x_1^{\alpha_1} \partial x_2^{\alpha_2} \ldots \partial x_n^{\alpha_n}},$$

where $|\alpha| = \alpha_1 + \alpha_2 + \cdots + \alpha_n$, $\alpha = (\alpha_1, \alpha_2, \ldots, \alpha_n)$.

1.30 Theorem. *Let G, W, Z be Banach spaces, $U \subset G$, $V \subset W$ open sets, $g \in C^r(U,V)$ and $f \in C^r(V,Z)$, $r \geq 1$. Then $f \circ g \in C^r$ and for all $x \in U$ it is $d_{g(x)} f \circ d_x g$ (Leibnitz formula).*

1.31 Taylor Theorem. *Let V, W be Banach spaces and $U \subset V$ be such an open set that for arbitrary x, $y \in U$ it is $\{x + ty : 0 \leq t \leq 1\} \subset U$. If*

$f \in C^r(U, W)$ then for $x, h \in U$ it holds that:

$$f(x+h) = f(x) + d_x f(h) + \frac{1}{2!} d_x^2 f(h^2) + \cdots + \frac{1}{r!} d_x^r f(h^r) + R(x, h) h^r$$

(Taylor expansion of the mapping f at x), where $h^k = (h, h, \ldots, h)$ (k-times), $R(x, h) \in L^r(V, W)$ and $R(x, 0) = 0$.

If $f \in C^\infty(V, W)$ then the infinite series

$$T(f)h = f(0) + \sum_{k=1}^{\infty} \frac{1}{k!} d_0^k f(h^k)$$

is called the formal Taylor series of the mapping f at $0 \in V$. The function $f : R \to R$, $f(x) = e^{-x^{-2}}$ for $x \neq 0$ and $f(x) = 0$ for $x = 0$ is of the class C^∞ and has all its derivatives at the point 0 equal to 0. This means that the formal Taylor series $T(f)(h)$ equals 0, and thus, the function f does not coincide with its formal Taylor series at the point 0 in any neighbourhood of the point 0.

The expression

$$T^n(f)(h) = \sum_{k=1}^{n} \frac{1}{k!} d_0^k f(h^k)$$

is called the Taylor polynomial of degree k of the mapping f at the point 0.

1.32 Borel Theorem. *For an arbitrary set of real numbers $\{c_k\}_{k \in M_n}$, where $M_n = N^n \cup \{(0, 0, \ldots, 0)\}$, there exists a function $f \in C^\infty(R^n, R)$ such that for arbitrary $\alpha = (\alpha_1, \alpha_2, \ldots, \alpha_n) \in M_n$ it is*

$$\frac{1}{\alpha!} \frac{\partial^{|\alpha|} f(0)}{\partial x_1^{\alpha_1} \partial x_2^{\alpha_2} \ldots \partial x_n^{\alpha_n}} = c_\alpha,$$

where $\alpha! = \alpha_1! \alpha_2! \ldots \alpha_n!$, $(0, 0, \ldots, 0)! \stackrel{def}{=} 1$.

1.33 Converse to Taylor Theorem. *Let V, W be Banach spaces, $U \subset V$ be an open and convex set, $f : U \to W$, $g_k : U \to L_s^k(V, W)$, $k = 0, 1, \ldots, r$ and let*

$$\varrho(x, h) = f(x + h) - \sum_{k=0}^{r} \frac{1}{k!} g_k(x) h^k$$

for $x \in U$, $h \in V$ for which $x + h \in U$. If every mapping g_k ($k = 0, 1, \ldots, r$) is continuous and

$$\lim_{(x, h) \to (x_0, 0)} \|h\|^{-r} \cdot \|\varrho(x, h)\| = 0,$$

then $f \in C^r(U, W)$ and $d^k f = g_k$ for $k = 0, 1, \ldots, r$.

1.34 Mean Value Theorem. *Let V, W be Banach spaces, $U \subset V$ be an open set, $f \in C^1(U, W)$, $x, y \in U$ and $P = \{(1 - \lambda)x + \lambda y : 0 \leq \lambda \leq 1\} \subset U$. Then $\|f(x) - f(y)\| \leq M\|x - y\|$, where $M = \max_{z \in P} \|d_z f\|_1$. If $V = R$, $W = R$, then there exists such $u \in P$ that $f(x) - f(y) = d_u f(x - y)$.*

Let V, W be Banach spaces and $U \subset V$. The mapping $f : U \to W$ is called a Lipschitz mapping if there exists such a number $L > 0$ that for arbitrary $x, y \in U$ it is $\|f(x) - f(y)\| \leq L\|x - y\|$. The number L is called the Lipschitz constant of the mapping f being denoted as $\text{Lip}(U, W)$. The set of all Lipschitz mappings from U into W is denoted as $\text{Lip}(U, W)$. The mapping $f : U \to W$ is called locally Lipschitz at the point $x \in U$ if there exists a neighbourhood $U' \subset U$ of the point x such that f/U' is a Lipschitz mapping. The mapping f is called a locally-Lipschitz mapping if it is locally-Lipschitz at every point $x \in U$.

1.35 Theorem. *If V, W are Banach spaces, $U \subset V$ is a compact set and the mapping $f : U \to W$ is a locally Lipschitz mapping then f is a Lipschitz mapping.*

Let (X, d) be a metric space. The mapping $f : X \to X$ is called a contractive mapping if there exists a number L such that $0 < L < 1$ and $d(f(x), f(y)) \leq L \cdot d(x, y)$ for all $x, y \in X$.

1.36 Banach Fixed Point Theorem. *If (X, d) is a complete metric space and $f : X \to X$ is a contractive mapping, then f has a unique fixed point in X, i.e., a point $x \in X$ such that $f(x) = x$.*

1.37 Inverse Mapping Theorem. *Let V, W be Banach spaces, $U \subset V$ be an open set and $f \in C^r(U, W)$, $1 \leq r < \infty$, $x \in U$ and the mapping $d_x f : V \to W$ be an isomorphism. Then there exists a neighbourhood $U' \subset U$ of the point x and a neighbourhood $W' \subset W$ of the point $y = f(x)$ such that the mapping $f/U' : U' \to W'$ is bijective, the inverse mapping f^{-1} to f/U' is of the class C^r and $d_y f^{-1} = (d_x f)^{-1}$. If $V = R^n$, $W = R^m$ and the mapping f is analytical then the mapping f^{-1} is analytical, as well.*

1.38 Implicit Function Theorem. *Let E_1, E_2, W be Banach spaces, $U_1 \subset E_1$, $U_2 \subset E_2$ be open sets, $f \in C^r(U_1 \times U_2, W)$, $0 \leq r < \infty$, $z_0 = (x_0, y_0) \in U_1 \times U_2$ be a point such that $f(z_0) = 0$, then the mapping $(\partial_2)_{z_0} f$ does exist and it is an isomorphism of the space E_2 onto the space W, where the mapping $H : U_1 \times U_2 \to L(E_2, W)$, $H(z) = (\partial_2)_z f$, $z = (z_1, z_2) \in U_1 \times U_2$ is continuous. Then there exists an open neighbourhood $U \subset U_1$ of the point x_0, an open neighbourhood $V \subset U_2$ of the point y_0 and the unique mapping $h \in C^r(U, V)$ such that $h(x_0) = y_0$ and $f(x, h(x)) = 0$ for all $x \in U$. If $r \geq 1$ then $d_x h = -((\partial_2)_u f)^{-1}(\partial_1)_u f$, where $u = (x, h(x))$.*

If $E_1 = R^n$, $E_2 = R^m$, $W = R^m$ and the mapping f is analytical, then the mapping h is also analytical.

Let $I \subset R$ be an interval, W be a Banach space and $f : I \to W$ be a continuous mapping. The mapping $g : I \to W$ is called a mapping primitive to the mapping f if it is continuous and there exists a countable set $D \subset I$ such that g is a differentiable mapping at every point $t \in I \setminus D$ and $\dot{g}(t) = f(t)$ for all $t \in I \setminus D$.

1.39 Theorem. *If g_1, g_2 are the mappings primitive to the mapping $f : I \to W$, $a, b \in I$ then $g_1(b) - g_1(a) = g_2(b) - g_2(a)$.*

If g is a mapping primitive to the mapping $f : I \to W$, $a, b \in I$ then the difference $g(b) - g(a)$ is called an integral of the mapping f from a to b and it is denoted as

$$\int_a^b f(t)\,dt.$$

Let $U_1 \times U_2 \subset F \times F^n$ ($F = R$ or $F = C$) is an open neighbourhood of the point $(0,0) \in F \times F^n$, $1 \leq r \leq \infty$. We say that the C^r-function $f : U_1 \times U_2 \to F$, $(t, x) \mapsto f(t, x)$ is a p-regular function ($p \leq r$) at $(0,0)$ if $f(0,0) = 0$,

$$\frac{\partial^k f(0,0)}{\partial t^k} = 0, \quad k = 1, 2, \ldots, p-1, \quad \frac{\partial^p f(0,0)}{\partial t^p} \neq 0.$$

1.40 Malgrange-Weierstrass Theorem. *Let $U_1 \times U_2 \subset F \times F^n$ ($F = R$, or $F = C$) is an open neighbourhood of $(0,0) \in F \times F^n$, $1 \leq r \leq \infty$ and $f : U_1 \times U_2 \to F$, $(t, x) \mapsto f(t, x)$, is a C^r-function (analytical function, $F = C$) p-regular at $(0,0)$. Then there exists an open neighbourhood $V_1 \times V_2 \subset U_1 \times U_2$ of the point $(0,0)$ and C^r-functions (analytical functions) $Q : V_1 \times V_2 \to F$, $(t, x) \mapsto Q(t, x)$, $u_j : V_2 \to F$, $x \mapsto u_j(x)$, $j = 1, 2, \ldots, p$, such that it holds that: $Q(0, 0) \neq 0$, $u_j(0) = 0$, $j = 1, 2, \ldots, p$ and*

$$f(t, x) = Q(t, x)(t^p + \sum_{j=1}^{p} u_j(x) t^{p-j})$$

for all $(t, x) \in V_1 \in V_2$.

Remark: For analytical functions of complex variables and smooth functions of both the real and complex variables the preceding theorem is proved in the books [24], [44], [74], [95]. For functions of the class C^r, $1 \leq r < \infty$ this theorem is proved in paper [90]. The theorem does not hold for the class of analytical functions of real variables (counter example in reference [155] p. 200).

References: [2, 24, 32, 44, 92, 95, 156].

Chapter 1

1.9 Differential Equations

Let V be a Banach space, $J \times U \times \subset R \times V$ be an open set and $f : J \times U \to V$ be a continuous mapping. The equation $\dot{x} = f(t,x)$ is called a differential equation in V. The differentiable mapping $u : I \to U$, where $I \subset J$ is an interval, is called the solution of the above differential equation on I if $\dot{u}(t) = f(t, u(t))$ for all $t \in I$. We say that this solution has its initial value $u_0 \in U$ at the point $t_0 \in I$ if it satisfies the so-called initial condition $u(t_0) = u_0$. The differential equation $\dot{x} = g(x)$, where $g : V \to V$ is called autonomous. If g is a linear mapping then this differential equation is called linear.

If $V = R^n$ and $(f_1, f_2, \ldots, f_n)$ are the coordinates of the mapping $f : J \times U \to R^n$ with respect to the basis of the unit vectors in R^n and $(u_1, u_2, \ldots, u_n)$ are the coordinates of the solution $u : I \to U$ with respect to the above basis then $\dot{u}(t) = f(t, u(t))$ iff $\dot{u}_i(t) = f_i(t, u_1(t), u_2(t), \ldots, u_n(t))$, $i = 1, 2, \ldots, n$, $t \in I$. We say that the system of mappings u_i, $i = 1, 2, \ldots, n$ represents a solution of the system of differential equations $\dot{x}_i = f_i(t, x_1, x_2, \ldots, x_n), i = 1, 2, \ldots, n$. If the functions f_i, $i = 1, 2, \ldots, n$ are linear in the variables $x_1, x_2, \ldots, x_n$ then the above system is called linear.

1.41 Existence and Uniqueness Theorem. *Let V be a Banach space, $J \times U \subset R \times V$ be an open set and $f : J \times U \to V$, $(t, x) \mapsto f(t, x)$ be a mapping continuous in (t, x) and it is locally Lipschitz in the variable x. Then to every point $(t_0, x_0) \in J \times U$ there exists an interval $I_0 \subset J$ such that $t_0 \in I_0$ and on I_0 there exists a unique solution u of the differential equation $\dot{x} = f(t, x)$ fulfilling the initial condition $u(t_0) = x_0$. If $f \in C^r$ then also $u \in C^r$.*

Let $u_i : I_i \to U$, $i = 1, 2$ be solutions of the differential equation $\dot{x} = f(t,x)$. We say that u_1 is a prolongation of the solution u_2 if $I_2 \subset I_1$ and $u_1(t) = u_2(t)$ for all $t \in I_2$. If $I_1 \neq I_2$ then we say that the above prolongation is proper. The solution $u : I \to U$ is called maximal if there is no its proper prolongation. The solution $u : I \to U$ is called a complete solution, if $I = (-\infty, \infty)$. The differential equation is called complete if every its solution is complete.

1.42 Theorem. *Let V be a Banach space $f : R \times V \to V$ be a continuous mapping and there exist such numbers $a, b \geq 0$ that $\|f(t,x)\| \leq a\|x\| + b$ for all $(t, x) \in R \times V$. Then the differential equation $\dot{x} = f(t, x)$ is complete.*

A more general theorem on the completeness of differential equations formulated for $V = R^n$ can be found, for example in [86].

1.43 Theorem. *Let V be a Banach space, $f \in C^r(V,V)$, $r \geq 1$ and $U \subset V$ is a compact set. If $u : I \to V$ is a maximal solution of the differential equation $\dot{x} = f(x)$ such that $u(t) \in U$ for all $t \in I$ then this solution is complete.*

Let V be a Banach space, $I \subset R$ is an interval and $T : I \to L(V)$ be a differentiable mapping. Then $T(t) \in L(V)$, and thus it is meaningful, except for the solutions of the linear differential equation $\dot{x} = T(t)x$ with the values at V, to look also for a differentiable mapping $S : I \to L(V)$ such that for every $t \in I$ the equality $\dot{S}(t) = T(t)S(t)$ is satisfied. The mapping S of such a property is called the operator solution of the above linear differential equation. If $V = E_n$ is an n-dimensional linear space with a given basis, then the matrix representation $\Phi(t) = [S(t)]$ of the mapping $S(t)$ with respect to the given basis satisfies the equality $\dot{\Phi}(t) = [T(t)]\Phi(t)$ for all $t \in I$. The mapping Φ is called a fundamental matrix of the differential equation $\dot{x} = [T(t)]x$, if $\det \Phi(t) \neq 0$ for all $t \in I$. The fundamental matrix Φ satisfying the condition $\Phi(t_0) = I_n$ is called normed at the point t_0, being denoted as $C(t, t_0)$. The mapping $C : I \times I \to M(n)$, $(t, t_0) \to C(t, t_0)$ is called a resolvent of the differential equation $\dot{x} = [T(t)]x$.

1.44 Theorem. *Let $\Phi : I \to M(n)$ be a fundamental matrix of the differential equation $\dot{x} = A(t)x$, where $A \in C^0(I, M(n))$. Then for arbitrary $t_0 \in I$*

$$\det \Phi(t) = \det \Phi(t_0) \cdot \exp \int_{t_0}^{t} \operatorname{Tr} A(s)\,ds$$

for all $t \in I$ (Liouville formula).

1.45 Theorem. *If $C(t,s)$ is the resolvent of the linear differential equation $\dot{x} = A(t)x$, $A \in C(R, M(n))$ then the solution u of the differential equation $\dot{x} = A(t)x + b(t)$, $b \in C(R, R^n)$ satisfying the intial condition $u(t_0) = x_0$ is of the form*

$$u(t) = C(t, t_0)x_0 + \int_{t_0}^{t} C(t, s)b(s)\,ds$$

(the variation constant formula).

1.46 Theorem. *If $\Phi : I \to M(n)$, $\Psi : I \to M(n)$ are fundamental matrices of the differential equation $\dot{x} = A(t)x$, $A \in C(I, M(n))$ then there exists a matrix $B \in \operatorname{GL}(n)$ such that $\Psi(t) = \Phi(t)B$ for all $t \in I$.*

The linear differential equation $\dot{x} = A(t)$ is called ω-periodic ($\omega > \infty$), if the mapping A is ω-periodic, i.e. $A(t + \omega) = A(t)$ for all $t \in R$.

1.47 Floquet Theorem. *Let the differential equation* $\dot{x} = A(t)x$, $A \in C(R, M(n))$, *be ω-periodic and Φ be its fundamental matrix normed at the point 0. Then there exists a unique ω-periodic mapping $G : R \to M_c(n)$ and the matrix $R \in M_c(n)$ such that $\Phi(t) = G(t)\exp tR$ for all $t \in R$.*

Let all the assumptions of the Floquet Theorem be satisfied. Then $\Psi : R \to M(n)$, $\Psi(t) = \Phi(t + \omega)$ is also a fundamental matrix, and thus, according to Theorem 1.45 there exists a matrix $B \in M(n)$ such that $\Phi(t + \omega) = \Phi(t)B$ for all $t \in R$. The matrix B is called a monodromy matrix. Obviously, $B = \Phi(\omega)$, and thus, from the Floquet Theorem we obtain that $B = \Phi(\omega) = \exp \omega R$. The eignvalues of the matrix B are called characteristic multipliers of the differential equation $\dot{x} = A(t)x$ and the eigenvalues of the matrix R are called characteristic exponents of the above differential equation.

The solution of the autonomous differential equation $\dot{x} = g(x)$ fulfilling the initial condition $u(0) = y$ will be denoted as $\varphi^g(y, t)$ or $\varphi(y, t)$.

Let B_1, B_2 be Banach spaces, $U \times V \subset B_1 \times B_2$ be an open set and $f : U \times V \to B_1$ be a continuous mapping. Let us consider the differential equation

$$\dot{x} = f(x, \varepsilon), \tag{1.1}$$

dependent on the parameter $\varepsilon \in V$. The present differential equation is called a parametrized differential equation. If for every $\varepsilon \in V$ we define the mapping $f_\varepsilon : U \to B_1$, $f_\varepsilon(x) = f(x, \varepsilon)$, then $\{f_\varepsilon\}_{\varepsilon \in V}$ is a system of mappings defining the system of differential equations

$$\dot{x} = f_\varepsilon(x), \quad \varepsilon \in V. \tag{1.2}$$

If $\dim B_2 = k < \infty$, then this system is also called a k-parametric system of differential equations.

If $s \in R$, $y \in U$ and $\varepsilon \in V$ then the solution u_ε of the differential equation (1.2) satisfying the initial condition $u_\varepsilon(s) = y$ is denoted as $\varphi^f(y, \varepsilon, s, t)$, or $\varphi(y, \varepsilon, s, t)$ whereby $\varphi(y, \varepsilon, 0, t)$ is denoted as $\varphi(y, \varepsilon, t)$.

1.48 Theorem. *Let B_1, B_2 be Banach spaces, $U \times V \subset B_1 \times B_2$ be an open set, $f \in C^r(U \times V, B_1)$, $1 \leq r \leq \infty$, the differential equation (1.2) is complete for every $\varepsilon \in V$ and let $(x_0, \varepsilon_0, t_0) \in U \times V \times R$. Then for arbitrary $T \in (0, \infty)$ there exists a number $\delta > 0$ such that it holds: If $I = \langle t_0 - T, t_0 + T \rangle$, $U_\delta = \{x \in U : \|x - x_0\| < \delta\}$, $V_\delta = \{\varepsilon \in V : \|\varepsilon - \varepsilon_0\| < \delta\}$, then the mapping $\varphi^f : U_\delta \times V_\delta \times I \times I \to B_1$, $(y, \varepsilon, s, t) \mapsto \varphi^f(y, \varepsilon, s, t)$ is C^r-differentiable. The mapping $\Phi : I \to L(B_1)$,*

$$\Phi(t) = \frac{\partial \varphi^f(y, \varepsilon, s, t)}{\partial y}$$

is an operator solution of the linear differential equation

$$\dot{z} = \frac{\partial f(\varphi^J(y,\varepsilon,s,t),\varepsilon)}{\partial x} z \qquad (1.3)$$

(the equation in variations) fulfilling the initial condition $\Phi(s) = \mathrm{id}$ (if $B_1 = R^n$, then $[\Phi(t)]$ is a fundamental matrix normed at the point $t = s$). The mapping $H : I \to L(B_2)$,

$$H(t) = \frac{\partial \varphi^J(y,\varepsilon,s,t)}{\partial \varepsilon}$$

is a solution of the differential equation

$$\dot{z} = \frac{\partial f(\varphi^J(y,\varepsilon,s,t),\varepsilon)}{\partial x} z + \frac{\partial f(\varphi^J(y,\varepsilon,s,t),\varepsilon)}{\partial \varepsilon} \qquad (1.4)$$

satisfying the initial condition $H(s) = 0$.

1.49 Gronwall Lemma. *Let $\mu > 0$, $a, b \in R$, $t_0 \in (a, b)$ and $u : (a, b) \to R$, $\varphi : (a, b) \to R$ be continuous non-negative functions fulfilling the inequality*

$$u(t) \leq \varphi(t) + \mu \left| \int_{t_0}^t u(s)\, ds \right| \quad \text{for } t \in (a, b). \qquad (1.5)$$

Then

$$u(t) \leq \varphi(t) + \mu \left| \int_{t_0}^t (\exp \mu |t - s|) \varphi(s)\, ds \right| \quad \text{for } t \in (a, b). \qquad (1.6)$$

If $\varphi(t) \equiv c$, then the inequality (1.6) is of the form

$$u(t) \leq c \exp \mu |t - t_0|. \qquad (1.7)$$

References: [13, 29, 32, 59, 86].

2 FOUNDATIONS OF THE THEORY OF DIFFERENTIABLE MANIFOLDS AND DIFFERENTIABLE MAPPINGS

This chapter presents the foundations of differential topology which are indispensable for providing a possible explanation of the essence of the contemporary qualitative theory of differential equations. Differential topology represents a mathematical discipline aimed at the study of the so-called differentiable manifolds and differentiable mappings defined on them. The notion of a manifold represents, in fact, a generalization of the notion of a curve, or a surface. There are several reasons for the study of differential equations in connection with differentiable manifolds. One of them is that the differential equations are very often being used for a description of the models of physical motion of bodies or of the course of chemical reactions and similar phenomena, that do not run freely, but are bound to some conditions. These conditions very often express the binding of the solutions of the differential equations to a certain surface, in general, to a differentiable manifold. Some other reasons of a rather methodological and technical character will be obvious only after having read the contents of the following chapters.

2.1 C^r-manifolds

2.1 Definition. *Let X be a Hausdorff topological space. If α is a homeomorphism of an open set U in X onto an open set in R^n, then the pair (U, α) is called a chart on X. We say that the system $S = \{(U_i, \alpha_i)\}_{i \in I}$ where (U_i, α_i) ($i \in I$) is a chart on X and I is an index set, is a C^r-atlas on X ($1 \leq r \leq \infty$) if $\cup_{i \in I} U_i = X$ and for arbitrary $i, j \in I$ such that $U_i \cap U_j \neq \emptyset$, the mapping $\alpha_{ij} = \alpha_j \circ \alpha_i^{-1} : \alpha_i(U_i \cap U_j) \to \alpha_j(U_i \cap U_j)$ is of class C^r. We also say that the charts (U_i, α_i) and (U_j, α_j) are C^r-compatible. Two C^r-atlases S_1, S_2 on X are called C^r-equivalent if $S_1 \cup S_2$ is a C^r-atlas on X. A differentiable structure of class C^r on X or a C^r-structure on X is a class $\mathscr{S}$ of C^r-equivalent atlases on X. The union of all the C^r-atlases on X belonging to $\mathscr{S}$ is the maximal C^r-atlas* on X, it is denoted as $A(\mathscr{S})$. The chart $(U,\alpha) \in A(\mathscr{S})$ is called an admissible chart on X and the mapping α is called coordinate mapping. A differentiable manifold of class C^r or a C^r-manifold (in the case that $r = \infty$ a smooth manifold) is a couple $(X, \mathscr{S})$ where X is the Hausdorff topological space with a countable basis and $\mathscr{S}$ is a C^r-structure on X. When it is obvious, where a C^r-structure is concerned, it will be simply written just as X. The space R^n is called a model space of the manifold X. The number n is called the dimension of the manifold X and we write $n = \dim X$. We say also that the manifold X is n-dimensional. If $\dim X = n$ and (U, α) is an admissible chart on X then $\alpha(x) = (\alpha_1(x), \alpha_2(x), \ldots, \alpha_n(x))$. The functions α_i, $i = 1, 2, \ldots, n$ are called coordinate functions or local coordinates on X. If there exists such a C^r-atlas that for two arbitrary charts (U_1, α_1), (U_2, α_2) of this atlas for which $U_1 \cap U_2 \neq \emptyset$ is $\det[d_x(\alpha_2 \circ \alpha_1^{-1})] > 0$ for arbitrary $x \in \alpha_1(U_1 \cap U_2)$ then the manifold X is called orientable.

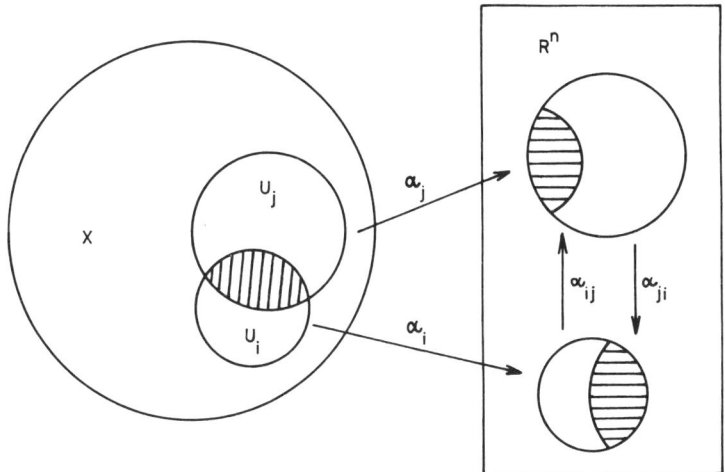

Fig. 1.

A more general concept of a differentiable manifold can be obtained so that instead of a model space R^n an arbitrary Banach space B (also one of infinite dimension is accepted) can be taken. Later in the book, when a manifold like this is mentioned a special note will be made, otherwise only C^r-manifolds with a finite dimensional space will be considered which moreover, will be orientable.

2.2 Example. Let $(X, \mathscr{S})$ be a C^r-manifold, $U \subset X$ be an open set and let the C^r-structure $\mathscr{S}$ be represented by the C^r-atlas $\mathscr{S} = \{(U_i, \alpha_i)\}_{i \in I}$. Then $S(U) = \{(U_i \cap U, \alpha_i/U_i \cap U)\}_{i \in I}$ is a C^r-atlas on U that represents the C^r-structure $\mathscr{S}(U)$ on U, and thus, $(U, \mathscr{S}(U))$ is a C^r-manifold where

$\dim U = \dim X$ (the topology on U is the relative topology induced by the topology on X).

2.3 Example. Let a topology be given on the set of matrices $M(m,n)$ induced by the Euclidean metric

$$\varrho(A, B) = \|A - B\| = \left(\sum_{i,j=1}^{n} |a_{ij} - b_{ij}|^2 \right)^{1/2},$$

$A = (a_{ij})$, $B = (b_{ij}) \in M(m,n)$. Let $U = M(m,n)$ and $\alpha: U \to R^{mn}$, $\alpha(A) = (a_{11}, a_{12}, \ldots, a_{1n}, \ldots, a_{m1}, a_{m2}, \ldots, a_{mn})$. Then $S = \{(U, \alpha)\}$ is a C^r-atlas on $M(m,n)$ that defines a C^∞-structure on $M(m,n)$. Throughout this book we shall often take $M(m,n)$ identical to R^{mn}.

2.4 Example. Let us consider an n-dimensional unit sphere

$$S^n = \left\{ x = (x_1, x_2, \ldots, x_{n+1}) \in R^{n+1} : \sum_{i=1}^{n+1} x_i^2 = 1 \right\}$$

with the relative topology induced by the topology on R^{n+1} that is induced by the Euclidean metric on R^{n+1}. Let $U_i = \{x \in S^n : x_i > 0\}$, $U_{i+n+1} = \{x \in S^n : x_i < 0\}$, $i = 1, 2, \ldots, n+1$ (x_i being the i-th coordinate of the point x), $U = \{y \in R^n : \|y\| < 1\}$, $\alpha_i: U_i \to U$, $\alpha_i(x_1, \ldots, x_{i-1}, x_i, x_{i+1}, \ldots, x_{n+1}) = (x_1, \ldots, x_{i-1}, x_{i+1}, \ldots, x_{n+1})$, $\alpha_{i+n+1}: U_{i+n+1} \to U$, $\alpha_{i+n+1}(x_1, \ldots, x_{i-1}, x_i, x_{i+1}, \ldots, x_{n+1}) = (x_1, \ldots, x_{i-1}, x_{i+1}, \ldots, x_{n+1})$, $i = 1, 2, \ldots, n+1$. The system $\{U_j\}_{j=1}^{2n+2}$ forms an open covering of the set S^n. The mappings α_j ($j = 1, 2, \ldots, 2n+2$) are bijective with corresponding inverse mappings of the form: $\alpha_i^{-1}(y) = (y_1, y_2, \ldots, y_i, h(y), y_{i+1}, \ldots, y_n)$, $\alpha_{i+n+1}^{-1}(y) = (y_1, y_2, \ldots, y_i, -h(y), y_{i+1}, \ldots, y_n)$, $i = 1, 2, \ldots, n+1$, where $y = (y_1, y_2, \ldots, y_n)$, $h(y) = (1 - \|y\|^2)^{1/2}$. The system $S = \{(U_j, \alpha_j)\}_{i=1}^{2n+2}$ is a C^∞-atlas on S^n representing a C^∞-structure on S^n, i.e. S^n is smooth manifold and $\dim S^n = n$.

2.5 Example. (Grassman manifolds). The example of a differentiable manifold that will be now introduced will be needed in Section 2.11 (in the proof of Theorem 2.113). Let us denote $G(n,k)(0 \leq k \leq n)$ as the set of all the linear k-dimensional subspaces in R^n. If $L \in G(n,k)$, then let us denote by $U(L)$ the set of all k-dimensional linear subspaces in R^n whose orthogonal projection into L is L. Let e_i, $1 \leq i \leq k$ be an orthonormal basis in L and v_j, $1 \leq j \leq n-k$ be vectors in R^n such that $e_1, e_2, \ldots, e_k, v_1, v_2, \ldots, v_{n-k}$ is an orthonormal basis in R^n. If $L' \in U(L)$ then for every vector e_i ($1 \leq i \leq k$) there exists a unique vector $e_i' \in L'$

such that its orthogonal projection into L is the vector e_i. This vector can be expressed through the orthonormal basis, and thus, we obtain

$$e'_i = e_i + \sum_{s=1}^{n-k} a_{is} v_s, \quad 1 \leq i \leq k, \quad a_{is} \in R. \tag{2.1}$$

Let us define the mapping $\varphi_L : U(L) \to M(k, n-k)$, $\varphi_L(L') = A = (a_{is})$. Obviously, the mapping φ_L is injective. Since for an arbitrary matrix $A = (a_{is}) \in M(k, n-k)$ the vectors e'_i, $1 \leq i \leq k$ defined by the expression (2.1) form a linear k-dimensional subspace belonging to $U(L)$, and thus, φ is also a surjective mapping. Let us define h, a topology on $G(n,k)$, by the requirement that every mapping φ_L was a homeomorphism U_L on $R^{k(n-k)}$. We shall show that the system $S = \{(U(L), \varphi_L)\}_{L \in G(n,k)}$ is a C^∞-atlas on $G(n,k)$. Let $L, K \in G(n,k)$, $L' \in U(L) \cap U(K)$, $\mathscr{B} = \{e_1, e_2, \ldots, e_k, v_1, v_2, \ldots, v_{n-k}\}$ is an orthonormal basis in R^n such that e_i, $1 \leq j \leq k$ is an orthonormal basis in L and $f_1, f_2, \ldots, f_k, w_1, w_2, \ldots, w_{n-k}$ is an orthonormal basis in R^n such that f_j, $1 \leq j \leq k$ is an orthonormal basis in K. Let $e'_j \in L'$ be the vector whose orthogonal projection into L is the vector e_j and $f'_j \in L'$ is the vector whose orthogonal projection into K is the vector f_j $(1 \leq j \leq k)$. Then

$$e'_j = e_j + \sum_{s=1}^{n-k} a_{js} v_s, \quad f'_j = f_j + \sum_{s=1}^{n-k} b_{js} w_s,$$

$1 \leq j \leq k$, i.e. $\varphi_L(L') = A = (a_{js})$, $\varphi_K(L') = B = (b_{mp})$. Expressing the vectors f'_j, f_j, w_s with the help of the basis $\mathscr{B}$ it can be shown that b_{mp} is a linear function of the variables a_{jr}, $1 \leq j \leq k$, $81 \leq r \leq n-k$. This means that the mapping $\varphi_K \circ \varphi_L^{-1} : R^{k(n-k)} \to R^{k(n-k)}$ is of class C^∞, and thus $G(n,k)$ is a smooth manifold, whereby $\dim G(n,k) = \dim M(k, n-k) = k(n-k)$. The manifold $G(n,k)$ is called a Grassman manifold. □

2.6 Proposition. *The Grassman manifold $G(n,k)$ is compact.*

Proof

Let $L, K \in G(n,k)$, $\mathscr{B} = \{e_1, e_2, \ldots, e_k, v_1, v_2, \ldots, v_{n-k}\}$ be an orthonormal basis in R^n such that the vectors $e_1, e_2, \ldots, e_k$ form L and $\mathscr{B}' = \{f_1, f_2, \ldots, f_k, w_1, w_2, \ldots, w_{n-k}\}$ is an orthonormal basis in R^n such that the vectors $f_1, f_2, \ldots, f_k$ form K. Let $C \in M(n)$ be a matrix such that $C\mathscr{B} = \mathscr{B}'$. Then obviously $CL = K$. Since the matrix C maps the orthonormal basis $\mathscr{B}$ to the orthonormal basis $\mathscr{B}'$, it must be orthogonal, i.e. $CC^* = I$. This mean that if $L_0 \in G(n,k)$ is fixed and $\Psi : O(n) \to G(n,k)$, $\Psi(C) = CL_0$, where $O(n) \subset M(n)$ is a set of orthogonal matrices, then $\Psi(O(n)) = G(n,k)$. The set $O(n)$ is a compact manifold (see Example 2.49) and since Ψ is a continuous mapping, from Theorem 1.15 it follows that the Grassman manifold $G(n,k)$ is compact. □

2.7 Proposition. *Let X_i, $i = 1, 2, \ldots, k$ be C^r-manifolds and $\dim X_i = n_i$. Then $X = X_1 \times X_2 \times \cdots \times X_k$ is a C^r-manifold of dimension $n = n_1 + n_2 + \cdots + n_k$.*

Proof

Let the product topology be given on X and (U_{i_s}, α_{i_s}), $i_s \in I_s$ be a C^r-atlas on X_s ($s = 1, 2, \ldots, k$). For every $i = (i_1, i_2, \ldots, i_k) \in I = I_1 \times I_2 \times \cdots \times I_k$ we defined $U_i = U_{i_1} \times U_{i_2} \times \cdots \times U_{i_k}$ and $\alpha_i = (\alpha_{i_1}, \alpha_{i_2}, \ldots, \alpha_{i_k})$. Then $\{U_i, \alpha_i)\}_{i \in I}$ is a C^r-atlas on X representing a C^r-structure S on X, i.e. (X, S) is a C^r-manifold of dimension n. □

2.8 Corollary. *$S^{n_1} \times S^{n_2} \times \cdots \times S^{n_k}$ is a smooth manifold of dimension $n = n_1 + n_2 + \cdots + n_k$.*

2.9 Definition. *$T^n = S^1 \times S^1 \times \cdots \times S^1$ (n-times) is called an n-dimensional torus.*

2.2 C^r-mappings

2.10 Definition. *Let X, Y be C^r-manifolds. The mapping $f : X \to Y$ is called a C^r-mapping ($1 \leq r \leq \infty$) or the differentiable mapping of the class C^r (smooth, if $r = \infty$), if for every $x \in X$ there exist admissible charts (U, α) on X, (V, B) on Y such that $x \in U$, $f(x) \in V$, $f(U) \subset V$ and the mapping $f_{\beta\alpha} = \beta \circ f \circ \alpha^{-1} : \alpha(U) \to \beta(V)$ is a C^r-mapping. The mapping $f_{\beta\alpha}$ is called a local representation of the mapping f (with respect to given admissible charts).*

2.11 Proposition. *Definition of a C^r-mapping is independent of the choice of admissible charts on X, and on Y, respectively.*

Proof

Let $(U', \alpha'), (V', \beta')$ be other admissible charts on X and Y, respectively as in the definition of C^r-mapping, where $x \in U'$, $f(x) \in V'$, $f(U') \subset V'$. It is to be proved that the mapping $f_{\beta'\alpha'} = \beta' \circ f \circ (\alpha')^{-1}$ is of the class C^r. Since $x \in U \cap U'$, it is sufficient to prove that the mapping $f' = f_{\beta'\alpha'}/\alpha'(U \cap U')$ is of the class C^r. But $f' = \delta \circ f_{\beta\alpha} \circ \gamma^{-1}$ where $\delta = \beta' \circ \beta^{-1}/\beta(V \cap V')$, $\gamma = \alpha \circ (\alpha')^{-1}/\alpha(U \cap U')$ and thus f' is of the class C^r. □

2.12 Definition. *Let X, Y be C^r-manifolds ($1 \leq r \leq \infty$). The mapping $f : X \to Y$ is called a local C^r-diffeomorphism at the point $x \in X$ if there*

Foundations of the Theory of Differentiable Manifolds and Differentiable Mappings

exists an open neighbourhood U of the point x and an open neighbourhood V of the point $y = f(x)$ such that $g = f/U : U \to V$ is a bijective mapping and both g and its inverse mapping g^{-1} are of the class C^r. The mapping f is called a local C^r-diffeomorphism if it is a local C^r-diffeomorphism at every point $x \in X$. The mapping f is called a C^r-diffeomorphism (global) iff f is a bijective mapping and both f and f^{-1} are C^r-mappings. The set of C^r-diffeomorphism from X onto Y is denoted as $\text{Diff}^r(X, Y)$ and $\text{Diff}^r(X) \stackrel{\text{def}}{=} \text{Diff}^r(X, X)$.

2.13 Example. Let $X = \{x \in R^n : \|x\|^2 < 1\}$, $Y = R^n$, $f : X \to Y$, $f(x) = (1 - \|x\|^2)^{-1/2}x$. Obviously, the mapping f is bijective, $f^{-1}(x) = (1 + \|x\|^2)^{1/2}x$, both f and f^{-1} are of the class C^∞, and thus, f is a C^∞-diffeomorphism.

2.14 Example. The mapping $f : R \to R$, $f(x) = x^3$ is of the class C^∞, bijective, but $f^{-1}(x) = x^{1/3}$ is not differentiable at the point $x = 0$. This means that f is not a local C^1-diffeomorphism at 0, though it maps R homeomorphically onto R and is of the class C^∞.

2.3 Tangent Space to a C^r-manifold

2.15 Definition. Let X be a C^r-manifold, $a, b \in R, a < b$. The mapping $\gamma : (a, b) \to X$ of class C^r is called a C^r-curve or a curve of the class C^r in X.

In case of a C^r-curve $\gamma : (a, b) \to R^n$, $r \geq 1$, a vector tangent to γ at the point $x = \gamma(s)$, $s \in (a, b)$ can be defined simply as an oriented segment with origin at the point x and the end at the point $x + v$, where $v = d_s\gamma(1)$. Similarly a tangent space to the surface of the class C^r in R^n can be defined at the point x lying at the plane tangent to the given surface at the point x. If X is not a subset of the vector space, but in general just a C^r-manifold, then the concept of a segment loses its sense. In spite of it, however, the concept of a "tangent space" to a C^r-manifold X at the point $x \in X$ can be defined so that it has the structure of a vector space over R and in the case that $X \subset R^n$ this space is isomorphic with the vector space of oriented segments having their origin at the point x lying in the plane tangent to X at the point x.

2.16 Definition. Let X be a C^r-manifold, $r \leq 1, x \in X$ and $C_x^r(X) = \{\gamma : (-\epsilon, \epsilon) \to X : \gamma$ is of the class C^r, $\gamma(0) = x$, $\varepsilon > 0\}$. We say that the curves $\gamma_1, \gamma_2 \in C_x^r(X)$ are equivalent $(\gamma_1 \sim \gamma_2)$ if there exists such an admissible chart (U, α) on X that $x \in U$ and $d_0(\alpha \circ \gamma_1)(1) = d_0(\alpha \circ \gamma_2)(1)$.

The definition of equivalence of curves does not depend upon the choice of an admissible chart. In fact, if (V, β) is a different admissible chart on X such that $x \in V$, then $d_0(\beta \circ \gamma_1)(1) = d_{\alpha(x)}(\beta \circ \alpha^{-1}) d_0(\alpha \circ \gamma_1)(1) = d_{\alpha(x)}(\beta \circ \alpha^{-1}) \circ d_0(\alpha \circ \gamma_2)(1) = d_0(\beta \circ \gamma_2)(1)$ (we have used the Leibnitz formula).

The relation $\gamma_1 \sim \gamma_2$ is obviously an equivalence relation on $C_x^r(X)$. The class $[\gamma]_x = \{\delta \in C_x^r(X) : \delta \sim \gamma\}$ is called a tangent vector to the manifold X at the point x. The set of all tangent vectors to X at the point x is called a tangent space to X at the point x, it is denoted as $T_x(X)$.

2.17 Definition. *Let $a_1 = [\gamma_1]_x$, $a_2 = [\gamma_2]_x \in T_x(X)$, $k_1, k_2 \in R$, (U, α) be an admissible chart on X such that $x \in U$. Then we define $k_1 a_1 + k_2 a_2 \stackrel{\text{def}}{=} a = [\gamma]_x$, where $\gamma : (-\varepsilon, \varepsilon) \to X, \gamma(t) = \alpha^{-1}(k_1 \alpha \circ \gamma_1(t) + k_2 \alpha \circ \gamma_2(t) + (1 - k_1 - k_2)\alpha(x))$ and ε is a small positive number such that $\gamma_1(t), \gamma_2(t), \gamma(t) \in U$ for all $t \in (-\varepsilon, \varepsilon)$.*

Using the Leibnitz formula the independence of the preceding definition of the choice of a chart can be proved (the proof is left to the reader).

2.18 Proposition. *If X is a C^r-manifold, $r \geq 1$, $\dim X = n$, $x \in X$ then the tangent space $T_x(X)$ is a vector space over R and it is isomorphic with R^n.*

Proof

The set $T_x(X)$ with the operations defined in the Definition 2.17 is a vector space over R. We shall prove that this space is isomorphic with R^n. Let (U, α) be an admissible chart on X, such that $x \in U$. We define the mapping $h_\alpha^x : T_x(X) \to R^n$, $h_\alpha^x([\gamma]_x) = d_0(\alpha \circ \gamma)(1)$. This mapping is obviously linear. If $h^x([\gamma_1]_x) = h^x([\gamma_2]_x)$, then $d_0(\alpha \circ \gamma_1)(1) = d_0(\alpha \circ \gamma_2)(1)$, i.e. $\gamma_1 \sim \gamma_2$, and thus, $[\gamma_1]_x = [\gamma_2]_x$. This means that h_α^x is an injective mapping. We define $\gamma : (-\varepsilon, \varepsilon) \to X$, $\gamma(t) = \alpha^{-1}(\alpha(x) + tv)$, $v \in R^n$. Since $\gamma(0) = x$, for $\varepsilon > 0$ sufficiently small it is $\gamma(t) \in U$ for all $t \in (-\varepsilon, \varepsilon)$, i.e. the curve γ is well defined and obviously, $d_0(\alpha \circ \gamma)(1) = v$, and thus, h_α^x is a surjective mapping. $\square$

2.19 Definition. *Let X, Y be C^r-manifolds, $r \geq 1$, $f : X \to Y$ is a C^r-mapping, $x \in X$ and $y = f(x)$. The mapping $D_x f : T_x(X) \to T_y(Y)$, $D_x f([\gamma]_x) = [f \circ \gamma]_y$ is called a derivative of the mapping f at the point x.*

2.20 Proposition. *The mapping $D_x f : T_x(X) \to T_y(Y)(y = f(x))$ is linear.*

Proof

According to Definition 2.17 $k_1[\gamma_1]_x + k_2[\gamma_2]_x = [\gamma]_x$ where $\gamma(t) = \alpha^{-1}(k_1\alpha \circ \gamma_1(t) + k_2\alpha \circ \gamma_2(t) + (1 - k_1 - k_2)\alpha(x))$, (U, α) is an admissible chart on X, $x \in U$ and according to Definition 2.19 it is $D_x f([\gamma]_x) = [f \circ \gamma]_y$, $k_1 D_x f([\gamma_1]_x) + k_2 D_x f([\gamma_2]_x) = k_1[f \circ \gamma_1]_y + k_2[f \circ \gamma_2]_y = [\delta]_y$, where $\delta(t) = \beta^{-1}(k_1\beta \circ f \circ \gamma_1(t) + k_2\beta \circ f \circ \gamma_2(t) + (1 - k_1 - k_2)\beta(y))$ where (V, β) is an admissible chart on Y, $y \in V$, $f(U) \subset V$. It is to be shown that $f \circ \gamma \sim \delta$. When using the Leibnitz formula we obtain $d_0(\beta \circ \delta)(1) = k_1 d_0(\beta \circ f \circ \gamma_1)(1) + k_2 d_0(\beta \circ f \circ \gamma_2)(1)$, $d_0(\beta \circ f \circ \gamma)(1) = d_{\alpha(x)}(\beta \circ f \circ \alpha^{-1})(k_1 d_0(\alpha \circ \gamma_1)(1) + k_2 d_0(\alpha \circ \gamma_2)(1) = k_1 d_{\alpha(x)}(\beta \circ f \circ \alpha^{-1}) \circ d_0(\alpha \circ \gamma_1)(1) + k_2 d_{\alpha(x)}(\beta \circ f \circ \alpha^{-1}) \circ d_0(\alpha \cdot \gamma_2)(1) = k_1 d_0(\beta \circ f \circ \gamma_1)(1) + k_2 d_0(\beta \circ f \circ \alpha_2)(1) = d_0(\beta \circ \delta)(1)$, i.e. $f \circ \gamma \sim \delta$. □

2.21 Proposition. *Let X be a C^r-manifold, $r \geq 1$, $\dim X = n$, $\mathscr{B} = \{v_i\}_{i=1}^n$ be a basis in R^n, $x \in X$ and (U, α) be such an admissible chart on X that $x \in U$. Then $(\partial/\partial\alpha_i)(x) \stackrel{\text{def}}{=} (h_\alpha^x)^{-1}(v_i)$, $i = 1, 2, \ldots, n$ (α_i is the i-th coordinate of the mapping α with respect to the basis $\mathscr{B}$) is the basis of the tangent space $T_x(X)$ where $h_\alpha^x : T_x(X) \to R^n$, $h_\alpha^x([\gamma]_x) = d_0(\alpha \circ \gamma)(1)$, i.e. every tangent vector $[\gamma]_x \in T_x(X)$ can be written in the form*

$$[\gamma]_x = \sum_{i=1}^n c_i (\partial/\partial\alpha_i)(x), \text{ where } c_i = \frac{d\tilde{\gamma}_i(0)}{dt}, \qquad (2.2)$$

$i = 1, 2, \ldots, n$ whereby $(\alpha \circ \gamma)(t) = (\tilde{\gamma}_1(t), \tilde{\gamma}_2(t), \ldots, \tilde{\gamma}_n(t))$.

Proof

The assertion of the proposition follows from the fact that the mapping h_α^x is an isomorphism (the proof of Proposition 2.18).

2.22 Definition. *If $[\gamma]_x \in T_x(X)$ is of the form (2.2) in a chart (U, α) then the vector $c = (c_1, c_2, \ldots, c_n) \in R^n$ is called the coordinate vector of the tangent vector $[\gamma]_x$ or the coordinates of the vector $[\gamma]_x$ in the chart (U, α).* □

2.23 Proposition. *Let (U, α), (V, β) be admissible charts on X, $x \in U \cap V$ and $c = (c_1, c_2, \ldots, c_n)$, $d = (d_1, d_2, \ldots, d_n)$ be coordinates of the tangent vector $\gamma_x \in T_x(X)$ in the chart (U, α) and (V, β), respectively. Then*

$$d^* = J_{\beta\alpha}(y) c^*, \qquad (2.3)$$

where $J_{\beta\alpha}(y) = [d_y(\beta \circ \alpha^{-1})]$, $y = \alpha(x)$.

Proof

Let $v_1, v_2, \ldots, v_n$ be a basis in R^n, $(\partial/\partial\alpha_i)(x) = (h_\alpha^x)^{-1}(v_i)$ and $(\partial/\partial\beta_i)(x) = (h_\beta^x)^{-1}(v_i)$, $i = 1, 2, \ldots, n$. Then

$$\sum_{i=1}^n d_i v_i = h_\beta^x([\gamma]_x) = d_0(\beta \circ \gamma)(1) = d_0(\beta \circ \alpha^{-1} \circ \alpha \circ \gamma)(1)$$

$$= d_{\alpha(x)}(\beta \circ \alpha^{-1}) \circ h_\alpha^x([\gamma]_x) = d_{\alpha(x)}(\beta \circ \alpha^{-1})\left(\sum_{i=1}^n c_i v_i\right),$$

and thus, $d^* = [d_{\alpha(x)}(\beta \circ \alpha^{-1})]c^*$. □

2.24 Proposition. *Let X, Y be C^r-manifolds, $r \geq 1$, $\dim X = n$, $\dim Y = m$, $f : X \to Y$ be a C^r-mapping, $x \in X$, $y = f(x)$, (U, α) be an admissible chart on X, $x \in U$, (V, β) be an admissible chart on Y, $y \in V$, $f(U) \subset V$. If $c = (c_1, c_2, \ldots, c_n) \in R^n$ are coordinates of the vector $[\gamma]_x \in T_x(X)$ and $d = (d_1, d_2, \ldots, d_m) \in R^m$ are coordinates of the vector $D_x f([\gamma]_x) \in T_y(Y)$, then*

$$d^* = [d_z f_{\beta\alpha}]c^*, \tag{2.4}$$

where $z = \alpha(x)$, $f_{\beta\alpha} = \beta \circ f \circ \alpha^{-1}$ is a local representation of the mapping f.

Proof

Let $v_1, v_2, \ldots, v_n$ be a basis in R^n, $w_1, w_2, \ldots, w_m$ be a basis in R^m, $(\partial/\partial\alpha_i)(x) = (h_\alpha^x)^{-1}(v_i)$, $i = 1, 2, \ldots, n$ $(\partial/\partial\beta_j)(y) = (h_\beta^y)^{-1}(w_j)$, $j = 1, 2, \ldots, m$. Since

$$[\gamma]_x = \sum_{i=1}^n c_i(\partial/\partial\alpha_i)(x), \; Df([\gamma]_x) = [f \circ \gamma]_y = \sum_{i=1}^m d_i(\partial/\partial\beta_i)(y),$$

then

$$\sum_{i=1}^m d_j w_j = h_\beta^y([f \circ \gamma]_y) = d_0(\beta \circ f \circ \gamma)(1) = d_0(\beta \circ f \circ \alpha^{-1} \circ \alpha \circ \gamma)(1)$$

$$= d_z f_{\beta\alpha} \circ d_0(\alpha \circ \gamma)(1) = d_z f_{\beta\alpha}\left(\sum_{i=1}^n c_i v_i\right).$$

Therefore $d^* = [d_z f_{\beta\alpha}]c^*$. □

2.25 Definition. *If X is a C^r-manifold, $r \geq 1$ then the set $T(X) = \cup_{x \in X} T_x(X)$ is called a tangent bundle to the manifold X. The mapping $\tau_X : T(X) \to X$, $\tau_X([\gamma]_x) = x$ is called a natural projection.*

2.26 Proposition. *If X is a C^r-manifold, $r \geq 2$ and $\dim X = n$, then $T(X)$ is a C^{r-1}-manifold of dimension $2n$.*

Proof

Topology on $T(X)$: Let $\{(U_i, \alpha_i)\}_{i \in I}$ be a C^r-atlas on X. For every $i \in I$ we define the mapping $T_i : W_i \to \tilde{U}_i \times R^n$, $T_i([\gamma]_x) = (\alpha_i(x), d_0(\alpha_i \circ \gamma)(1))$, where $W_i = \tau_X^{-1}(U_i)$, $\tilde{U}_i = \alpha_i(U_i)$, $\tau_X : T(X) \to X$ is the natural projection. The mapping T_i is obviously bijective. First let us define a topology on W_i. We say that the set $A \subset W_i$ is open in W_i if $A = T_i^{-1}(V)$, where V is an open set in $\tilde{U}_i \times R^n$. The system of all such sets forms a topology S_i on W_i with respect to which the mapping T_i is a homeomorphism. The system $\{S_i\}_{i \in I}$ forms a basis of a topology S on $T(X)$ with respect to which all the mappings T_i, $i \in I$ are homeomorphisms. It can be shown that the topology S is independent of the choice of the C^r-atlas on X (proof is left to the reader). Since X satisfies the second axiom of countability, from the definition of the topology S it can easily be seen that it is satisfied also by the topological space $(T(X), S)$. We shall show that this space is a Hausdorff space. Let $[\gamma_1]_x, [\gamma_2]_y \in T(X)$, where $[\gamma_1]_x \neq [\gamma_2]_y$. If $x \neq y$ then from the Hausdorff property of X the existence of disjoint neighbourhoods of the points $[\gamma_1]_x$ and $[\gamma_2]_y$ follows. If $x = y \in U_i$, for some $i \in I$ and $T_i([\gamma_1]_x) = (u, v_1)$, $T_i([\gamma_2]_x) = (u, v_2)$, then $v_1 \neq v_2$. Obviously, there exist disjoint neighbourhoods V_1, V_2 of the points v_1 or v_2 in R^n. Since the mapping T_i is a homeomorphism, $T_i^{-1}(\tilde{U}_i \times V_1)$, $T_i^{-1}(\tilde{U}_i \times V_2)$ are disjoint neighbourhoods of the points $[\gamma_1]_x$ and $[\gamma_2]_x$, respectively and thus, the space $(T(X), S)$ is a Hausdorff space. Let us define a C^{r-1}-structure on $T(X)$ in the following way: Let $\{(U_i, \alpha_i)\}_{i \in I}$ be a C^r-atlas on X and $\{T_i\}_{i \in I}$ be a system of mappings defined in the preceding part of the proof. We shall prove that $\{(\tau_X^{-1}(U_i), T_i)\}_{i \in I}$ is a C^{r-1}-atlas on $T(X)$. Obviously, $T(X) = \cup_{i \in I} \tau_X^{-1}(U_i)$. If $\tau_X^{-1}(U_i) \cap \tau_X^{-1}(U_j) \neq \emptyset$, then $U_i \cap U_j \neq \emptyset$ and $T_i \circ T_j^{-1} : \alpha_j(U_i \cap U_j) \times R^n \to \alpha_i(U_i \cap U_j) \times R^n$, $(T_i) \circ (T_j)^{-1}(y, v) = (\alpha_i \circ \alpha_j^{-1}(y), d_y(\alpha_i \circ \alpha_j^{-1})(v))$ is a C^{r-1}-diffeomorphism. $\square$

2.27 Definition. *If X is a C^r-manifold, $r \geq k+1$, then we define $T^k(X) \stackrel{\text{def}}{=} T(T^{k-1}(X))$ $(k = 1, 2, \ldots)$, where $T^0(X) \stackrel{\text{def}}{=} X$.*

2.28 Corollary. *If X is an n-dimensional C^r-manifold, $r \geq k+1$, then $T^k(X)$ is a C^{r-k}-manifold of dimension $2^k n$.*

2.29 Definition. *Let X, Y be C^r-manifold, $r \geq 2$ and $f : X \to Y$ be a C^r-mapping. The mapping $Df : T(X) \to T(Y)$, $Df([\gamma]_x) = D_x f([\gamma]_x)$ is called the derivative of the mapping f. If $r \geq k+1$, then the mapping $D^k f : T^k(X) \to T^k(Y)$, $D^k f = D(D^{k-1} f)$ $(k = 1, 2, \ldots)$ where $D_0 f \stackrel{\text{def}}{=} f$ is called the k-th derivative of the mapping f.*

2.30 Proposition. *Let X, Y be C^r-manifolds, $r \geq 1$, $f : X \to Y$ is a C^r-mapping, $x \in X$, $y = f(x)$, (U, α) is an admissible chart on X, $x \in U$ and (V, β) is an admissible chart on Y, $y \in V$, $f(U) \subset V$. Then the local representation of the mapping $Df : T(X) \to T(Y)$ with respect to the charts $(\tau_X^{-1}(U), T_\alpha), (\tau_Y^{-1}(V), T_\beta)$ is of the form $(Df)_{\beta\alpha} \stackrel{\text{def}}{=} T_\beta \circ Df \circ T_\alpha^{-1} : \alpha(U) \times R^n \to \beta(V) \times R^n$, $(Df_{\beta\alpha})((z,e)) = (f_{\beta\alpha}(z), d_z f_{\beta\alpha}(e))$, where $f_{\beta\alpha} = \beta \circ f \circ \alpha^{-1}$.*

Proof

Since $z = \alpha(u)$ for some $u \in U$ then $T_\alpha^{-1}(z,e) = [\gamma]_u \in T_u(X)$, where $d_0(\alpha \circ \gamma)(1) = e$ and $D_u f([\gamma]_u) = D_u f([\gamma]_u) = [f \circ \gamma]_v$, where $v = f(u)$. Therefore $T_\beta \circ Df([\gamma]_u) = (\beta \circ f(u), d_0(\beta \circ f \circ \gamma)(1)) = (\beta \circ f \circ \alpha^{-1}(z), d_0(\beta \circ f \circ \alpha^{-1} \circ \alpha \circ \gamma)(1)) = (f_{\beta\alpha}(z), d_z f_{\beta\alpha}(e)) = T_\beta \circ Df \circ T_\alpha^{-1}(z,e)$. □

2.31 Theorem. *Let X, Y, Z be C^r-manifolds, $r \geq 1$, $U \subset X$, $V \subset U$ be open sets, $g \in C^r(U, V)$ and $f \in C^r(V, Z)$. Then $f \circ g \in C^r$ and for all $x \in U$ it is $D_x(f \circ g) = D_{g(x)}f \circ D_x g$.*

Proof

If $[\gamma]_x \in T_x(X)$, then $D_x g([\gamma]_x) = [g \circ \gamma]_{g(x)}$, $D_x(f \circ g)([\gamma]_x) = [f \circ g \circ \gamma]_{f \circ g(x)}$ and $D_{g(x)}f \circ d_x g([\gamma]_x) = [f \circ g \circ \gamma]_{f \circ g(x)}$. □

2.32 Corollary. *If the assumptions of Theorem 2.31 are fulfilled, then $D(f \circ g) = Df \circ Dg$.*

Proof

$D(f \circ g)([\gamma]_x) = D_x(f \circ g)([\gamma]_x) = D_{g(x)}f \circ D_x g([\gamma]_x)$ and $Df \circ Dg([\gamma]_x) = Df \circ D_x g([\gamma]_x) = Df([g \circ \gamma]_{g(x)}) = D_{g(x)}f \circ D_x g([\gamma]_x)$. □

2.4 C^r-submanifolds

2.33 Definition. *Let X, Y be C^r-manifolds, $r \geq 1$. The mapping $f : X \to Y$ of the class C^r is called a C^r-immersion (C^r-submersion) at the point $x \in X$ if the mapping $D_x f : T_x(X) \to T_y(Y)$ where $y = f(x)$ is injective (surjective). The mapping f is called a C^r-immersion (C^r-submersion) if it is C^r-immersion (C^r-submersion) at every point $x \in X$. The mapping f is called a C^r-embedding if it is a C^r-immersion and it homeomorphically maps X onto $f(X)$. The mapping f is called proper if $f^{-1}(K)$ is a compact set in X for an arbitrary compact set K in Y.*

2.34 Example. Let $X = (-\pi, \pi)$, $Y = R^2$ and $f : X \to Y$, $f(t) = (\sin t, (\sin t) \cos t)$. If $x = \sin t, y = \sin t \cdot \cos t$ then $x \cos t = y$ and that is

Foundations of the Theory of Differentiable Manifolds and Differentiable Mappings

why $y^2 + x^4 = x^2$, i.e. $y^2 = x^2(1-x^2)$. The mapping f is a C^∞-immersion but, it does not map X homeomorphically onto $f(X)$, and thus, it is not an embedding mapping (Fig. 2).

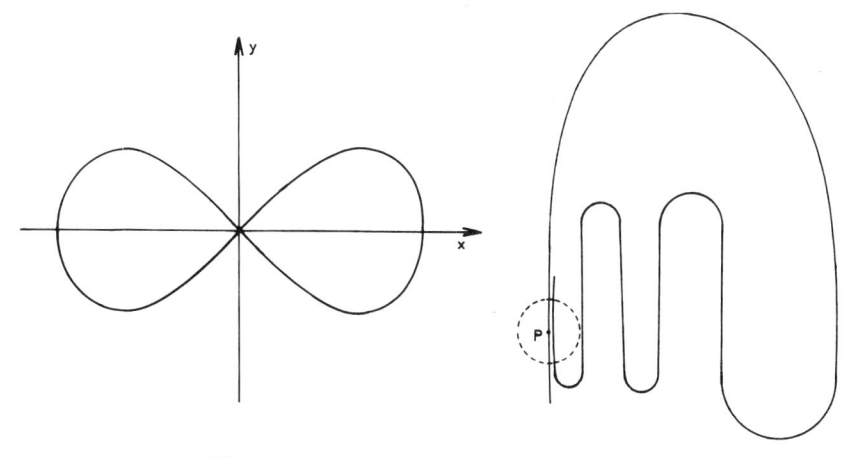

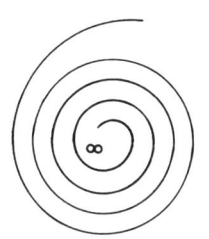

Fig. 2.   Fig. 3.

Fig. 4.

2.35 Example. Let us consider a C^1-mapping from R to R^2 that maps a set R to the curves shown in Figs. 3, 4. In case of the curve shown in Fig. 3 an immersion is concerned that does not homeomorphically map R onto its image (see the neighbourhood of the point P) though this mapping is bijective. An embedding that is not proper is shown in Fig. 4.

2.36 Definition. *Let X be a C^r-manifold, $1 \leq r \leq \infty$ and $\dim X = n$. A subset $Z \subset X$ is called a C^r-submanifold in X (smooth if $r = \infty$) if there exists $k \in N$, such that for every $x \in Z$ there exists such an admissible chart (U, α) on X that $\alpha(X \cap Z) = (R^k \times \{0\}) \cap V = \{x = (x_1, x_2, \ldots, x_n) \in V : x_{k+1} = \cdots = X_n = 0\}$, where $V \subset R^n$ is an open set. The number k is* called a dimension of a submanifold Z ($k = \dim Z$) and the number $n - k$ is called a codimension of the submanifold Z ($n - k = \operatorname{codim} Z$).

2.37 Example. If $X = R^n = R^m \times R^k$, then $Z = \{x = (x_1, x_2, \ldots, x_{m+k}) \in R^{m+k} : x_{m+1} = x_{m+2} = \cdots = x_{m+k} = 0\}$ is a smooth submanifold in X of codimension k. The mapping $f : R^n \to R^k$, $f(x_1, x_2, \ldots, x_n) = (x_{m+1}, x_{m+2}, \ldots, x_{m+k})$ is a C^∞-submersion and $Z = f^{-1}(0)$. (Later on we shall show that if $g : X \to Y$ is a C^r-submersion and $y \in Y$, then $g^{-1}(y)$ is a C^r-submanifold in X; see Theorem 2.48).

2.38 Proposition. *If X is a C^r-manifold ($r \geq 1$) of dimension n and Z is C^r-submanifold in X of dimension k, then Z is a C^r-manifold of dimension k.*

Proof

Let a relative topology be given on Z induced by a topology on X and let $\{(U_i, \alpha_i)\}_{i \in I}$ be the maximal C^r-atlas on X representing a C^r-structure on X. Then there exists a set $J \subset I$ such that for arbitrary $x \in Z$ there exists at least one $j \in J$ such that $x \in U_j$ and $\alpha_j(U_j \cap Z) = (R^k \times \{0\}) \cap V_j$, where $V_j \subset R^n$ is an open set. If $p_1 : R^k \times R^{n-k} \to R^k$ is a projection, then $\{(U_j \cap Z, p_1 \circ \alpha_j)\}_{j \in J}$ is obviously a C^r-atlas on Z representing a C^r-structure on Z, and thus, Z is a C^r-manifold of dimension k. □

2.39 Definition. *The subset Z of the C^r-manifold Y ($r \geq 1$) is called a C^r-immersive submanifold in Y if there exists a C^r-manifold X and an injective C^r-immersion $f : X \to Y$ such that $Z = f(X)$.*

The curve shown in Fig. 3 is an immersive submanifold in R^2. This example also shows that $Z = f(X)$ may not be a submanifold.

2.40 Theorem. *If X is a C^r-manifold ($1 \leq r \leq \infty$) and the mapping $f : X \to R^N$ is a C^r-immersion, injective and proper, then $f(X)$ is a C^r-submanifold in R^N.*

Proof of this theorem can be found, e.g. in [44, 112].

2.41 Theorem. *If X is a compact C^r-manifold, $1 \leq r \leq \infty$, $\dim X = n < \infty$, then there exists a C^r-embedding $f : X \to R^N$ where N is a large enough natural number.*

The above theorem was proved by H. Whitney (for $N = 2n + 1$) and it is referred to as the Whitney Embedding Theorem. The proof of this theorem can be found for example in [44, 51, 65].

From the above two theorems it follows that it is sufficient to consider the compact C^r-manifolds as subsets in R^N. This is advantageous not only because the notions and the results of differential topology become very illustrative but also the technique for the proofs can be simplified to a considerable extent. The next section will be devoted to C^r-manifolds in R^N.

2.5 C^r-manifolds in R^N.

A very important class is formed by differentiable manifolds such that they are given as submanifolds in R^N. They will be called C^r-manifolds in R^N. Very often they are defined by a system of equations dependent on the coordinates on R^N and with the help of the Implicit Function Theorem admissible charts can be defined on them forming their differentiable structure. It is also advantageous to define a smooth manifold in R^N of dimension 0.

2.42 Definition. *The set $X \subset R^N$ is called a smooth manifold of dimension 0 if for every $x \in X$ there exists an open set U in R^N such that $U \cap X = \{x\}$.*

For differentiable manifolds in R^N their tangent spaces can also be defined in a way different from that defined in Section 2.1. The definition of the tangent space from Section 2.1 is geometrically illustrative but sometimes it is not as convenient to use as that introduced in the following text.

2.43 Definition. *Let X be a C^r-manifold in R^N, $r \geq 1$, $\dim X = n$, $x \in X$ and (U, α) is an admissible map on X such that $x \in U$, $\alpha(x) = u$. Let us define a set*

$$\hat{T}_x(X) = \text{Image}\,(d_u \Phi), \qquad (2.5)$$

where $\Phi = \alpha^{-1} : R^n \to R^N$.

2.44 Proposition. *The definition of the set $\hat{T}_x(X)$ is independent of the choice of admissible chart.*

Proof

Let (U_1, α_1), (U_2, α_2) be admissible charts on X, $x \in U_1 \cap U_2$, $\Phi_i = \alpha_i^{-1}$, $i = 1, 2$ and $\Phi_1(u) = \Phi_2(v) = x$. Let W be an open neighbourhood of the point u in R^n ($n = \dim X$) such that $\Phi_2(W) \subset U_1$. Let us define a mapping $F = \Phi_1^{-1} \circ G$ where $G = \Phi_2/W$. Since $\Phi_1 \circ F = G$ then $d_v \Phi_2 = d_v G = d_v(\Phi_1 \circ F) = d_u \Phi_1 \circ d_v F$, i.e. $\text{Image}\,(d_v \Phi_2) \subset \text{Image}\,(d_u \Phi_1)$. If in the present consideration we change Φ_1 for Φ the opposite inclusion can be proved. $\square$

2.45 Definition. *Let X be a C^r-manifold in R^N, $r \leq 1$. Let us define a mapping $G_x : T_x(X) \to \hat{T}_x(X)$, $G_x([\gamma]_x) = d_0\gamma(1)$.*

If $\gamma : (-\varepsilon, \varepsilon) \to X \subset R^N$ is a C^r-curve, (U, α) is an admissible chart on X, $x \in U$, then $d_0\gamma(1) = d_0(\alpha^{-1} \circ \alpha \circ \gamma)(1) = d_u\Phi \circ d_0(\alpha \circ \gamma)(1)$, where $u = \alpha(x)$, $\Phi = \alpha^{-1}$. This means that $d_0\gamma(1) \in \hat{T}_x(X) = \text{Image}\,(d_u\Phi)$, and thus the mapping G_x from the above definition is well defined.

2.46 Proposition. *The mapping $G_x : T_x(X) \to \hat{T}_x(X)$ is an isomorphism.*

Proof

If $G_x([\gamma_1]_x) = G_x([\gamma_2]_x)$ then $d_u\Phi \circ d_0(\alpha \circ \gamma_1)(1) = d_u\Phi \circ d_0(\alpha \circ \gamma_2)(1)$, and thus, $d_0(\alpha \circ \gamma_1)(1) = d_0(\alpha \circ \gamma_2)(1)$, i.e. $[\gamma_1]_x = [\gamma_2]_x$, and G_x is an injective mapping. Let $v \in \text{Image}\,(d_u\Phi) = \hat{T}_x(X)$. Then there exists $w \in R^n$ ($n = \dim X$) such that $v = d_u\Phi(w)$. According to Proposition 2.18 the vector space $T_x(X)$ is isomorphic with R^n, and thus, there exists a C^r-curve γ in X such that $\gamma(0) = x$ and $d_0(\alpha \circ \gamma)(1) = w$. That is why $v = d_u\Phi \circ d_0(\alpha \circ \gamma)(1) = d_0\gamma(1) = G_x([\gamma]_x)$, and thus, G_x is also the surjective mapping. $\square$

2.6 Immersion and Submersion Theorems

2.47 Theorem. *Let X, Y be C^r-manifolds, $r \geq 1$, $\dim X = n$, $\dim Y = m$ and $f : X \to Y$ be a C^r-mapping. Then the following assertions hold.*

1. If f is a C^r-immersion at the point $x \in X$, then there exists an admissible chart (U, α) on X and an admissible chart (V, β) on Y such that $x \in U$, $y = f(x) \in V$, $0 \in U' = \alpha(U)$, $0 \in V' = \beta(V)$, $f(U) \subset V$ and the mapping $g = \beta \circ f \circ \alpha^{-1} : U' \to V$ H is of the form $g(u) = (u, 0) \in R^n \times R^{n-m}$ if $n < m$ and $g(u) = u$, if $n = m$.

2. If f is C^r-submersion at the point $x \in X$ then there exist admissible charts (U, α), (V, β) on X and Y, respectively that have the same properties as said in the assertion 1, where the mapping $g = \beta \circ f \circ \alpha^{-1}$ has the form $g(u_1, u_2, \ldots, u_m, \ldots, u_n) = (u_1, u_2, \ldots, u_m)$.

Proof

Let (U, α), (V, β) be some admissible charts on X and Y, respectively, such that $x \in U$, $y = f(x) \in V$, $\alpha : U \to W$, $\beta : V \to W'$. Without loss of generality it can be assumed that $0 \in W$, $0 \in W'$, $\alpha(x) = 0$, $\beta(y) = 0$. A small neighbourhood U of the point x can be chosen, such that $f(U) \subset V$. The mapping $f_{\beta\alpha} = \beta \circ f \circ \alpha^{-1} : W \to W'$ is a local representation of the mapping f. From Proposition 2.24 it follows that the mapping f is an immersion (submersion) at x iff the mapping $d_0 f_{\beta\alpha} : R^n \to R^m$ is injective (surjective). It means that it is sufficient to prove the theorem for the mapping $f : R^n \to R^m$, $x = 0$ and $y = 0$. At first we shall prove the assertion 1 for $n = m$. Since the mapping $d_0 f$ is injective, the matrix $[d_0 f]$ is regular and from the Inverse Mapping Theorem it follows that there exist neighbourhoods O_1, O_2 of the point 0 in R^n and a C^r-mapping $h : O_1 \to O_2$ such that $h \circ f(u) = u$ for all $u \in O_1$. Thus, assertion 1 for $n = m$ is proved ($\beta = h, \alpha = \text{id}$). Let $n < m$. Since the mapping $d_0 f$ is injective then rank $[d_0 f] = n$. This means that there exists a non-zero minor of order n of the matrix $[d_0 f]$. We can assume that the matrix formed by the first n rows of the matrix $[d_0 f]$ is regular. In fact, if another n-tuple of the rows of this matrix is linearly independent, such as of the k_1-th, k_2-th, ..., k_n-th row then there exists a permutation $\sigma : \{1, 2, \ldots, m\} \to \{1, 2, \ldots, m\}$ such that $\sigma(i) = k_i$, $i = 1, 2, \ldots, n$. The mapping $\varrho : R^m \to R^m$, $\varrho(v_1, v_2, \ldots, v_n, \ldots, v_m) = (v_{\sigma(1)}, v_{\sigma(2)}, \ldots, v_{\sigma(n)}, \ldots, v_{\sigma(m)})$ is a C^∞-diffeomorphism and the mapping $f' = \varrho \circ f$ is such that the matrix formed by the first n rows of the matrix $[d_0 f']$ is regular. Thus, let us suppose that this property still has the matrix $[d_0 f]$ and let $F : U \times R^{m-n} \to R^m$, $F(u, v) = f(u) + (0, v)$. Then

$$[d_{0,0} F] = \begin{bmatrix} [d_0 \tilde{f}] & 0 \\ C & I \end{bmatrix},$$

where $0 = 0_{n, m-n}$, $I = I_{m-n}$ and $C \in M(n, m-n)$. The matrix $[d_{(0,0)} F]$ is obviously regular, and that is why according to the Inverse Mapping Theorem there exists such open neighbourhoods W_1, $W_2 = V_1 \times V_2 \subset R^n \times R^{m-n}$ of the point 0 in R^n and a C^r-mapping $G : W_1 \to R^m$ such that $G \circ F(u, v) = (u, v)$ for all $(u, v) \in W_2$. Therefore $g(u) = G \circ F(u, 0) = G \circ f(u) = (u, 0)$ for all $u \in V_1$. Thus, proposition 1 is proved also for $n < m$ ($\beta = G, \alpha = id$).

For the reasons analogous to those in the proof of proposition 1 without loss of generality we can assume that the matrix formed by the first m columns of the matrix $[d_0, f]$ is regular. Let $F : U_1 \times U_2 \to V \times R^{n-m}$, $F(u, v) = (f(u, v), v)$, where $U_1 \times U_2$ is an open neighourhood of the point 0 in R^n, $U_1 \subset R^m$, $U_2 \subset R^{n-m}$ and V is an open neighourhood of the point 0 in R^m. Then

$$[d_{(0,0)} F] = \begin{bmatrix} [d_0 \varphi] & C \\ 0 & I \end{bmatrix},$$

where $\varphi : U_1 \to R^m$, $\varphi(u) = f(u, 0)$, $0 = 0_{n-m, m}$, $I = I_{n-m}$. Since rank $[d_o \varphi] = [d_o f] = m$, the matrix $[d_{(0,0)} F]$ is regular and thus, according to the Inverse Mapping Theorem there exists an open neighbourhood W of the point 0 in R^n and the C^r-mapping $G : W \to U_1 \times U_2$ such that

$F \circ G(u) = (f \circ G(u), G_{m+1}(u), \ldots, G_n(u)) = u$ for all $u \in W$, where $G = (G_1, G_2, \ldots, G_n)$. From this equation, however, proposition 2. follows immediately ($\beta = \text{id}, \alpha^{-1} = G$). $\square$

2.48 Theorem. *Let X, Y be C^r-manifolds, $r \geq 1$, $\dim X = n$, $\dim Y = m$, $f : X \to Y$ be a C^r-submersion and $y \in Y$. Then $f^{-1}(y)$ is a C^r-submanifold in X and $\text{codim} f^{-1}(y) = \dim Y = m$.*

Proof

If $n = m$ then according to Theorem 2.47 f is the local C^r-diffeomorphism, and thus, $f^{-1}(y)$ is the only point, i.e. $f^{-1}(y)$ is a C^r-submanifold in X of dimension 0. Let $n > m$ and $f(x) = y$. According to Theorem 2.47 there exist admissible charts (U, α), (V, β) on X and Y, respectively, such that $x \in U$, $y \in V\alpha(x) = 0$, $\beta(y) = 0$ and the mapping $g = \beta \circ f \circ \alpha^{-1}$ is of the form $g(u_1, u_2, \ldots, u_n) = (u_1, u_2, \ldots, u_m)$ for all $(u_1, u_2, \ldots, u_n) \in U' = \alpha(U)$. Since $f^{-1}(y) \cap U = \alpha^{-1}(g^{-1}(0) \cap U')$ and $g^{-1}(0) = (\{0\} \times R^{n-m}) \cap U' \subset R^n$, then $f^{-1}(y)$ is the C^r-submanifold in X, where $\text{codim} f^{-1}(y) = \text{codim} g^{-1}(0) = m$. $\square$

2.49 Example. Based on the above theorem it will be proved that the set of orthogonal matrices $O(n) = \{A \in M(n) : AA^* = I_n\}$ is the smooth, compact submanifold in $M(n)$, where $\dim S_n = 1/2n(n+1)$. The mapping $f : M(n) \to S_n$, $f(A) = AA^*$ is smooth and $O(n) = f^{-1}(I_n)$. We shall prove that the mapping f is a submersion on $O(n)$. If $A \in M(n)$, then $d_A f(B) = \lim_{s \to 0}[f(a + sB) - f(A)]s^{-1} = \lim_{s \to 0}[(A + sB)(A + sB)^* - AA^*]s^{-1} = \lim_{s \to 0}(BA^* + AB^* + sBB^*) = BA^* + AB^*$. It is to be shown that for an arbitrary matrix $C \in S_n$ there exists a matrix $B \in M(n)$ such that $BA^* + AB^* = C$. Since the matrix C is symmetrical, $C = (1/2)(C + C^*)$ which means that it is sufficient to solve the equation $BA^* = (1/2)C$. After multiplying this equation from the right by the matrix A we obtain the solution $B = (1/2)CA$. Thus, we have shown that the mapping f is a submersion and that is why according to Theorem 2.48 it holds that $O(n) = f^{-1}(I)$ is a smooth submanifold in $M(n)$, whereby $\dim O(n) = \text{codim} S_n = (1/2)n(n-1)$. From the equation defining the set $O(n)$ it follows that this set is both closed and bounded, and thus, compact.

2.50 Proposition. *Let X be a C^r-manifold, $r \geq 1$, $\dim X = n$, Z be a C^r-submanifold in X, $\text{codim} Z = k$ and $x \in Z$. Then there exists a neighbourhood U of the point x in X and a C^r-submersion $h : U \to R^k$ such that $Z \cap U = h^{-1}(0)$.*

Proof

According to the definition of the C^r-submanifold there exists an admissible chart (U, α) on X such that $x \in U$, $\alpha(x) = (y, 0) \in R^n$, $\alpha(U) = V_1 \times V_2$

where V_1 is an open neighbourhood of the point y in R^{n-k}, V_2 is an open neighbourhood of the point 0 in R^k and $\alpha(U \cap Z) = V_1 \times \{0\}$. If $\pi_2 : V_1 \times V_2 \to V_2$ is a projection, then $h = \pi_2 \circ \alpha : U \to R^k$ is the C^r-submersion and $h^{-1}(0) = Z \cap U$. $\square$

2.7 Regular and Critical Values of Mappings

2.51 Definition. *Let X, Y be the C^r-manifolds, $r \geq 1$. The point $y \in Y$ is called the regular value of the C^r-mapping $f : X \to Y$, if the mapping $D_x f : T_x(X) \to T_y(Y)$ is surjective for every $x \in f^{-1}(y)$. The point $y \in Y$ is called the critical value of the mapping f if it is not the regular value of the mapping f. The point $x \in X$ is called the critical point of the mapping f if there exists a critical value $y \in Y$ of the mapping f such that $x \in f^{-1}(y)$. The set of all regular values (critical points) of the mappings f is denoted as $R(f)(K(f))$.*

2.52 Example. If $f : R \to R$, $f(u) = u(u^2 - 3)$, then the mapping $d_u f : R \to R$ is not surjective (and thus, neither is the mapping $d_u f : T_u(R) \to T_{f(u)}(R)$) for values $u \in R$ for which $d_u f = 0$, i.e. for $u = -1$, $u = +1$. This means that the values $y_1 = f(-1) = 2$, $y_2 = f(1) = -2$ are critical values and $x_1 = -1$, $x_2 = +1$ are critical points of the mapping f. The mapping f has no other critical values and critical points.

2.53 Example. The mapping $f : R \to R$, $f(x) = x^3$ only has the critical value $y = 0$ and the critical point $x = 0$.

2.54 Example. The mapping $f : R^n \to R^m$, $f(x) = c$ for all $x \in R^n$ only has the critical value $y = c$ but the set of the critical points $K(f)$ is represented by the the whole space R^n.

2.55 Example. Let $f : R \to R$ be the C^r-mapping, $r \geq 1$. If y is a regular value of the mapping f then for every $x \in f^{-1}(y)$ it is $d_x f \neq 0$, i.e. the function f is strictly monotonous in a certain neighbourhood U of the point x. That is why $f^{-1}(y) \cap U = \{x\}$, and thus, $f^{-1}(y)$ is a submanifold in R of dimension 0.

From the above examples it can be seen that the set of regular values is "large" compared with the difference of the set of critical values of a mapping.

2.56 Definition. *Let $a = (a_1, a_2, \ldots, a_n)$, $b = (b_1, b_2, \ldots, b_n) \in R^n$, $a_i < b_i$, $i = 1, 2, \ldots, n$. The set $K(a, b) = \{(x_1, x_2, \ldots, x_n) \in R^n : a_i < x_i < b_i, i = 1, 2, \ldots, n\}$ is called the open cube in R^n. The number $V(K(a, b)) = (b_1 - a_1)(b_2 - a_2) \ldots (b_n - a_n)$ is called the volume of the cube $K(a, b)$.*

2.57 Definition. *We say that the set $A \subset R^n$ has the measure 0 (written as $m(A) = 0$), if for arbitrary $\varepsilon > 0$ there exists a sequence $\{K_i\}_{i=1}^{\infty}$ of open cubes in R^n such that $A \subset \cup_{i=1}^{\infty} K_i$ and*

$$\sum_{i=1}^{\infty} V(K_i) < \varepsilon.$$

2.58 Definition. *Let X be a C^r-manifold. We say that the set $A \subset X$ has the measure 0 ($m(A) = 0$) if there exists a sequence $\{(U_i, \alpha_i)\}_{i=1}^{\infty}$ of admissible charts on X such that $m(\alpha_i(U_i \cap A)) = 0$ for every $i \in N$, where $A \subset \cup_{i=1}^{\infty} U_i$.*

2.59 Sard Theorem. *Let X, Y be C^r-manifolds, $r \geq 1$, $\dim X = n$, $\dim Y = m$, $f : X \to Y$ be a C^s-mapping, $s \leqq r$ and $s > \max(0, n - m)$. Then the set of critical values of the mapping f has the measure 0.*

Proof of the Sard Theorem can be found, e.g. in [44, 65, 145]. Let us note that the greater the difference $n - m$ is, the greater differentiability degree of the mapping f is needed so that the assumptions of the Sard Theorem are fulfilled. H. Whitney [160] has constructed an example of a C^1-function $f : R^2 \to R$ for which the set of its critical values has positive measure.

2.60 Definition. *The property P of the elements of the Baire topological space T is called generic in T if the set of the elements of T having the property O is a massive set in T. The classification of the elements of the topological space T according to their generic properties is called generic classification. If this proposition holds for all elements of a certain massive subset in T then we say that this proposition holds generically.*

2.61 Remark. In the literature on dynamical systems very often the concept of a massive set is mixed with that of a residual set. This notion, however is not in agreement with the notion of a residual set accepted in the topology (see, e.g., [148] 2.5). The notion of a massive set is also used in [44, 65].

2.62 Theorem. *If the assumptions of the Sard Theorem are fulfilled, then set $R(f)$ of regular values of the mapping f is massive (and thus, dense) in Y.*

Proof of Theorem 2.62 can be found, for example, in [2]. From the above theorem it follows that if the assumptions of the Sard Theorem are fulfilled, then the regularity property of the values of the mapping f is the generic property in Y.

2.63 Remark. According to Theorem 1.14 the set of the elements of a complete metric space possessing a certain generic property is dense, and thus, "large" enough. For that reason the generic property will play a very important role in the following chapters everywhere where the topological space will be specified on the space of C^r-mappings, the space of matrices, or the space of differential equations, and sim. However, why do we focus our attention on "large" subsets and not whole spaces? In fact, a classification of the whole spaces, mostly contains too many equivalence classes, thus becoming of no use. Both in the classification of differentiable mappings and of the differential equations that will be studied in the next chapters a very substantial role is played by the set of critical points of the mappings. According to the Whitney Theorem that is proved in [24] every closed subset M in R^n represents the set of zero points of a certain differentiable function f. If $n = 1$ then the set M is the set of the critical points of the function

$$\int f(x)\,dx$$

which means that to classify the differentiable functions from the viewpoint of the properties of their critical points represents the task that is no less complicated than that to classify all the closed subsets in R^n. By confining ourselves to "large" subsets in the above sense the chance that at least for these subsets the problem of classification will be simplified is increased. In many cases, as well as in those that will be studied in the following chapters, the generic classification is really simple. However, in general this is not the case. For example (see [24], Theorem 10.6 or [44], Assertion 6.1) the set of the structurally stable differentiable mappings from R^{n^2} to R^{n^2} ($n \geq 3$) is not dense in the space of all such mappings (see definition of structural stability in Chapter 3).

2.8 Topology on the Space of C^r-mappings

Let X be a C^r-manifold, Y be a C^r-manifold in R^M, $r \geq 1$. If the manifold X is compact of dimension n, then there exists a finite number of admissible charts $\{(U_i, \alpha_i)\}_{i=1}^s$ on X such that $\cup_{i=1}^s U_i = X$, where the set $\overline{\alpha_i(U_i)}$ is compact for any i. For $f, g \in C^r(X, Y)$ we define

$$d^k(f, g, U_i) = \max_{x \in \overline{\alpha_i(U_i)}} \|d_x^k(f \circ \alpha_i^{-1}) - d_x^k(g \circ \alpha_i^{-1})\|_k,$$

$k = 0, 1, \ldots, r;\ k < \infty,\ i = 1, 2, \ldots, s$. We define

$$d(f, g, U_i) = \sum_{k=0}^{r} d^k(f, g, U_i),$$

if $r < \infty$ and

$$d(f, g, U_i) = \sum_{k=0}^{\infty} 2^{-k} d^k(f, g, U_i)(1 + d^k(f, g, U_i))^{-1},$$

if $r = \infty$ $(i = 1, 2, \ldots, s)$.

2.64 Definition. *If $f, g \in C^r(X, Y)$, $0 \leq r \leq \infty$ then we define*

$$d(f, g) = \max_{1 \leq i \leq s} d(f, g, U_i).$$

2.65 Proposition. *If X is a compact C^r-manifold, $0 \leq r \leq \infty$ then it holds that*
 1. the function $d : C^r(X, Y) \times C^r(X, Y) \to R$, $(f, g) \mapsto d(f, g)$ is a metric on $C^r(X, Y)$;
 2. the topology on $C^r(X, Y)$ induced by the metric d is independent of the choice of admissible charts on X;
 3. the metric space $(C^r(X, Y), d)$ is complete.

Proof

Since $d(f, g) = d(f, g, U_i)$ is the metric on $C^r(U_i, R^M)$ for every i, $1 \leq i \leq s$ and $\cup_{i=1}^{s} U_i = X$ then d is obviously a metric on $C^r(X, Y)$. Assertion 2 will be proved for $r = 1$ only. For $r > 1$ the proof is analogous, but we will not present it here because of its formal complexity. Let $d(\varrho)$ be a metric on $C^1(X, Y)$ defined as in Definition 2.65 with the help of the system of admissible charts $\{(U_i, \alpha_i)\}_{i=1}^{s}$, $(\{(V_j, \beta_j)\}_{j=1}^{p})$. Let $f_0 \in C^1(X, Y)$ and $U_\varrho(f_0) = \{g \in C^1(X, Y) : \varrho(g, f_0) < \varepsilon\}$, $\varepsilon > 0$. We shall prove that there exists $\delta > 0$ such that $U_d(f_0) = \{g \in C^1(X, Y) : d(g, f_0) < \delta\} \subset U_\varrho(f_0)$. Let us suppose that $\delta > 0$ does not exist. Then, for every $n \in N$ there exist $g_n \in C^1(X, Y)$, $i_n \in \{1, 2, \ldots, s\}$, $j_n \in \{1, 2, \ldots, p\}$, $y_n \in U_{i_n}$, $u_n \in \alpha_{i_n}(U_{i_n})$, $v_n \in \beta_{j_n}(V_{j_n})$ such that $y_n = \alpha_{i_n}^{-1}(u_n) = \beta_{j_n}^{-1}(v_n)$ and it holds that

$$\left\| g_n \circ \alpha_{i_n}^{-1}(u_n) - f_0 \circ \alpha_{i_n}^{-1}(u_n) \right\|_0 < \frac{1}{n}, \tag{M_0}$$

$$\left\| d_{u_n}(g_n \circ \alpha_{i_n}^{-1}) - d_{u_n}(f_0 \circ \alpha_{i_n}^{-1}) \right\|_1 < \frac{1}{n} \tag{M_1}$$

Foundations of the Theory of Differentiable Manifolds and Differentiable Mappings

for all $n \in N$ and also either

$$\left\| g_n \circ \beta_{j_n}^{-1}(v_n) - f_0 \circ \beta_{j_n}^{-1}(v_n) \right\|_0 > \varepsilon \qquad (N_0)$$

or

$$\left\| d_{v_n}(g_n \circ \beta_{j_n}^{-1}) - d_{v_n}(f_0 \circ \beta_{j_n}^{-1}) \right\|_1 > \varepsilon \qquad (N_1)$$

for all $n \in N$. It can be seen immediately, that the inequality (N_0) cannot be satisfied for large enough n since otherwise it would be in contradiction with the inequality (M_0). Since

$$\| d_{v_n}(g_n \circ \beta_{j_n})^{-1} - d_{v_n}(f_0 \circ \beta_{j_n}^{-1}) \|_1$$
$$= \| d_{u_n}(g_n \circ \alpha_{i_n}^{-1}) \circ d_{v_n}(\alpha_{i_n} \circ \beta_{j_n}^{-1}) - d_{u_n}(f_0 \circ \alpha_{i_n}^{-1}) \circ d_{v_n}(\alpha_{i_n} \circ \beta_{j_n}^{-1}) \|_1$$
$$\leqq K \| d_{u_n}(g_n \circ \alpha_{i_n}^{-1}) - d_{u_n}(f_0 \circ \alpha_{i_n}^{-1}) \|_1 < K \frac{1}{n}$$

for all $n \in N$, where

$$K = \sup_{n \in N} \| d_{v_n}(\alpha_{i_n} \circ \beta_{j_n}^{-1}) \|_1$$

and the number of charts is finite, $K < \infty$, and thus, for large enough n, the inequality (N_1) cannot be satisfied. We have proved that every open set in the metric space $M_d = (C^1(X,Y), d)$. If in the above argumentation we replace ϱ by d, it can be proved that every open set in M_d is open also in M_ϱ, i.e. the metric d, ϱ are equivalent to each other. Assertion 3 is a consequence of the completeness of the metric spaces $(C^r(\overline{U}_i, R^M), d_i)$, $i = 1, 2, \ldots, n$, where $d_i = d/C^r(\overline{U}_i, R^M) \times C^r(\overline{U}_i, R^M)$. □

2.66 Remark. According to Theorem 2.41 every C^r-manifold ($r \geqq 1$) of finite dimension can be embedded into R^N for a certain natural number N and thus, in case of compact C^r-manifolds X, Y Proposition 2.65 is general enough. If X, Y are arbitrary, finite-dimensional C^r-manifolds, where X is compact, then there exists a metric d on $C^r(X,Y)$ for which the assertions 1—3 of Proposition 2.65 hold. The metrics like this can be defined analogously to that defined in case of $Y \subset R^N$, however, also the admissible charts on Y must be taken into account. Let us note that the topology on $C^r(X,Y)$ induced by this metrics is called the topology of uniform convergence on $C^r(X,Y)$.

Now we define the so-called weak and strong topology on the space of C^r-mappings. Let X, Y be the C^r-manifolds (in general, non compact ones), $1 \leqq r \leqq \infty$ and $f \in C^r(X,Y)$. Let us denote the set of all 4-tuples of the form $[(U, \alpha), (V, \beta), K, \varepsilon]$ as J, where (U, α), (V, β) are admissible charts on X and Y, respectively, $K \subset U$ there exists a compact set, such that $f(K) \subset V$ and $0 < \varepsilon \leq \infty$. For $\eta = [(U, \alpha), (V, \beta), K, \varepsilon] \in J$ we define the set

$$U^r(f, \eta) = \left\{ g \in C^r(X, Y) : g(K) \subset V, \max_{x \in \alpha(K)} \|d_x^k g_{\alpha\beta} - d_x^k f_{\alpha\beta}\|_k < \varepsilon, \right.$$
$$\left. k = 0, 1, \ldots, r \right\},$$

where $g_{\alpha\beta} = \beta \circ g \circ \alpha^{-1}$, $f_{\alpha\beta} = \beta \circ f \circ \alpha^{-1}$. Let T_f be a system of subsets of the space $C^r(X, Y)$ such that for arbitrary $A \in T_f$ there exists $\eta \in J$ such that $U^r(f, \eta) \subset A$. The system T_f has the properties 1–5 from Theorem 1.6. According to this theorem there exists the only topology W on $C^r(X, Y)$ at which T_f is the set of all neighbourhoods of the mapping f for every $f \in C^r(X, Y)$. This topology is called the *weak topology* on $C^r(X, Y)$. The topological space $(C^r(X, Y), W)$ is denoted as $C_W^r(X, Y)$. It can be verified that if X is a compact manifold, then the topology W coincides with the topology of uniform convergence on $C^r(X, Y)$. The construction of the weak topology W on $C^r(X, Y)$ can be modified in the following way. Let $f \in C^r(X, Y)$. Let us consider the 4-tuple $[\Phi, \Psi, K, \varepsilon]$, where $\Phi = \{(U_i, \alpha_i)\}_{i \in \Lambda}$, Λ being a certain index set $(U_i, \alpha_i)(i \in I)$ being an admissible chart on X, whereby for every $x \in X$ there exists its neighbourhood such that there is a non-empty intersection with a finite number of sets of the system $\{U_i\}_{i \in \Lambda}$, $\Psi = \{(V_i, \beta_i)\}_{i \in x}$, $(V_i, \beta_i)(i \in \Lambda)$ is an admissible chart on Y, $K = \{K_i\}_{i \in \Lambda}$, K_i is such a compact set that $K_i \subset U_i$, $f(K_i) \subset V_i$, $\varepsilon = \{\varepsilon_i\}_{i \in \Lambda}$, $0 < \varepsilon_i \leq \infty$. The set of all such 4-tuples will be denoted as L. For $\eta = [\Phi, \Psi, K, \varepsilon] \in L$ we define the set

$$U^r(f, \eta) = \{ g \in C^r(X, Y) : g(K_i) \subset V_i, \max_{x \in \alpha_i(K_i)} \|d_x^k g_i - d_x^k f_i\|_k < \varepsilon_i$$
$$\text{for all } i \in \Lambda, \ k = 0, 1, \ldots, r \}$$

where $g_i = \beta_i \circ g \circ \alpha_i^{-1}$, $f_i = \beta_i \circ f \circ \alpha_i^{-1}$. Let $\tilde{T}$ be a system of subsets of the space $C^r(X, Y)$ such that for arbitrary $A \in \tilde{T}_f$ there exists $\eta \in L$ such that $U^r(f, \eta) \subset A$. The system $\tilde{T}_f$ possesses the properties 1–5 from Theorem 1.6 and according to the above theorem there only exists the topology S on $C^r(X, Y)$ at which $\tilde{T}_f$ is the set of all neighbourhoods of the mapping f for all $f \in C^r(X, Y)$. This topology is called the strong topology on $C^r(X, Y)$ or the Whitney C^r-topology. The topological space $(C^r(X, Y), S)$ will be denoted as $C_S^r(X, Y)$.

2.67 Theorem. *If X, Y are C^r-manifolds, $1 \leq r \leq \infty$ then the topological space $C_S^r(X, Y)$ is a Baire space.*

Proof of the above theorem can be found for example in [44, 65].

Foundations of the Theory of Differentiable Manifolds and Differentiable Mappings

2.9 Jets

2.68 Definition. *Let X, Y be C^r-manifolds, $1 \leq r \leq \infty$. We say that the mappings $f, g \in C^r(X, Y)$ are germ equivalent at the point $x \in X$ (we write $f \sim_x g$) if there exists such an open neighbourhood U of the point x that $f/U = g/U$. The class of equivalence represented by the mapping f is called the germ of the mapping f at the point x and we denote it as $[f]_x$. When it is clear, where the above germ is defined, we shall write just $\tilde{f}$. The set of all the germs of the mappings from $C^r(X, Y)$ at the point x is denoted as $C_x^r(X, Y)$.*

2.69 Definition. *Let X, Y be C^r-manifolds, $1 \leq r \leq \infty$ and $k \in N$, $k \leq r$. We say that the couple $(x, f) \in X \times C^r(X, Y)$ is j^k-equivalent to the couple $(x', g) \in X \times C^r(X, Y)$, if $x = x'$, $f(x) = g(x)$ and there exist admissible charts (U, α), (V, β) on X and Y, respectively, such that $f(U) \subset Y$, $g(U) \subset V$, $d_z^i f_{\alpha\beta} = d_z^i g_{\alpha\beta}$, $i = 1, 2, \ldots, k$, where $z = \alpha(x)$, $f_{\alpha\beta} = \beta \circ f \circ \alpha^{-1}$, $g_{\alpha\beta} = \beta \circ g \circ \alpha^{-1}$. The class of this equivalence represented by the germ $\tilde{f} = [f]_x$ is called the k-jet of the germ $\tilde{f}$, or the k-jet of the mapping f at the point x denoted as $j^k \tilde{f}(x)$ or $j^k f(x)$. The set of all such k-jets is denoted as J_x^k, i.e. $J_x^k = \{j^k \tilde{f}(x) : \tilde{f} \in C_x^r(X, Y)\}$ and we define $J^k(X, Y) = \cup_{x \in X} J_x^k(X, Y)$. The mapping $j^k f : X \to J^k(X, Y)$, $x \mapsto j^k f(x)$, is called the k-jet extension of the mapping f.*

Let $P_n^k = \{p \in R[X_1, X_2, \ldots, X_n] : \deg(p) \leq k, p(0) = 0\}$, $H_n^k : P_n^k \to R^L$,

$$H_n^k(\sum \alpha_{i_1 \ldots i_n} x_1^{i_1} x_2^{i_2} \ldots x_n^{i_n}) = (a_{10\ldots 0}, \ldots, a_{i_1 i_2 \ldots i_n}, \ldots, a_{00\ldots 01}) \in R^L,$$

where $L = L(n, k)$ is the maximum number of coefficients of the polynomial in n variables of degree k without an absolute term. The above notations will be used to prove the following proposition.

2.70 Proposition. *The set $B_{n,m}^k = P_n^k \times P_n^k \times \cdots \times P_n^k$ (m-times) is a smooth manifold.*

Proof

Let us define a topology on P_n^k. The set $W \subset P_n^k$ is called open in P_n^k if there exists an open set V in R^L ($L = L(n, k)$, such that $W = (H_n^k)^{-1}(V)$. If $\{W_i\}_{i \in I}$ is a system of open sets in P_n^K covering P_n^k, then $\{(W_i, \gamma_i)\}_{i \in I}$, where $\gamma_i = H_n^k/W_i$ is a C^∞-atlas on P_n^k representing a C^∞-structure $\mathscr{S}$ on P_n^k. This means that $(P_n^k, \mathscr{S})$ is a smooth manifold and from Proposition 2.7 it follows that $B_{n,m}^k$ is also a smooth manifold. $\square$

2.71 Theorem. *Let X, Y be C^r-manifolds, $k \in N$, $1 \leq k \leq r \leq \infty$, $\dim X = n$ and $\dim Y = m$. Then $J^k(X,Y)$ is the C^{r-k}-manifold and $\dim J^k(X,Y) = n + m + \dim(B_{n,m}^k)$.*

Outline of proof (for more details see [44]).

Let T_1, T_2 be topologies on X and Y, respectively. Then the system of the sets $T = \{J^k(U,V)\}_{(U,V) \in T_1 \times T_2}$ is a topology on $J^k(X,Y)$. Let (U,α), (V,β) be admissible charts on X and Y, respectively. Let us define the mapping $\Theta_1 : J^k(U,V) \to J^k(\alpha(U), \beta(V))$, $\Theta_1(j^k f(x)) = j^k(\beta \circ f \circ \alpha^{-1})(\alpha(x))$ and mapping $\Theta_2 : J^k(\alpha(U), \beta(V)) \to \alpha(U) \times \beta(V) \times B_{n,m}^k$, $\Theta_2(j^k g(u)) = (u, g(u)), H_n^k T_k g_1(u), \ldots, H_n^k T_k g_m(u))$, where $g = (g_1, g_2, \ldots, g_m)$, $T_k g_1 : R^n \to P_n^k$, $x \mapsto T_k g_j(x)$ $(j = 1, 2, \ldots, m)$, $T_k g_j(x)$ is a polynomial of degree k that is based on the Taylor expansion of the function g_j at x omitting the terms of the order higher than k and omitting the absolute term $g_j(x)$. The mapping $\Theta = \Theta_2 \circ \Theta_1 : J^k(U,V) \to \alpha(U) \times \beta(V) \times B_{n,m}^k$ is a coordinate mapping on $J^k(U,V)$ induced by the charts (U,α), (V,β), i.e. $(J^k(U,V), \Theta)$ is an admissible chart on $J^k(X,Y)$. The set of the charts generated like this, through all the admissible charts on X, and on Y, respectively, forms the C^{r-k}-atlas on $J^k(X,Y)$ defining the C^{r-k}-structure and obviously $\dim J^k(X,Y) = n + m + \dim(B_{n,m}^k)$. □

2.10 Transversality

2.72 Definition. *Let X, Y be C^r-manifolds, $1 \leq r \leq \infty$, Z be a C^r-submanifold in Y and $f : X \to Y$ be C^s-mapping, $1 \leq s \leq r$. We say that the mapping f transversally intersects Z at the point $x \in X$ (we write $f\overline{\pitchfork}_x Z$) if either $y = f(x) \notin Z$ or if $y \in Z$, then*

$$\text{Image}(D_x f) + T_y(Z) = T_y(Y). \qquad (2.6)$$

We say that the mapping f transversally intersects Z on the set $M \subset X$ (we write $f\overline{\pitchfork}_M Z$; instead of $f\overline{\pitchfork}_x Z$ we write $f\overline{\pitchfork} Z$), if $f\overline{\pitchfork}_x Z$ for all $x \in M$. If X is a C^r-submanifold in Y, $f : X \to Y$, $f(u) = u$ for $u \in X$, $x \in X \cap Z$ and $f\overline{\pitchfork}_x Z$ then we say that the submanifolds X and Z transversally intersect each other at x (writing $X\overline{\pitchfork} Z$) if $X\overline{\pitchfork}_x Z$ for every $x \in X \cap Z$.

If $G_y : T_y(Y) \to \hat{T}_y(Y)$ is the isomorphism from Definition 2.45 then the equality (2.6) is equivalent to the equality

$$\text{Image}(G_y \circ D_x f) + \hat{T}_y(Z) = \hat{T}_y(Y). \qquad (2.7)$$

2.73 Proposition. *Let X be a C^r-manifold in R^M, $1 \leq r \leq \infty$, Z be a C^r-submanifold in Y, $f : X \to Y$ be a C^r-mapping, $x \in X$ and $y = f(x) \in Z$. Let (U, α) be an admissible chart on X, $x \in U$ and (V, β) be an admissible chart on Y, $y \in V$. Then $f \overline{\cap}_x Z$ iff*

$$\text{Image}(d_v \Psi \circ d_u f_{\alpha\beta}) + \hat{T}_y(Z) = \hat{T}_y(Y), \tag{2.8}$$

where $\Psi = \beta^{-1}$, $v = \beta(y)$, $u = \alpha(x)$, $f_{\alpha,\beta} = \beta \circ f \circ \alpha^{-1}$.

Proof

$G_y(D_x f([\gamma]_x)) = G_y([f \circ \gamma]_y) = d_0(f \circ \gamma)(1) = d_0(\beta^{-1} \circ \beta \circ f \circ \alpha^{-1} \circ \alpha \circ \gamma)(1) = d_v \Psi \circ d_u f_{\alpha\beta} \circ d_0(\alpha \circ \gamma)(1)$, and thus, the equality (2.8) follows from the equality (2.8). □

2.74 Example. Let $X = R^n$, $Y = R^{2n}$, $Z = \{(u_1, u_2) \in R^{2n} : u_1 \in R^n, u_2 = 0\}$, $f : X \to Y$, $f(u) = (u, g(u))$, where $g \in C^r(R^n, R^n)$, $1 \leq r \leq \infty$ and $g(x) = 0$, $x \in X$, i.e. $y = f(x) \in Z$. Let U, V be open sets in X and Y, respectively, whereby $x \in U$, $y \in V$. Then (U, α), (V, β), where $\alpha = \text{id}$, $\beta = \text{id}$, are admissible charts on X and Y, respectively, and $f_{\alpha\beta} = \beta \circ f \circ \alpha^{-1} = f$. Therefore $\text{Image}(G_y \circ d_x f) = \text{Image}(d_x f)$. Since $\hat{T}_y(Y) = R^{2n}$, $\hat{T}_y(Z) = Z$ and $d_x f(v) = (v, d_x g(v))$ for $v \in R^n$, the equality (2.8) is fulfilled iff for arbitrary $w_1, w_2 \in R^n$ there exist $v_1, v_2 \in R^n$ such that $(v_1, d_x g(v_1)) + (v_2, 0) = (w_1, w_2)$ and this is possible iff $\text{Image}(d_x g) = R^n$. If for example $n = 1$ and $g(u) = 2u$ then $f \overline{\cap}_0 Z$ (Fig. 5). If, however, $g(u) = u^2$, then $\text{Image}(d_0 g) = 0$, and thus, f does not intersect Z transversally at 0 (Fig. 6).

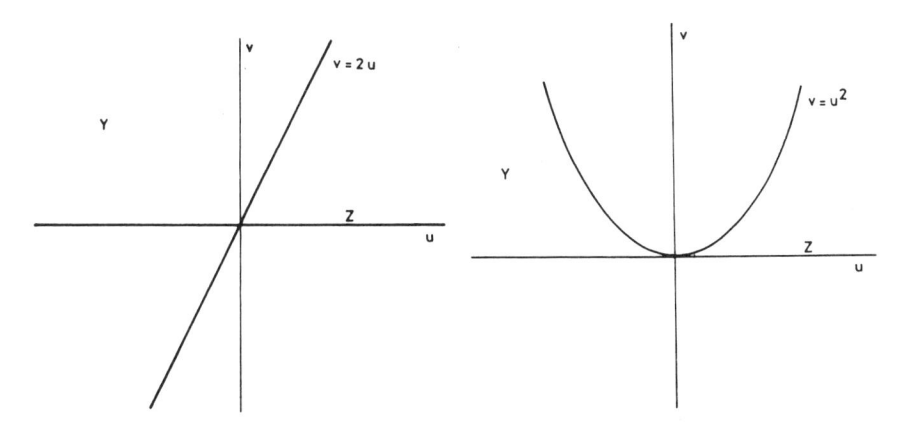

Fig. 5.   Fig. 6.

2.75 Proposition. Let X, Y be C^r-manifolds, $1 \leq r \leq \infty$, Z be a C^r-submanifold in Y and $\dim X + \dim Z < \dim Y$, i.e. $\operatorname{codim} Z > \dim X$. Then if $f \in C^r(X,Y)$ and $f \overline{\pitchfork} Z$ then $f(X) \cap Z = \emptyset$.

Proof

Let $y = f(x) \in Z$. Since $D_x f(T_x(X)) + T_y(Z) = T_y(Y)$, from Proposition 2.18 it follows that $\dim(D_x f(T_x(X)) + T_y Y)) \leq \dim T_x(X) + \dim T_y(Z) = \dim X + \dim Z < \dim Y = \dim T_y(Y)$ and it is a contradiction with the transversality condition which means that $f(x) \notin Z$ for every $x \in X$. □

2.76 Proposition. Let X, Y be C^r-manifolds, $1 \leq r \leq \infty$, Z be a C^r-submanifold in Y, $\operatorname{codim} Z = k$, $x \in X$, $y = f(x) \in Z$. Let V be an open neighbourhood of the point y in Y and $h : V \to R^k$ be such a C^r-submersion that $Z \cap V = h^{-1}(0)$. Then

1. $T_y(Z) = \operatorname{Ker}(D_y h)$;
2. $f \overline{\pitchfork}_x Z$ iff the mapping $h \circ f$ is the submersion at the point x;
3. if $f \overline{\pitchfork} Z$ then $f^{-1}(Z)$ is a C^r-submanifold in X and $\operatorname{codim} f^{-1}(Z) = \operatorname{codim} Z$.

Proof

Since $h(z) = 0$ for all $z \in Z \cap V$ so $D_y h(v) = 0$ for all $v \in T_y(Z)$, i.e., $T_y(Z) \subset \operatorname{Ker}(D_y h)$. According to the above assumptions, the mapping $D_y h$ is surjective, and thus, according to Theorem 2.48 is $\operatorname{codim}(D_y h)^{-1}(0) = k$, i.e., $\operatorname{codim} \operatorname{Ker}(D_y h) = k$. We obtain that $T_y(Z)$ is the vector subspace in $\operatorname{Ker}(D_y h)$, $\dim T_y(Z) = \dim \operatorname{Ker}(D_y h)$, and thus, $T_y(Z) = \operatorname{Ker}(D_y h)$, i.e., assertion 1 holds. The mapping $D_x(h \circ f) = D_y h \circ D_x f : T_x(X) \to T_0(R^k)(h \circ f(x) = 0, y = f(x))$ is surjective iff

$$D_y h(\operatorname{Image} D_x f) = T_0(R^k). \tag{2.9}$$

Since the mapping $D_y h : T_y(Y) \to T_0(R^k)$ according to the assumption is surjective, so the equality (2.9) is fulfilled iff

$$\operatorname{Image}(D_x f) + \operatorname{Ker}(D_y h) = T_y(Y). \tag{2.10}$$

According to assertion 1, however $T_y(Z) = \operatorname{Ker}(D_y h)$, and thus, the equality (2.10) can be written in the form $\operatorname{Image}(D_x f) + T_y(Z) = T_y(Y)$, i.e., $f \overline{\pitchfork}_x Z$, and thus it also holds for assertion 2. Assertion 3 is a consequence of Proposition 2.50 and assertion 2. □

2.77 Example. Let $f = (f_1, f_2, \ldots, f_n) \in C^r(R^m, R^n)$, $1 \leq r \leq \infty$, $m \geq n - k$, $Z = \{u = (u_1, u_2, \ldots, u_k, \ldots, u_n) : u_{k+1} = u_{k+2} = \cdots = u_n = 0\}$. Then, obviously $Z = h^{-1}(0)$, where $h(u_1, u_2, \ldots, u_k, \ldots, u_n) = (u_{k+1}, u_{k+2}, \ldots, u_n)$. According to Proposition 2.76 $f \overline{\pitchfork}_x Z$ iff the mapping $F = h \circ f = (f_{k+1}, f_{k+2}, \ldots, f_n) : R^m \to R^{n-k}$ is the submersion at point x, i.e., if rank $[d_x F] = n - k$.

2.78 Example. Let $f \in C^r(U, R^n)$, $1 \leq r \leq \infty$, $U \subset R^m$ is an open set, $m = n - k$, $0 \leq k < n$, $V \subset R^n$ is an open set, $M = \{y \in V : F_{k+1}(y) = F_{k+2}(y) = \cdots = F_n(y) = 0\}$, where $F = (F_{k+1}, F_{k+2}, \ldots, F_n) \in C^r(V, R^{n-k})$ is a mapping, such that $d_y F : R^n \to R^{n-k}$ is a submersion for every $y \in V$. According to Theorem 2.48 $M = F^{-1}(0)$ is a C^r-submanifold in R^n and codim $M = n - k = m$, i.e., dim $M = k$. Let us define the mapping $\tilde{F} : V \to R^n$, $\tilde{F}(y) = (y_1, y_2, \ldots, y_k, F_{k+1}(y), F_{k+2}(y), \ldots, F_n(y))$. Since for every $y \in V$, $d_y \tilde{F} : R^n \to R^n$ is a linear isomorphism, then from the Inverse Mapping Theorem it follows that F is the local C^r-diffeomorphism. This means that for every $v \in V$ there exists its open neighbourhood $\tilde{V} \subset V$ such that $\tilde{F}/\tilde{V}$ is the C^r-diffeomorphism $\tilde{V}$ on $\tilde{F}(\tilde{V})$, and thus, in this neighbourhood we can introduce new coordinates $w_1 = y_1$, $w_2 = y_2$, $\ldots$, $w_k = y_k$, $w_{k+1} = F_{k+1}(y), \ldots$, $w_n = F_n(y)$. In these new coordinates the set $M \cap \tilde{V}$ can be expressed as follows: $M \cap \tilde{V} = \{(w_1, w_2, \ldots, w_k, w_{k+1}, \ldots, w_n) : w_{k+1} = w_{k+2} = \cdots = w_n = 0\}$. According to the assertion in Example 2.77 $f \overline{\pitchfork}_u M$ iff rank $[d_u(F \circ f)] = n - k = m$.

2.79 Weak Transversality Theorem. *Let X, Y be C^r-manifolds, $1 \leq r \leq \infty$, Z be a C^r-submanifold in Y. Then the following assertions hold:*

1. If $K \subset X$ is a compact set and Z is closed in Y, then the set $T_{K,Z}^r(X,Y) = \{f \in C^r(X,Y) : f \overline{\pitchfork}_K Z\}$ is open in $C_S^r(X,Y)$

2. The set $T_Z^r = \{f \in C^r(X,Y) : f \overline{\pitchfork} Z\}$ (X need not be compact and Z need not be closed) is massive (and thus, dense) in $C_S^r(X,Y)$.

The above theorem will be proved only under the assumption that X is the compact manifold. Proof of the above theorem for arbitrary C^r-manifolds can be found for example in [44] and [65]. To be able to prove this theorem, we shall need several lemmas.

2.80 Lemma. *If $m \geq k$, then the set $F = \{g \in L(R^m, R^k) : g$ is not surjective mapping$\}$ is closed in $L(R^m, R^k)$.*

Proof

Since the space $L(R^m, R^k)$ is isomorphic with the space of the matrices $M(m, k)$ it is sufficient to prove that the set $G = \{A = (a_{ij} \in M(m,k) : \text{rank } A < k\}$ is closed in $M(m,k)$. Let $G(s) = \{A = (a_{ij}) \in M(m,n); \text{rank } A = k-s\}$. Since $\overline{G}(s) = \cup_{s=1}^{k-s} G(s+i)$ the set $G = \cup_{s=1}^{k} G(s)$ is closed. $\square$

2.81 Lemma. *There exists a function $\chi : R^k \to R$ of the class C^∞ such that $0 \leq \chi(x) \leq 1$ for all $x \in R^k$, $\chi(x) = 1$ if $\|x\| \leq 1/2\varrho$ and $\chi(x) = 0$, if $\|x\| \geq \varrho$, where $\varrho > 0$.*

Proof

Let us define

$$\chi : R^k \to R, \; \chi(x) = \Psi(\varrho - \|x\|)\left(\Psi(\varrho - \|x\|) + \Psi\left(\|x\| - \frac{1}{2}\varrho\right)\right)^{-1},$$

where $\Psi : R \to R$, $\Psi(s) = \exp(s^{-2})$, if $s > 0$ and $\Psi(s) = 0$ if $s \leq 0$. The function χ has all the required properties. $\square$

2.82 Lemma. *Let X be a C^r-manifold of dimension k, $1 \leq r \leq \infty$, $U \subset X$ be an open set and $K \subset U$ be an open set with a compact closure, where $\overline{K} \subset U$. Then there exists a function $\varphi : X \to R$ of class C^r such that $0 \leq \varphi(x) \leq 1$, for all $x \in X$, $\varphi(x) = 1$, if $x \in K$ and $\varphi(x) = 0$, if $x \notin U$.*

Proof

Let $\{(U_i, \alpha_i)\}_{i=1}^s$ be a system of admissible charts on X such that $\alpha_i(U_i) = B_{r_i} = \{x \in R^k : \|x\| < r_i\}$, $r_i > 0$, $K \subset \alpha_i^{-1}(B_{r_i/2}) \subset U$. According to Lemma 2.81 there exist functions $\chi_i : R^k \to R$, $i = 1, 2, \ldots, s$ of class C^∞ such that $\chi(x) \geq 0$ for all $x \in R^k$, $\chi_i(x) = 1$, if $x \in R_{r_i/2}$ and $\chi_i(x) = 0$ if $x \notin B(r_i)$. Let us define the function $\varphi_i : X \to R$, $\varphi_i(x) = \chi_i \circ \alpha_i(x)$ if $x \in U_i$ and $\varphi_i(x) = 0$ if $x \in U_i$; $(i = 1, 2, \ldots, s)$. Then the function $\varphi : X \to R$, $\varphi(x) = 1 - (1 - \varphi_1(x))(1 - \varphi_2(x))\ldots(1 - \varphi_s(x))$ has all the required properties. $\square$

2.83 Definition. *Let M be a set, $f : M \to R^n$ and $N(f) = \{x \in M : f(x) \neq 0\}$. Then the closure of the set $N(f)$ is called the support of the mapping f and it is denoted as $\operatorname{supp} f$.*

2.84 Definition. *Let X be a compact C^r-manifold, $1 \leq r \leq \infty$ and $S = \{U_i\}_{i=1}^m$ be its finite open covering. The system $\{\chi_i\}_{i=1}^m$, $\chi_i : X \to R$ $(i = 1, 2, \ldots, m)$ is called the partition of unity on X subordinated to the covering S if it holds:*

1. $\chi_i \in C^r$, $i = 1, 2, \ldots, m$;
2. $\operatorname{supp} \chi_i \subset U_i$, $i = 1, 2, \ldots, m$, $\bigcup_{i=1}^m \operatorname{supp} \chi_i = X$;
3. $\chi_i(x) \geq 0$ for all $x \in X$, $i = 1, 2, \ldots, m$;
4. $\sum_{i=1}^m \chi_i(x) = 1$ for all $x \in X$.

2.85 Lemma. *If X is a compact C^r-manifold, $1 \leq r \leq \infty$ and $S = \{U_i\}_{i=1}^m$ is its finite open covering then there exists the partition of unity on X subordinated to this covering.*

Proof

Obviously, there exist compact sets $K_i \subset X$, $i = 1, 2, \ldots, m$ such that $K_i \subset U_i$ and $\cup_{i=1}^m K_i = X$. According to Lemma 2.82 for every $i = 1, 2, \ldots, m$ there exists a C^r-function $\varphi_i : X \to R$ such that $\varphi_i(x) \geq 0$ for all $x \in X$, $\varphi_i(x) = 1$, if $x \in K_i$ and $\varphi_i(x) = 0$, if $x \notin U_i$. Let us define the function

$$\chi_i : X \to R, \quad \chi_i(x) = \left(\sum_{i=1}^m \varphi_i(x)\right)^{-1} \cdot \varphi_i(x), \tag{2.11}$$

$i = 1, 2, \ldots, m$. Obviously, the system of functions $\{\chi_i\}_{i=1}^m$ has the properties 1–4 from Definition 2.84. □

Proof of Theorem 2.79. (We suppose that X is a compact manifold). At first, we shall prove the assertion 1. Let d be a metric on $C^r(X,Y)$ defined through the system $\{(U_i, \alpha_i)\}_{i=1}^s$ of admissible charts on X and some countable system of admissible charts on Y (see Section 2.8). We shall show that the set $W = C^r(X,Y) \setminus T^r_{K,Z}(X,Y)$ is closed in $C^r(X,Y)$. Let $\{f_n\}_{n=1}^\infty$ be a sequence of mappings from W converging to the mapping $f \in C^r(X,Y)$ with respect to the metric d. We shall prove that $f \in W$. Since $f_n \notin T^r_{K,Z}(X,Y)$ for every $n \in N$ there exist $x_n \in K$, $y_n = f_n(x_n) \in Z$ such that f_n does not transversally intersect Z at the point x_n. From the closeness of the manifold Z and compactness of K it follows that there exist sub-sequences $\{y_{n_k}\}_{k=1}^\infty$, $\{x_{n_k}\}_{k=1}^\infty$ of the sequences $\{y_n\}_{n=1}^\infty$ and $\{x_n\}_{n=1}^\infty$, respectively, such that $y = \lim_{k \to \infty} y_{n_k} \in Z$ and $x = \lim_{k \to \infty} x_{n_k} \in K$, i.e., $f(x) = y \in Z$. It can be shown that the mapping f does not intersect Z transversally, at the point x. According to Proposition 2.50 there exists such an open neighbourhood V of the point y in Y and a submersion $h : V \to R^k$ ($k = \text{codim}\, Z$) such that $V \cap Z = h^{-1}(0)$. There exists a number $N > 0$, such that for all $n > N$ it is $y_n \in V \cap Z$. From Proposition 2.76 it follows that the mapping $h \circ f_n$ is not a submersion at the point x_n. Without loss of generality we can assume that $V \subset f_n(U_i)$ for some i, $1 \leq i \leq s$ and all $n > N$. The mapping $g_n = h \circ f_n \circ \alpha_i^{-1} : \alpha_i(U_i) \to R^k$ is not a submersion at the point $u_n = \alpha_i(x_n)$ for $n > N$, i.e. the mapping $d_{u_n} g_n = d_{y_n} h \circ d_{u_n}(f_n \circ \alpha_i^{-1}) : R^m \to R^k$ ($m = \dim X$) is not surjective. This means that $d_{u_n} g_n \in F$ (see Lemma 2.80). Since according to Lemma 2.8 F is a closed set in $L(R^m, R^k)$ then $d_y h \circ d_u(f \circ \alpha_i^{-1}) = \lim_{n \to \infty} d_{y_n} h \circ d_{u_n}(f \circ \alpha_i^{-1}) \in F$, where $u = \alpha_i(x)$. However, this means that the mapping $d_u(h \circ f \circ \alpha_i^{-1})$ is not surjective, thus, the mapping $h \circ f$ is not a submersion at the point x. This is why, according to Proposition 2.76, the mapping f

does not intersect Z transversally at the point x, and thus, the openness of the set $T^r_{K,Z}(X,Y)$ is proved. Now, we shall prove assertion 2. Let $f \in C^r(X,Y), (V, \beta)$ be an admissible chart on Y and $h : V \to R^k$ is such a C^r-submersion that $V \cap Z = h^{-1}(0)$. Let W, W_1 be such open sets in X that $\overline{W} \subset W_1$ and $f(W_1) \subset V$. According to Lemma 2.81 there exists a function $\varphi \in C^r(X, R)$ such that $\varphi(x) \geq 0$ for all $x \in X$, $\varphi(x) = 1$, if $x \in W$ and $\varphi(x) = 0$, if $x \in W_1$. Obviously, there exists a number $\delta > 0$ such that for $v \in R^n (n = \dim Y)$, $\|v\| < \delta$ is $\beta \circ f(x) + \varphi(x)v \in \beta(V)$ for all $x \in W_1$. Therefore for every such $v \in R^n$ the mapping $g_v \in C^r(W_1, V)$, $g_v(x) = \beta^{-1}(\beta \circ f(x) + \varphi(x)v)$ is defined. Let us define the mapping $g : X \to Y$ as follows: $g(x) = f(x)$, if $x \notin W_1$ and $g(x) = g_v(x)$, if $x \in W_1$ where $\|v\| < \delta$. From the construction of the mapping g it follows that $g \in C^r(X,Y)$. Moreover, for arbitrary $\varepsilon > 0$ there exists a number $\delta > 0$, such that if $\|v\| < \delta$, then $d(g, f) < \varepsilon$. According to Theorem 2.62 the set of regular values of the mapping $h \circ f$ is dense in R^k, and thus, there exists $v \in R^n$ such that $\|v\| < \delta$ and $0 \in R^k$ is a regular value of the mapping $h \circ g$, i.e. $h \circ g$ is a submersion at every point $x \in (g^{-1}(Z \cap V)) \cap W$. From Proposition 2.76 it follows that $g \in T^r_{\overline{W},Z}(X,Y)$. We have proved that the set $T^r_{\overline{W},Z}(X,Y)$ is dense in $C^r_S(X,Y)$. The topological space Y fulfils the second axiom of countability and thus, there exists such a countable system $\{(V_i, \beta_i)\}_{i=1}^{\infty}$ of admissible charts on Y such that it holds:

1. $\cup_{i=1}^{\infty} V_i$ contains the submanifold Z;

2. for every $i \in N$ there exists a C^r-submersion $h_i : V_i \to R^k$ such that $V_i \cap Z = h_i^{-1}(0)$;

3. there exists a system $\{W_i\}_{i=1}^{\infty}$ of open sets in X such that $\cup_{i=1}^{\infty} W_i = X$ and $f(W_i) \subset V_i$ for every $i \in N$.

With the use of the assertion just proved we obtain that for every $i \in N$ the set $T^r_{\overline{W}_i,Z}(X,Y)$ is dense in $C^r_S(X,Y)$. Since $T^r_Z(X,Y) = T^r_{X,Z}(X,Y) = \cap_{i=1}^{\infty} T^r_{\overline{W}_i,Z}(X,Y)$ and each of the sets $T^r_{\overline{W}_i,Z}(X,Y)(i \in N)$ according to assertion 1 is also open, then $T^r_Z(X,Y)$ is a massive set in $C^r_S(X,Y)$. □

2.86 Remark. Assertion 1 of Theorem 2.79 does not hold, if it is not supposed that the set Z is closed. Example: $X = S^1 = \{(x,y) \in R^2 : x^2 + y^2 = 1\}$, $Z = \{(t,0,0) : 0 < t < 1\} \subset Y = R^3$, $f : X \to Y$, $f(x,y) = (x+1, y, 0)$. Since $\dim X + \dim Z < \dim Y$, according to Proposition 2.75 the mapping $g \in C^r(X,Y)$ transversally intersects Z iff $g(X) \cap Z = \emptyset$. Since $f(X) \cap (R \times \{(0,0)\}) = \{(0,0,0),(2,0,0)\}$ then $f(X) \cap Z = \emptyset$, and thus, $f \overline{\pitchfork} Z$. If, however, ε is a number, such that $0 < \varepsilon < 1$ and $f_\varepsilon : X \to Y, f_\varepsilon(x,y) = (x+1+\varepsilon, y, 0)$ then $f_\varepsilon(-1,0) = (\varepsilon, 0, 0) \in Z$, and thus, the mapping f_ε does not intersect Z transversally.

2.87 Definition. Let X, Y, A be C^r-manifolds, $1 \leq r \leq \infty$ and $G : G \times A \to Y$ is a C^r-mapping. The system of mappings $\{G_a\}_{a \in A}$, where

Foundations of the Theory of Differentiable Manifolds and Differentiable Mappings

$G_a : X \to Y$, $G_a(x) = G(x, a)$ is called an A-parametric system of C^r-mappings from $C^r(X, Y)$ (dependent on $a \in A$).

2.88 Parametric Transversality Theorem. *Let X, Y, A be C^r-manifolds, $1 \leq r \leq \infty$, Z is a C^r-submanifold in Y, $\{G_a\}_{a \in A}$ is an A-parametric system of mappings from $C^r(X, Y)$, $\dim X = n$ and $\operatorname{codim} Z = k$. Then the following assertions hold:*

1. *If $K \subset X$ is a compact set and Z is closed in Y then the set $A_{K,Z} = \{a \in A : G_a \overline{\pitchfork}_K Z\}$ is open in A*

2. *If $r > \max(0, n - k)$ and $G \overline{\pitchfork} Z (G : X \times A \to Y, G(x, a) = G_a(x))$ then the set $A_Z = \{a \in A : G_a \overline{\pitchfork} Z\}$ is massive in A.*

Proof

Assertion 1 follows directly from assertion 1 of Theorem 2.79. Since $G \overline{\pitchfork} Z$, according to Proposition 2.76 $W = G^{-1}(Z)$ is a C^r-submanifold in $X \times A$. Let $\pi_1 : X \times A \to A$ is the projection and $\pi = \pi_1/W$. We shall prove that if $a \in A$ is a regular value of the mapping π then $G_a \overline{\pitchfork} Z$. Thus, the theorem will be proved since according to the Sard Theorem (here the assumption is used that the assumption $r > \max(0, n - k)$ of the Sard Theorem is used) is the set of regular values of the mapping π massive in A. Let $a \in A$ be a regular value of the mapping π and $y = G(x, a) \in Z$ for some $x \in X$. Since $G \overline{\pitchfork} Z$, then

$$D_{(x,a)} G(T_{(x,a)}(X \times A)) + T_y(Z) = T_y(Y), \qquad (2.12)$$

i.e. for arbitrary $w \in T_y(Y)$ there exist $u \in T_{(x,a)}(X \times A), v \in T_y(Z)$ such that

$$D_{(x,a)} G(u) + v = w. \qquad (2.13)$$

Since $T_{(x,a)}(X \times A) = T_x(X) \times T_a(A)$, then $u = (u_1, u_2)$ where $u_1 \in T_x(X)$ and $u_2 \in T_a(A)$. According to the assumption a is the regular value of the mapping π and thus, the mapping $D_{(x,a)} \pi : T_{(x,a)}(W) \to T_a(A)$ is surjective. Thus, there exists $b \in T_x(X)$, such that $(b, u_2) \in T_{(x,a)}(W)$ and $D_{(x,a)} \pi(b, u_2) = u_2$. Since $G/W : W \to Z$, then $D_{(x,a)} G(T_{(x,a)}(W)) \subset T_y(Z)$ and this means that

$$D_{(x,a)} G(b, u_2) \in T_y(Z) \qquad (2.14)$$

$e \stackrel{\text{def}}{=} D_x G_a(u_1 - b) - w = D_{(x,a)} G(u_1 - b, 0)) - w = D_{(x,a)} G((u_1, u_2) - (b, u_2)) - w = e_1 - e_2$, where $e_1 = D_{(x,a)} G(u) - w$, $u = (u_1, u_2)$ and $e_2 = D_{(x,a)} G(b, u_2)$. From the equality (2.13) it follows that $e_1 \in T_y(Z)$ and from the equality (2.14) it follows that $e_2 \in T_y(Z)$. We obtain that for arbitrary $w \in T_y(Y)$ there exist $v_1 \in T_x(X)$ $(v_1 = u_1 - b)$ and $v_2 \in T_y(Z)$ $(v_2 = e_1)$ such that $D_x G_a(v_1) + v_2 = w$, i.e. $G_a \bar{\pitchfork} Z$. □

2.89 Remark. In Theorem 2.88 we have supposed that the manifold A is finite dimensional. Very often an analogous theorem will be needed under the assumption that A is an infinite dimensional Banach space, e.g. $A = C^r(X, Y)$, where X is a compact manifold. In literature (e.g. in [2]) a theorem like this is formulated under the assumption that A is a C^r-manifold, whose model space is represented by an infinite dimensional Banach space. The assumption on differentiability of the mapping G is too strong for applications of the parametric Transversability Theorem but it can be made weaker as will be introduced in the following transversality theorems.

2.90 Definition. Let A be a Banach space, X, Y be C^r-manifolds, $1 \leq r \leq \infty$ and $F : A \to C^r(X, Y)$. For every $a \in A$ we write F_a instead of $F(a)$, i.e. $F_a : X \to Y$. The mapping $\mathrm{ev}_F : A \times X \times Y$, $\mathrm{ev}_F(a, x) = F_a(x)$ is called the evaluation mapping of the mapping F. The mapping F is called the C^s-representation ($0 \leq s \leq r$), if the evaluation mapping ev_F is the C^s-mapping. The mapping F is called the C^s-pseudorepresentation, if the mapping $F^{(k)} : A \to C^{r-k}(T^k(X), T^k(Y))$, $F^{(k)}(a) = D^k F_a$ is the C^0-representation for every $k = 0, 1, \ldots, s$.

2.91 Abraham Transversality Theorem. Let A be a Banach space, X, Y be C^r-manifolds, Z be a C^r-submanifold in Y, $1 \leq r \leq \infty$, $\dim X = n$, $\dim Y = m$ and $\mathrm{codim}\, Z = k$. Then the following assertions hold.

1. If $K \in X$ is a compact set, Z is closed in Y and $F : A \to C^r(X, Y)$ is the C^r-pseudorepresentation, then the set $A_{K,Z} = \{a \in A : F_a \bar{\pitchfork}_K Z\}$ is open in A.

2. Let us assume that

    a) A satisfies the second axiom of countability,

    b) $r > \max(0, n - k)$,

    c) $F : A \to C^r(X, Y)$ is the C^r-representation,

    d) $\mathrm{ev}_F \bar{\pitchfork} Z$.

    Then the set $A_Z = \{a \in A : F_a \bar{\pitchfork} Z\}$ is massive (and thus, dense) in A.

Proof of Theorem 2.91 is, in fact the same as the proof of Theorem 2.88, however, some techniques specific to infinite-dimensional Banach spaces are to be used in it. For a better understanding of the next chapters, however, it is not necessary to go into such details. Theorem 2.92 can be found in [2] in the form of two assertions (18.2 and 19.1) along with a detailed proof.

2.92 Thom Transversality Theorem. *Let X, Y be C^{r+k}-manifolds, $1 \leq r \leq \infty$, Z be a C^{r+k}-submanifold in $J^k(X,Y)$, $\dim X = n$, $\dim Y = m$, $\text{codim } Z = q$. Then, the following assertions hold:*

1. if $K \subset X$ is a compact set and Z is closed in $J^k(X,Y)$, then the set $T_{K,Z}^{(k)} = \{f \in C^{r+k}(X,Y) : j^k f \pitchfork_K K\}$ is open in $C_S^{r+k}(X,Y)$;

2. the set $T_Z^{(k)} = \{f \in C^{r+k}(X,Y) : j^k f \pitchfork Z\}$ is massive (and thus, dense) in $C_S^{r+k}(X,Y)$.

Proof of Theorem 2.92 can be found in the proceedings [155] and also in [44] and [65].

2.93 Remark. Throughout the entire chapter when formulating theorems and propositions we paid special attention to the differentiability degree of manifolds. Now, when having defined the notion of diffeomorphism and knowing that many notions we have introduced are independent of the system of coordinates, the fact that there is no difference between the C^1-manifolds and C^∞-manifolds, will be easier to understand. More precisely: the differentiable structure $\mathscr{S}$ of class C^s on X is called compatible with the differentiable structure $\mathscr{T}$ of class C^r on X ($s > r$) if every chart of the atlas $\mathscr{S}$ is a chart of the atlas $\mathscr{T}$ as well. According to [65] (Lemma 2.9) it holds that if $\mathscr{S}$ is a differentiable structure of class C^r on X ($1 \leq r \leq \infty$) then for every s, $1 \leq s \leq \infty$ there exists a differentiable structure $\mathscr{T}$ of class C^s on X compatible with $\mathscr{S}$. According to [64] (Theorem 2.10) every C^r-manifold ($1 \leq r \leq \infty$) is C^r-diffeomorphic with some C^∞-manifold. Therefore in the following chapters C^r-manifolds with $1 \leq r \leq \infty$ need not be considered and the only manifolds to be taken into account will be C^∞-manifolds.

The reader who might be interested in broadening his knowledge of differential topology is recommended to go through some exercises of [51, 65].

2.11 Stratification of Algebraic and Semi-Algebraic Manifolds

Let the polynomial of n variables $p(x) \in R([X])$. Let us denote $V_n(p) = \{x = (x_1, x_2, \ldots, x_n) : p(x) = 0\}$. For example, $V_1(x_1^2 + 1) = \emptyset$, $V_2(x_1^2 + x_2^2 - 1)$ is the circle in R^2, $V_3(x_1^2 + x_2^2 - 1)$ is the surface of a cylinder, $V_3(x_1 - x_2)$ is a plane. All of the above sets are smooth manifold. In general the set $V_n(p)$, however, need not be a smooth manifold. For example, $V_2(x_1^2 - x_2^2) = V_2((x_1 - x_2)(x_1 + x_2)) = V_2(x_1 - x_2) \cup V_2(x_1 + x_2)$, i.e., the set $V_2(x_1^2 - x_2^2)$ is composed of two lines crossing each other at the point $P = (0,0)$. Each of the sets $V_2(x_1 - x_2) \setminus P$ and $V_2(x_1 + x_2) \setminus P$

is composed of two connected components representing smooth manifolds, and thus, $V_2(x_1^2 - x_2^2) = A_1 \cup A_2$, where A_1 is a smooth one-dimensional manifold composed of four connected components, and $A_2 = \{P\}$ which is a smooth manifold of dimension 0.

2.94 Definition. *The set $A \subset R^n$ is called algebraic if there exist polynomials $p_i \in R[X_1, X_2, \ldots, X_n]$, $i = 1, 2, \ldots, k$, such that $A = V_n(p_1) \cap V_n(p_2) \cap \cdots \cap V_n(p_k)$.*

2.95 Definition. *The set $A \in R^n$ is called semi-algebraic if there exist polynomials $p_i, q_j \in R[X_i, X_2, \ldots X_n]$, $1 \leq i \leq k$, $1 \leq j \leq s$, such that $A = \{x \in R^n : p_i(x) = 0, i = 1, 2, \ldots, k, q_i(x) \geq 0, j = 1, 2, \ldots, s\}$.*

2.96 Definition. *Let X, Y be smooth submanifolds in R^n, $\dim Y = k$ and $x \in X$. We say that the submanifold Y is (a)-regular over X at the point x if it holds that if the sequence $\{y_n\}_{n=1}^{\infty}$ of the points of Y converges to the point $x \in X$ and the sequence $\{T_{y_n}(Y)\}_{n=1}^{\infty}$ of linear spaces converges to the k-dimensional linear subspace K in R^n in the topology of the Grassman manifold $G(k, n)$, (see, e.g. Example 2.5; $T_{y_n}(Y)$ is identified with the k-dimensional linear subspace in R^n parallel with the tangential hyperplane to X passing throug the point y_n) then $T_x(X) \subset K$. The manifold Y is called (a)-regular over X if it is (a)-regular over X at every point $x \in X$.*

If $x, y \in R^n$ then $\overline{xy}$ will denote a one-dimensional linear subspace in R^n generated by the vector $x - y$.

2.97 Definition. *Let X, Y be smooth submanifolds in R^n, $\dim Y = k$ and $x \in X$. We say that the submanifold Y is (b)-regular over X at the point x if it holds that: for an arbitrary sequence $\{x_n\}_{n=1}^{\infty}$ of the points of X and an arbitrary sequence $\{y_n\}_{n=1}^{\infty}$ of the points of Y such that*

1. *$x = \lim_{n \to \infty} x_n = \lim_{n \to \infty} y_n$, $x_n \neq y_n$ for all $n \in N$,*

2. *the sequence $\{\overline{x_n y_n}\}_{n=1}^{\infty}$ converges to one-dimensional linear subspace L (in the topology of the Grassman manifold $G(1, n)$),*

3. *the sequence $\{T_{y_n}(Y)\}_{n=1}^{\infty}$ converges to a k-dimensional linear subspace K (in the topology of the Grassman manifold $G(k, n)$) the inclusion $L \subset K$ is satisfied. The manifold Y is called (b)-regular over X if it is (b)-regular over X at every point $x \in X$.*

2.98 Remark. If the submanifold Y is (a)-regular or (b)-regular over X at the point $x \in X$ then it is also said that the triple (Y, X, x) satisfies the Whitney condition (a) and the Whitney condition (b), respectively (see, e.g., [99]).

Foundations of the Theory of Differentiable Manifolds and Differentiable Mappings

2.99 Proposition. *Let X, Y be smooth submanifolds in R^n. If the submanifold Y is (b)-regular over X, then Y is (a)-regular over X.*

Proof of the above proposition can be found in [100].

2.100 Definition. *Let A be a subset in R^n. The decomposition $A = \cup_{i \in I} A_i$ (I be an index set) is called Whitney stratification of the set A if the following conditions hold.*

1. *For every $i \in I$, A_i is a smooth submanifold in R^n (A_i is called the stratum of the set A).*

2. $A_i \cap A_j = \emptyset$, *if $i \neq j$.*

3. (Local finiteness condition): *For arbitrary $x \in A$ there exists its neighbourhood U and just a finite number of the indices $i_1, i_2, \ldots, i_s \in I$ such that $U \cap A_{i_m} \neq \emptyset$, $m = 1, 2, \ldots, s$.*

4. (Boundary condition): *For every $i \in I$ there exists a set $J \subset I$ such that $\overline{A_i} = \cup_{j \in J} A_j$.*

5. (Regularity condition): *If $A_i \subset \overline{A_j}$ then the submanifold A_j is (b)-regular over A_i.*

6. $A_i \cap \overline{A_j} \neq 0 \implies A_i \subset \overline{A_j}$.

2.101 Proposition. *If $A = \cup_{i \in I} A_i$ is the Whitney stratification of the set $A \subset R^n$, $A_i \subset \overline{A_j}$, then $\dim A_i < \dim A_j$.*

Proof of the above proposition can be found in [42], (see Theorem 1.1). If $I = N$ and for $i < j$ with $\dim A_i > \dim A_j$, then the Whitney stratification from Definition 2.100 is called ordered.

2.102 Example. Let $A = V_2(x_1(x_1 - x_2^2)) = \cup_{i=1}^5 A_i$, where $A_1(A_2)$ is the positive (negative) part of the x_2-axis, $A_3 = \{(x_1, x_2) : x_2 = x_1^{1/2}, x_1 > 0\}$, $A_4 = \{(x_1, x_2) : x_2 = -(x_1)^{1/2}, x_1 > 0\}$, $A_5 = \{(0,0)\}$. The above decomposition of the set A is the Whitney stratification (Fig. 7).

2.103 Example. Let $A = V_3(x_1(x_1 - x_2^2 - x_3^2)) = \cup_{i=1}^3 A_i$, where $A_1 = V_3(x_1) \setminus \{(0,0,0)\}$, $A_2 = V_3(x_1 - x_2^2 - x_3^2) \setminus \{(0,0,0)\}$, $A_3 = \{(0,0,0)\}$. The above decomposition of the set A is the Whitney stratification (Fig. 8).

2.104 Example. Let $A = V_3(x_1^2 - x_3 x_2^2) = \cup_{i=1}^3 A_i$, where $A_1 = \{x \in A : x_2 > 0\}$, $A_2 = \{x \in A : x_2 < 0\}$, $A_3 = \{x \in A; x_1 = x_2 = 0\}$. (The set A is called the Whitney umbrella). If $\{y_n\}_{n=1}^\infty$ is an arbitrary sequence of the points of A_1 or of A_2 converging to the point $P = (0,0,0)$, then the sequence $\{T_{y_n}\}_{n=1}^\infty$ converges to the two-dimensional linear subspace $T_P(A_3)$, and thus, neither A_1 nor A_2 is (a)-regular (and thus, not (b)-regular) submanifold over A_3. This means that the above decomposition is not the Whitney stratification (Fig. 9). Let $B_1 = A_1$, $B_2 = A_2$, $B_3 = \{x \in$

57

Chapter 2

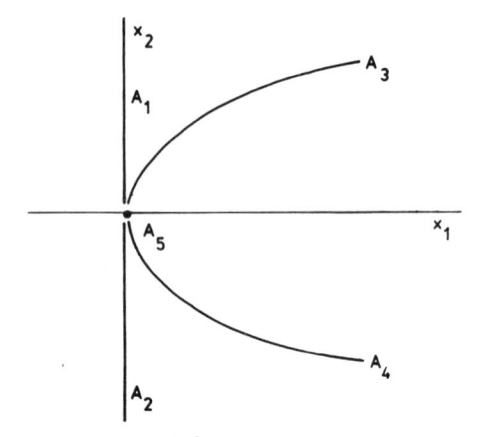

Fig. 7.

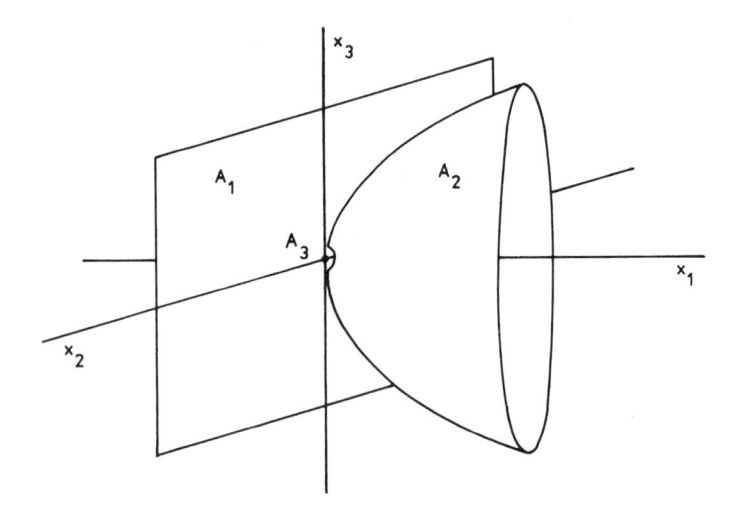

Fig. 8.

$A_3 : x_3 > 0\}$, $B_4 = \{x \in A_3 : x_3 < 0\}$, $B_5 = \{(0,0,0)\}$. The decomposition $A = \cup_{i=1}^{5} B_i$ is the Whitney stratification. Obviously, $B_3 \subset \overline{B}_1$, $B_3 \subset \overline{B}_2$, $B_5 \subset \overline{B}_3$, $B_5 \subset \overline{B}_4$. (Fig. 10). The submanifolds B_1, B_2 are (a)-regular over B_3 and the submanifolds B_3, B_4 are (a)-regular over B_5 and also the regularity condition from Definition 2.100 holds.

2.105 Example. Let $A = V_3(x_2^2 - x_3^2 x_1^2 - x_1^3) = \cup_{i=1}^{5} A_i$, where $A_1 = \{x \in A : x_3 > 0, x_1 < 0\}$, $A_2 = \{x \in A : x_3 < 0, x_1 < 0\}$, $A_3 = \{x \in A : x_1 > 0, x_2 < 0\}$, $A_4 = \{x \in A : x_1 > 0, x_2 > 0\}$, $A_5 = \{x \in A : x_1 = x_2 = 0\}$. The submanifolds A_1, A_2 are (a)-regular over A_5 but not (b)-regular over

58

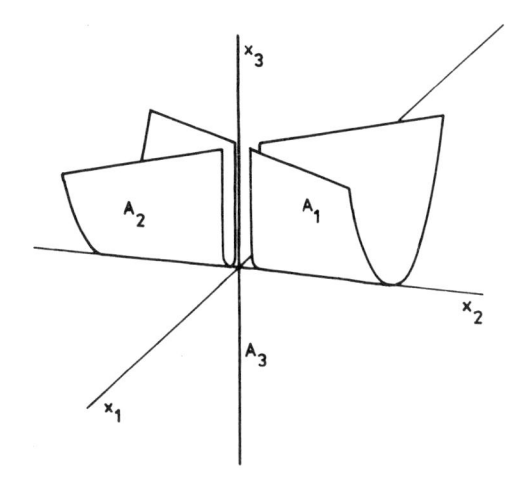

Foundations of the Theory of Differentiable Manifolds and Differentiable Mappings

Fig. 9.

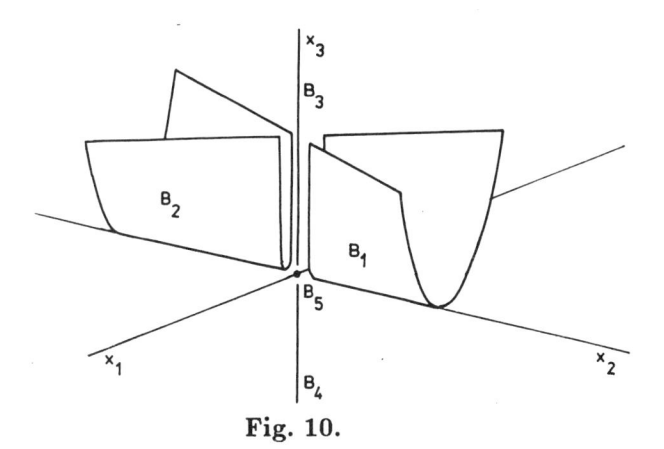

Fig. 10.

A_5, and thus, the above decomposition is not the Whitney stratification (Fig. 11). Let $B_1 = A_1$, $B_2 = A_2$, $B_3 = \{x \in A_3 : x_1 > 0\}$, $B_4 = \{x \in A_3 : x_3 < 0\}$, $B_5 = \{x \in A_4 : x_3 > 0\}$, $B_6 = \{x \in A_4 : x_3 < 0\}$, $B_7 = \{x \in A_5 : x_3 > 0\}$, $B_8 = \{x \in A_5 : x_3 < 0\}$, $B_9 = \{x \in A_3 : x_3 = 0\}$, $B_{10} = \{x \in A_4 : x_3 = 0\}$, $B_{11} = \{(0,0,0)\}$. The decomposition $A = \cup_{i=1}^{11} B_i$ is still the Whitney stratification. (Fig. 12)

2.106 Whitney Stratification Theorem. *For every algebraic set there exists its ordered Whitney stratification with a finite number of strata.*

Proof of the above theorem can be found in Whitney papers [161], [162]. In [162] a similar theorem is proved for the so-called analytical sets, i.e. the sets that are locally defined by a finite number of equations of the form

59

Fig. 11.   Fig. 12.

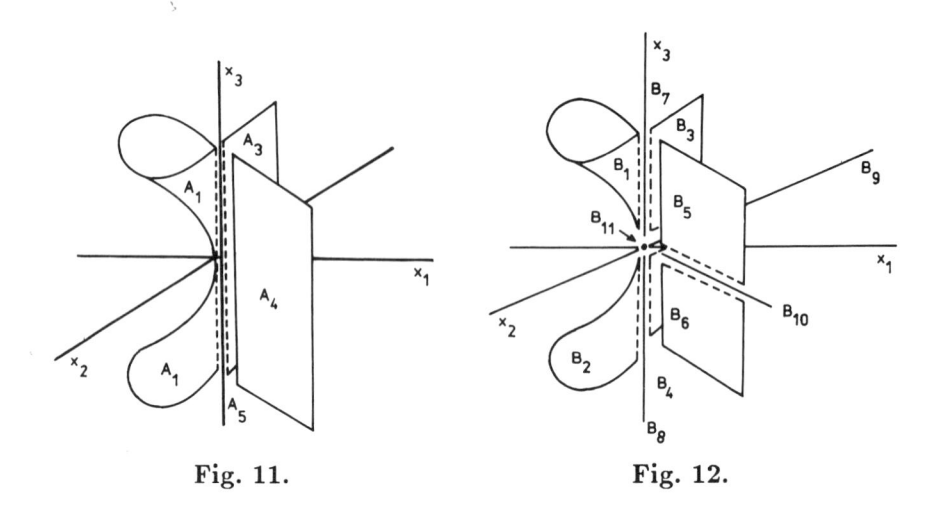

$f_i(x) = 0$, $i = 1, 2, \ldots, n$, $x \in R^m$, f_i are analytical functions.

2.107 Theorem. *For every semi-algebraic set there exists its ordered Whitney stratification with a finite number of strata.*

In [93] a theorem similar to Theorem 2.107 is proved for the so-called local semi-analytical sets, i.e., the sets that are locally defined by a finite number of equations and inequalities of the form: $f_i(x) = 0$, $i = 1, 2, \ldots, k$, $f_j(x) \geq 0$, $j = 1, 2, \ldots, m$, $x \in R^n$, f_i, f_j are analytical functions. Proof of Theorem 2.107 is outlined in [42].

2.108 Definition. *The mapping $f : R^n \to R^m$ is called algebraic, if $f = (f_1, f_2, \ldots, f_m)$, where f_i, $i = 1, 2, \ldots, m$ are polynomials.*

2.109 Tarski–Seidenberg Theorem. *If $A \subset R^n$ is a semi-algebraic set and $f : R^n \to R^m$ is an algebraic mapping then $f(A) \subset R^m$ is a semi-algebraic set.*

The proof of the above theorem can be found in [38, 135] while its more general formulation can be found in [2].

2.110 Corollary. *Let $A \subset R^n$ be a semi-algebraic set, $1 \leq m \leq n$ and $\pi_m : R^n \to R^m$ be the projection. Then $\pi_m(A) \subset R^m$ is a semi-algebraic set.*

2.111 Remark. The theory of algebraic and semi-algebraic sets represents the subject of algebraic geometry. As to a convenient introduction to algebraic geometry the reader is referred to [80]. The topic of stratification of sets (not confining ourselves to subsets of R^n) is dealt with in more detail in [42], [94], [97], [99], [111], [159].

2.12 Transversality to Stratification

2.112 Definition. *Let X be a smooth manifold, $A = \cup_{i \in I} A_i \subset R^n$ be the Whitney stratification and $r \geq 1$. We say that the mapping $f : X \to R^n$ transversally intersects the Whitney stratification of the set A at the point $x \in X$ (we write $f \overline{\pitchfork}_x A$), if $f \overline{\pitchfork}_x A_i$ for all $i \in I$. We say that f transversally intersects the Whitney stratification of the set A on the set $K \subset X$ (we write $f \overline{\pitchfork}_K A$) if $f \overline{\pitchfork}_K A_i$ for all $i \in I$. Instead of $f \overline{\pitchfork}_X A$ we write $f \overline{\pitchfork} A$.*

We define the sets: $T_{K,A}^r(X, R^n) = \{f \in C^r(X, R^n) : f \overline{\pitchfork}_K A\}$, $T_A^r(X, R^n) = \{f \in C^r(X, R^n) : f \overline{\pitchfork} A\}$.

2.113 Transversality to Stratification Theorem. *Let X be a smooth manifold, $A = \cup_{i \in I} A_i \subset R^n$ be the Whitney stratification and $1 \leq r \leq \infty$. Then it holds that:*
1. if $K \subset X$ is a compact set and the set A is closed, then the set $T_{K,A}^r(X, R^n)$ is open in $C_S^r(X, R^n)$;
2. if the index set I is countable then the set $T_A^r(X, R^n)$ is massive (and thus, dense) in $C_S^r(X, R^n)$.

Proof

First, assertion 1 will be proved. Let $f \in T_{K,A}^r(X, R^n)$. If $f(K) \cap A = \emptyset$, then as the consequence of closeness of the set A and compactness of the set K there exists a neighbourhood $U(f)$ of the mapping f in $C_S^r(X, R^n)$ such that for arbitrary $g \in U(f)$ it is $g(K) \cap A = \emptyset$. If $y = f(x) \in A$ for some $x \in K$ and $f \overline{\pitchfork}_x A$ then there exists such a stratum A_s that $y \in A_s$ and $f \overline{\pitchfork}_x A_s$. Let us suppose that $f \notin \text{int } T_{K,A}^r(X, R^n)$. Then there exist sequences $\{f_m\}_{m=1}^\infty, f_m \in C^r(X, R^n) \setminus T_{K,A}^r(X, R^n)$, $m \in N$, $\{x_m\}_{m=1}^\infty$, $x_m \in K$, $m \in N$ such that $\lim_{m \to \infty} f_m = f$, $y_m = f(x_m) \in A$ for all $m \in N$, $\lim_{m \to \infty} y_m = y \in A$, where the mapping f_m does not intersect the stratification of the set A transversally at the point x_m. According to condition 4 from the definition of Whitney stratification there exists a set $J \subset I$ such that $\overline{A}_s = \cup_{j \in J} A_j$. From local finiteness of Whitney stratification it follows that there exists $p \in J$ such that for large enough $L > 0$ it is $y_m \in A_p$ for all $m > L$. From Proposition 2.6 on compactness of the Grassman manifolds it follows that some partial sequence of the sequences $\{T_{y_m}(A_p)\}_{m>L}^\infty$ converges on a linear subspace $T \subset T_y(R^n)$. Without loss of generality we can suppose that the original sequence still converges to T. Since $A_s \subset \overline{A}_p$, from the regularity condition of the

61

Whitney stratification it follows immediately that the submanifold A_p is (b)-regular over A_s. This means that $T_y(A_s) \subset T$. Since $f \bar\pitchfork_{x_m} A_s$, then Image $(D_x f) + T_y(A_s) = T_y(R^n)$, and then using the above inclusion we obtain the equality Image $(D_x f) + T_y(A_s) = T_y(R^n)$. Since $\lim_{m\to\infty} f_m = f$, $\lim_{m\to\infty} y_m = y$ and $\lim_{m\to\infty} T_{y_n}(A_p) = T$ from the above equality it follows that for large enough m Image $(D_{x_m} f_m) + T_{y_m}(A_p) = T_{y_m}(R^n)$. This means, however, that $f_m \bar\pitchfork_{x_m} A_p$, which is a contradiction. Thus, assertion 1 is proved. From the second assertion of the Weak Transversality Theorem it follows that for every $i \in I$ the set $T^r_{A_i}(X, R^n) = \{f \in C^r(X, R^n) : f \pitchfork A_i\}$ is massive in $C^r_S(X, R^n)$. Since the set I is, according to the assumption, countable and $T^r_A(X, R^n) = \cap_{i \in I} T^r_{A_i}(X, R^n)$, the set $T^r_A(X, R^n)$ is massive in $C^r_S(X, R^n)$. □

Analogously to the proof of Theorem 2.113, using the Abraham Transversality Theorem we shall prove the following theorem.

2.114 Abraham Theorem on Transversality to Stratification. *Let A be a Banach space, X be a smooth manifold, $B = \cup_{i \in I} B_i \subset R^n$ be the Whitney stratification and $\dim Y = n$. Then the following assertions hold.*

1. If $K \subset X$ is the compact set, A is closed and $F : A \to C^r(X, R^n)$ is the C^1-pseudo-representation ($1 \leq r \leq \infty$) then the set $A_{K,B} = \{a \in A : F_a \bar\pitchfork_K B\}$ is open A.

2. We suppose that

a) A satisfies the second axiom of countability
b) $r > \max(0, n - k)$, where $k = \min_{i \in I} \operatorname{codim} B_i$
c) $F : A \to C^r(X, R^n)$ is the C^r-representation
d) $\operatorname{ev}_F \bar\pitchfork B$.

Then the set $A_B = \{a \in A : F_a \bar\pitchfork B\}$ is massive (and thus, dense) in A.

3 VECTOR FIELDS AND DYNAMICAL SYSTEMS

3.1 Vector Fields on Differentiable Manifolds

3.1 Definition. *Let X be a smooth manifold of dimension n, $0 \leq r \leq \infty$ and $\pi_X : T(X) \to X$ be the natural projection. The mapping $F : X \to T(X)$ of the class C^r is called the C^r-vector field on X, or the vector field of the class C^r on X if*

$$\pi_X \circ F = \mathrm{id}_X. \tag{3.1}$$

The set of all C^r-vector fields on X will be denoted as $V^r(X)$ considering it as a topological space with the relative topology induced by the Whitney C^r-topology on $C^r(X, T(X))$.

In fact, the vector fields on X form a special class of mapping from X into the tangent bundle $T(X)$ whose structure can be studied best with the use of their local representations. In Section 2.3 (see proof of Proposition 2.26) we have proved that if (U, α) is an admissible chart on X, then $(\pi_X^{-1}(U), T_\alpha)$, where $T_\alpha : \pi_X^{-1}(U) \to R^n \times R^n$, $T_\alpha([\gamma]_x) = (\alpha(x), d_0(\alpha \circ \gamma)(1))$, is an admissible chart on $T(X)$. From the form of the mapping T_α it follows that the local representation $F_\alpha = T_\alpha \circ F \circ \alpha^{-1}$ of the vector field $F \in V^r(X)$ is of the form

$$F_\alpha(y) = (y, f_\alpha(y)), \quad y \in \alpha(U), \tag{3.2}$$

where $f_\alpha(U) \to R^n$ is a mapping of the class C^r. The mapping f_α is called the main part of the local representation of the vector field F.

If $U \subset R^n$ is an open set, $1 \leq r \leq \infty$ and $f \in C^r(U, R^n)$, then according to the theorem on the existence and uniqueness of solutions of differential equations, for every $x \in U$ there exists a number $\varepsilon > 0$ and a unique C^r-mapping $c_x : (-\varepsilon, \varepsilon) \to R^n$ such that $c_x(0) = x$ and $\dot{c}_x(t) = d_t c_x(1) = f(c_x(t))$ for all $t \in (-\varepsilon, \varepsilon)$. The mapping $F : U \to T(U)$, $F(x) = [c_x]_x \in$

63

$T_x(U)$ is the C^r-vector field on U, whose local representation with respect to the admissible chart (U, α), $\alpha = \mathrm{id}_U$, the mapping $F_\alpha : U \to R^n \times R^n$, $F_\alpha(y) = (y, f(y))$. We say that the vector field F is generated by the differential equation $\dot{y} = f(y)$.

3.2 Definition. C^r-curve $c : I \to X$, where $1 \leq r \leq \infty$ and $I \subset R$ is an open interval containing the point 0 is called the integral curve of the vector field $F \in V^r(X)$, passing through the point $x \in X$, if $c(0) = x$ and

$$D_t c(e_t) = F(c(t)) \quad \text{for all} \quad t \in I, \tag{3.3}$$

where e_t is the unit vector in $T_t(R)$; see the remark as follows.

3.3 Remark. The tangent space to the interval I at the point $t \in I$ is the set $T_t(I) = T_t(R) = \{k e_t : k \in R\}$, where $e^t = [\gamma]_t$, $\gamma : (t - \varepsilon, t + \varepsilon) \to R$, $\gamma(s) = t + s$, $\varepsilon > 0$. Since

$$d_0 \gamma(1) = \frac{d\gamma(t)}{ds} = 1$$

and $\gamma(0) = t$, the tangent vector e_t is called the unit vector.

3.4 Lemma. Let $(\pi_X^{-1}(U), T_\alpha)$ be an admissible chart on $T(X)$ derived from the admissible chart (U, α) on X. Then $T_\alpha D_t c(e_t) = (\Psi(t), \dot{\Psi}(t))$, where $\Psi = \alpha \circ c$, $c : I \to X$, $\dot{\Psi}(t) = d_t \Psi(1)$.

Proof

$T_\alpha D_t(e_t) = T_\alpha([c \circ \gamma]_t) = (\alpha \circ c \circ \gamma(0), d_0(\alpha \circ c \circ \gamma)(1)) = ((\alpha \circ c)(t), d_t(\alpha \circ c)(1)) = (\Psi(t), \dot{\Psi}(t))$, where $e_t = [\gamma]_t$. $\square$

3.5 Proposition. Let X be a smooth manifold, $c : I \to X$ be a C^r-curve $(1 \leq r \leq \infty)$ and (U, α) be an admissible chart on X such that $c(t) \in U$ for all $t \in I$. Then c is the integral curve of the vector field $F \in V^r(X)$ iff the C^r-curve $\Psi : I \to R^n$, $\Psi(t) = \alpha \circ c(t)$ is the solution of the differential equation

$$\dot{x} = f_\alpha(x), \tag{3.4}$$

where f_α is the main part of the local representation of the vector field F with respect to the chart (U, α).

Proof

From Lemma 3.4 we obtain that if $c : I \to X$ is the integral curve of the vector field F then $(\Psi(t), \dot{\Psi}(t)) = T_\alpha Dc(e_t) = T_\alpha F(c(t)) = T_\alpha F(\alpha^{-1} \circ$

$\alpha \circ c(t)) = F_\alpha(\Psi(t)) = (\Psi(t), f_\alpha(\Psi(t))$ for all $t \in I$. This means that $\dot{\Psi}(t) = f_\alpha(\Psi(t))$ for all $t \in I$. Let $\Psi = \alpha \circ c$ be the solution of the differential equation (3.4). Since $c(t) \in U$ for $t \in I$ it is suffices to prove that $T_\alpha D_t(e_t) = T_\alpha F(c(t))$ for $t \in I$. From Lemma 3.4 we get

$$T_\alpha F(c(t)) = T_\alpha F(\alpha^{-1} \circ \Psi(t)) = F_\alpha(\Psi(t)) = (\Psi(t), f_\alpha(\Psi(t)))$$
$$= (\Psi(t), \dot{\Psi}(t)) = T_\alpha D_t c(e_t). \quad \square$$

3.6 Definition. *Let X be a smooth manifold, $F \in V^r(x)$, $0 \leq r \leq \infty$, $U \subset X$ be an open set, $x_0 \in U$ and $I \subset R$ be an open interval containing the point 0. A C^r-mapping $\varphi : U \times I \to X$ is called the local C^r-flow of the vector field F at the point x_0 if:*

1. for every $x \in U$ is $\varphi_x : I \to X$, $\varphi_x(t) = \varphi(x, t)$, the integral curve of the vector field F passing through the point x;

2. for every $t \in I$ the mapping $\varphi_t : U \to X$, $\varphi_t(x) = \varphi(x, t)$, is the C^r-diffeomorphism (homeomorphism, if $r = 0$) of the set U onto the set $\varphi_t(U)$.

If $I = (-\infty, \infty)$, $U = X$ and the mapping φ satisfies the conditions 1-2, then φ is called the global C^r-flow, C^r-flow or simply the flow of the vector field F. If for the vector field F there exists a global C^r-flow, then this vector field is called complete.

3.7 Definition. *Let $c_i : I_i \to X$, $i = 1, 2$ be integral curves of the vector field $F \in V^r(X)$. We say that c_1 is the extension of the integral curve c_2 to the interval I_1 if $I_2 \subset I_1$ and $c_2(t) = c_1(t)$ for all $t \in I_2$. If, moreover, $I_1 \neq I_2$ then this extension is called the proper extension. The integral curve $c : I \to X$ of the vector field F is called maximal if it has no proper extension.*

3.8 Lemma. *If X is a smooth manifold, $F \in V^r(X)$, $1 \leq r \leq \infty$, $x \in X$, then there exists a unique maximal integral curve of the vector curve F passing through the point x.*

Proof

Let $C(x)$ be the set of all integral curves of the vector field F passing through the point x. Let us denote by I_c the domain of the integral curve $c \in C(x)$. From Theorem 1.41 it follows that the set $C(x)$ is non-empty. Let us put $I = \cup_{c \in C(x)} I_c$. From Theorem 1.41 it also follows that if $t \in I_{c_1} \cap I_{c_2}$, where $c_1, c_2 \in C(x)$, then $c_1(t) = c_2(t)$. Therefore the mapping $c : I \to X$, $c(t) = u(t)$ can be defined, where $u \in C(x)$ is an integral curve such that $t \in I_u$. Obviously, the mapping c is an integral curve of the vectorfield F passing through the point x. From the definition of the interval I it follows that the integral curve c is maximal and from Theorem 1.41 it follows that it is the only one. $\quad \square$

3.9 Lemma. *Let X be a smooth manifold, $1 \leq r \leq \infty$, U be an open neighbourhood of the point x_0, $I \subset R$ be an open interval containing the point 0 and $\varphi : U \times I$ be a C^r-mapping having the property 1 of the local C^r-flow (Definition 3.6) of the vector field $F \in V^r(X)$. Then there exists an open neighbourhood $U_1 \subset U$ of the point x_0 and an open interval $J \subset I$ containing the point 0, such that it holds that:*

$$\varphi(\varphi(x,s),t) = \varphi(x, s+t), \quad t, s \in I_1, \quad x \in U_1, \tag{3.5}$$

where $I_1 \subset J$ is an open interval containing the point 0 such that for all t, $s \in I_1$ and $t + s \in J$.

Proof

The existence of an open interval J and of the neighbourhood U_1 of the point x_0 such that $\varphi(x,t) \in U$ for all $(x,t) \in U_1 \times J$ follows from the continuity of the mapping φ and from the property 1 of the local C^r-flow. Let I_1 be an interval having the properties listed in Lemma 3.9, $s \in I_1, x \in U_1$ and $y = \varphi(x,s)$. Then $c_1 : I_1 \to X$, $c_1(t) = \varphi(y,t)$, $c_2 : I_1 \to X$, $c_2(t) = \varphi(x, t+s)$ are integral curves of the vector field F fulfilling the condition $c_1(0) = c_2(0) = y$. From Proposition 3.5 and Theorem 1.41 it follows that $c_1(t) = c_2(t)$ for all $t \in I_1$, and thus, the equality (3.5) holds. □

3.10 Theorem. *Let X be a smooth manifold, $1 \leq r \leq \infty$ and $F \in V^r(X)$. Then it holds that:*

1. for every $x \in X$ there exists a local C^r-flow of the vector field F at the point x;

2. if X is a compact manifold, then there exists a unique global C^r-flow of the vector field F.

Proof

Since assertion 1 is of local character, on the basis of Proposition 3.5 it can be assumed that $X = R^n$. The existence of the C^r-mapping $\varphi : U \times I \to R^n$ having property 1 from Definition 3.6 follows from Theorem 1.48. Let $U_1 \subset U$, $I_1 \subset J \subset I$ be as in Lemma 3.9 and moreover let I_1 be an interval such that if $t \in I_1$, then also $-t \in I_1$. From the equality (3.5) it follows that for $t \in I_1$ it is $\varphi_{-t} \circ \varphi_t = \mathrm{id}_{U_1}$, and thus, the mapping φ_t is injective. From the Liouville formula it follows that for $x \in U_1$, $t \in R$ the mapping $d_x\varphi_t : R^n \mapsto R^n$ is an isomorphism, and thus, from the Inverse Mapping Theorem it follows that φ_t is the C^r-diffeomorphism of the set U_1 onto the set $\varphi_t(U_1)$. Now we shall prove assertion 2. According to assertion 1, for every $x \in X$ there exists a local C^r-flow $\tilde{\varphi} : U \times I \to X$ at the point x. Let $c : (a, b) \to X$ be the maximal integral curve of the vector field F passing through the point $y \in U$ which according to Lemma 3.8 exists and is the only one. We shall prove that $a = -\infty$ and $b = +\infty$. It is sufficient to prove that $b = +\infty$ since the assertion saying that $a = -\infty$ can be proved in an analogous way. Let us assume that $b < \infty$. Then there exists a sequence $\{t_n\}_{n=1}^{\infty}$ of real numbers from the interval (a, b) such that $\lim_{n \to \infty} t_n = b$. From compactness of the manifold X it follows that the sequence $\{c(t_n)\}_{n=1}^{\infty}$ contains a convergent subsequence $\{c(t_{n_k})\}_{k=1}^{\infty}$ converging to a point $z \in X$. Let (V, β) be an admissible chart on X such that $z \in V$. From Proposition 3.5 and from Theorem 1.41 it follows that if f_β is the main part of the local representation $F_\beta = T_\beta \circ F \circ \beta^{-1}$ of the vector field F, then there exist $\alpha \in (a, b)$, $\varepsilon > 0$ and a solution $c_1 : (\alpha, b + \varepsilon) \to R^n$ of the autonomous differential equation $\dot{v} = f_\beta(v)$ such that $c_1(t) \in \beta(V)$ for all $t \in (\alpha, b + \varepsilon)$, $c_1(t) = \beta \circ c(t)$ for $t \in (a, b)$. Then, however, $\tilde{c} : (a, b, +\varepsilon) \to X$, $\tilde{c}(t) = c(t)$ for $t \in (a, a >$ and $\tilde{c}(t) = \beta^{-1} \circ c_1(t)$ for $t \in (\alpha, b + \varepsilon)$ is the integral curve of the vector field F, that represents the extension of the integral curve c, which is in contradiction with the maximality of the integral curve c. Let $\{(U_i, \alpha_i)\}_{i=1}^{n}$ be a system of admissible charts on X such that $\cup_{i=1}^{m} U_i = X$ and for every $x \in U_i$ ($i = 1, 2, \ldots, m$) there exists a unique local C^r-flow $\varphi : U_1 \times R \to X$. From the compactness of X and the previous part of the proof it follows that such a system really exists. Let us define $\varphi : X \times R \to X$, $\varphi(x, t) = \varphi_i(x, t)$ for $x \in U_i$ ($i = 1, 2, \ldots, m$), $t \in R$. From the properties of the local C^r-flows φ_i ($i = 1, 2, \ldots, m$) and their uniqueness it follows that the mapping φ is well defined and has all the properties of the global C^r-flow of the vector field F.

3.11 Remark. If $X = R^n$, then (V, β), where $V = R^n$, $\beta = \operatorname{id}_{R^n}$ is the chart on R^n and the main part f_β of the local representation of the vector field $F \in V^r(R^n)$ determines its global C^r-flow if the latter exists. For the proof of the existence of global C^r-flow ($r \geq 1$) of the vector field $F \in V^r(R^n)$ it is sufficient to prove that some of the conditions of extensibility of every solution of the autonomous differential equation $\dot{v} = f_\beta(v)$ on the interval $(-\infty, \infty)$ were satisfied, such as, for example, those mentioned in Theorem 1.42.

In the rest of the present section as well as in the following sections and chapters we shall always suppose completeness of the vector fields on manifolds even in the case where the compactness of manifolds is not assumed.

3.12 Proposition. *Let X be a smooth manifold, $1 \leq r \leq \infty$ and φ be a global C^r-flow of the vector field $F \in V^r(X)$. Then*

$$\varphi(\varphi(x, s), t) = \varphi(x, s + t) \tag{3.6}$$

for all $t, s \in R$, $x \in X$.

Proof

Let $I = (-\infty, \infty)$, $s \in I$, $x \in X$ and $y = \varphi(x, s)$. Then $c_1 : I \to X$, $c_1(t) = \varphi(y, t)$, $c_2 : I \to X$, $c_2(t) = \varphi(x, t + s)$ are the integral curves of the vector field F satisfying the condition $c_1(0) = c_2(0) = y$. From Theorem 3.10 it follows that $c_1(t) = c_2(t)$ for all $t \in I$, and thus the equality (3.6) holds. $\square$

3.13 Definition. *Let X be a smooth manifold. The mappings $\varphi : X \times R \to X$ ($\varphi : X \times Z \to X$, where Z is the set of integers) is called the continuous (discrete) C^k-dynamical system or also the continuous (discrete) dynamical system of the class C^k on X if the following conditions are satisfied:*

1.
$$\varphi_0 = \mathrm{id}_X; \tag{3.7}$$

2. *the mapping*

$$\varphi_t : X \to X, \varphi_t(x) = \varphi(x, t) \tag{3.8}$$

is the C^k-diffeomorphism for every $t \in R$ ($t \in Z$);

3.
$$\varphi_{t+s} = \varphi_t \circ \varphi_s \tag{3.9}$$

for all $t, s \in R$ ($t, s \in Z$).

The system $\{\varphi_t\}_{t \in R}(\{\varphi_t\}_{t \in Z})$ fulfilling the conditions (3.7)–(3.9) is called the one-parameter (discrete) group of C^k-diffeomorphisms on X, X being called the phase space of the dynamical system φ. The points of the phase space X are called the states of the dynamical system φ. Let R_0^+ (Z_0^+) be the set of non-negative real (integer) numbers. The mapping $\varphi : X \times R_0^+ \to X$ ($\varphi : X \times Z_0^+ \to X$) is called the continuous (discrete) semi-dynamical system of class C^k on X, if $\varphi_{t+s} = \varphi_t \circ \varphi_s$ for all $t, s \in R_0^+$ ($t, s \in Z_0^+$) and the mapping $\varphi_t : X \to X$, $\varphi_t(x) = \varphi(x, t)$ is of the class C^k for all $t \in R_0^+$ ($t \in Z_0^+$).

3.14 Example. Let X be a smooth manifold, $1 \leq r \leq \infty$ and F be a complete vector field on X. Then, according to Proposition 3.12 the global C^r-flow φ of the vector field F is a continuous C^r-dynamical system on X and $\{\varphi_t\}_{t \in R}$ is a one-parametric group of C^r-diffeomorphisms on X. The restriction $\tilde{\varphi} = \varphi / X \times Z$ is obviously a discrete C^r-dynamical system on X. The dynamical system φ will also be called the flow of the vector field F.

3.15 Example. Let $f : X \to X$ be a C^r-diffeomorphism, where X is a smooth manifold. Let us define the mapping $\varphi : X \times Z \to X$, $\varphi(x, k) = f^k(x)$, where f^k is the k-th iteration of the diffeomorphism f ($f^0 = \mathrm{id}_X$, $f^m = f \circ f^{m-1}$ for $m > 0$, f^{-m} is the m-th iteration of the inverse f^{-1} of f). The mapping φ is obviously a discrete dynamical system of the class C^r on X. This dynamical system is called the dynamical system generated by the diffeomorphism f.

3.16 Example. Let X be a smooth manifold. If $f : X \to X$ is a C^k-mapping (it need not be a diffeomorphism) then the mapping $\varphi : X \times Z_0^+ \to X$, $\varphi(x,n) = f^n(x)$ ($f_0(x) \stackrel{\text{def}}{=} f(x)$, $f^m(x) \stackrel{\text{def}}{=} f \circ f^{m-1}(x)$ for $n > 0$) is a discrete semi-dynamical system of class C^r on X. In [141] the reader can find an introduction to the theory of discrete semi-dynamical systems on R. The above semi-dynamical systems have a very complicated topological structure of semi-trajectories (a semi-trajectory is defined in the same way as a positive semi-trajectory of a dynamical system; see Definition 3.17) and their study is important, amongst other reasons, for the study of global properties of trajectories of dynamical systems on the so-called strange attractors (see Chapter 6).

3.17 Definition. *Let $\varphi : X \times G \to X$ be a C^k-dynamical system on the smooth manifold X, where $G = R$ or $G = Z$. The set $\gamma_x = \{\varphi(x,t) : t \in G\}$ is called the trajectory of the dynamical system φ passing through the point x. The set $\gamma_x^+ = \{\varphi(x,t) : t \in G, t > 0\}$ ($\gamma_x^- = \{\varphi(x,t) : t \in G, t < 0\}$) is called a positive (negative) semi-trajectory of the dynamical system φ starting from the point x. The point $x \in X$ is called the singular point of the dynamical system φ if $\gamma_x = \{x\}$, i.e., if $\varphi(x,t) = x$ for all $t \in G$. The point $x \in X$ is called the regular point of the dynamical system φ if it is not its singular point. The trajectory γ_x is called T-periodic, if $\varphi(x,t+T) = \varphi(x,t)$ for all $t \in G$ and $T \in G$ is the smallest positive number satisfying this equality. The number T is called the period (prime period) of the trajectory γ_x. If φ is a continuous dynamical system, then the T-periodic trajectory is also called the closed trajectory. If the dynamical system φ is generated by the diffeomorphism $f : X \to X$, then the trajectory γ_x is also called the trajectory of the diffeomorphism f. In this case the singular point of the dynamical system φ is the fixed point of the diffeomorphism f. If γ_x is a T-periodic trajectory of this dynamical system, then the point x is called the periodic point of the diffeomorphism f. The set of all periodic trajectories of the dynamical system φ is denoted as $\text{Per}(\varphi)$ or $\text{Per}(v)$, if φ is the flow of the vector field v. The set of all T-periodic trajectories (points) of the dynamical system φ (of the diffeomorpism f) is denoted as $\text{Per}_T(\varphi)(\text{Per } f)$. The trajectory γ_x of the dynamical system φ is called oriented if it has ordering as follows: if $y = \varphi(x,t_1)$, $z = \varphi(x,t_2)$, where $x \in X$, $t_1, t_2 \in G$, then $y \leq z$ iff $t_1 \leq t_2$. This ordering will be depicted graphically as an arrow oriented in the direction of this ordering.*

3.18 Proposition. *If φ is a continuous C^{r+1}-dynamical system ($1 \leq r \leq \infty$) on the smooth manifold X, then there exists a vector field $F \in V^r(X)$ such that φ is a C^r-flow of this vector field.*

Proof

If $x \in X$ then the curve $c_x : R \to X$, $c_x(s) = \varphi_s(x)$ is of the class C^{r+1}, $c_x(0) = x$, and thus, it defines the class $[c_x]_x \in T_x(X)$. The mapping $F : X \to T(X)$, $F(x) = [c_x]_x$ is obviously a C^r-vector field on X. It is to be proved that for arbitrary $x \in X$, c_x is the integral curve of the vector field F. Let $t \in R$ and $e_t = [\gamma]_t$ be the unit vector in $T_t(R)$, i.e. $\gamma : (t-\varepsilon, t+\varepsilon) \to X$, $\gamma(s) = t+s$, $\varepsilon > 0$. Since for arbitrary $s \in (t-\varepsilon, t+\varepsilon)$ is $C_{\varphi_t(x)}(s) = \varphi_s \circ \varphi_t(x) = \varphi_{s+t}(x) = c_x \circ \gamma(s)$, then

$$F(c_x(t)) = F(\varphi_t(x)) = [c_{\varphi_t}(x)]_{\varphi_t(x)} = [c_x \circ \gamma]_{\varphi_t(x)} = D_t c_x(e_t). \quad \square$$

3.19 Definition. *Let X, Y be smooth manifolds. We say that the C^r-diffeomorphisms $f : X \to X$, $g : Y \to Y$ are C^k-conjugate, $0 \leq k \leq r$ (if $k = 0$, then also topologically conjugate), if there exists a C^k-diffeomorphism (homeomorphism, if $k = 0$) $h : X \to Y$ such that $g \circ h = h \circ f$ or also $g = h \circ f \circ h^{-1}$. The mapping h is called the conjugating C^k-diffeomorphism (conjugating homeomorphism).*

3.20 Proposition. *If C^r-diffeomorphism $f : X \to X$, $g : X \to Y$ are C^k-conjugate, then the conjugating C^k-diffeomorphism h maps the trajectories of the diffeomorphism f onto the trajectories of the diffeomorphism g.*

Proof

Let $\gamma_x^f = \{f^n(x) : n \in Z\}$, $x \in X$ and $y = h(x)$. Since $h \circ f^n \circ h^{-1} = (h \circ f \circ h^{-1}) \circ (h \circ f \circ h^{-1}) \circ \cdots \circ (h \circ f \circ h^{-1}) = g^n$, then

$$h(\gamma_x^f) = \{h \circ f^n(x) : n \in Z\} = \{h \circ f^n \circ h^{-1}(y) : n \in Z\}$$
$$= \{g^n(y) : n \in Z\} = \gamma_y^g.$$

Analogously, it can be proved that $h^{-1}(\gamma_y^g) = \gamma_x^f$. $\quad \square$

Since every continuous C^r-dynamical system $\varphi : X \times R \to X$ defines the discrete C^r-dynamical system $\tilde{\varphi} = \varphi/X \times Z$, naturally, the question arises, whether to every discrete C^r-dynamical, system $\tilde{\Psi} : X \times Z \to X$ there exists a continuous C^k-dynamical system $\Psi : X \times R \to X$ that $\tilde{\Psi} = \Psi/X \times Z$. This task can also be reformulated in the following way. If the C^r-diffeomorphism $f : X \to X$ is given then does there exist a continuous C^k-dynamical system $\varphi : X \times R \to X$ such that $\varphi_1 = f$? If such a continuous C^k-dynamical system exists, then we say that the C^r-diffeomorphism f can be embedded into a continuous C^r-dynamical system. We shall now introduce the result of J. Palis according to which in the set $\text{Diff}^1(X)$ there are only a few C^1-diffeomorphisms which can be embedded into a continuous C^0-dynamical system.

3.21 Theorem. *There exists a massive subset G of the set $\mathrm{Diff}^1(X)$ (X being a smooth manifold) such that if $f \in G$ then f cannot be embedded into a C^0-dynamical system on X.*

Proof of the above theorem can be found in the paper [121]. The problem of embedding the diffeomorphism into a continuous dynamical system is also solved in paper [21].

Theorem 3.21 seemingly proves that the class of discrete dynamical system is with respect to the structure of their trajectories broader than the class of continuous dynamical systems. We shall show that it is not exactly the case. In fact, we shall prove that the structure of trajectories of an arbitrary C^r-diffeomorphism $f : X \to X$ is fully characterized by the structure of trajectories of a certain C^r-diffeomorphism $g : \Sigma \to \Sigma$ which is C^r-conjugate with f, where Σ is a C^r-submanifold of some C^r-manifold $\tilde{X}$, $\dim \Sigma = \dim X$, $\dim \tilde{X} = \dim X + 1$, whereby g is defined with the use of the intersection points of trajectories of some continuous C^r-dynamical system φ on $\tilde{X}$ with the submanifold Σ.

3.22 Definition. *Let $\varphi : X \times R \to X$ be a continuous C^r-dynamical system on the smooth manifold X and $\Sigma \subset X$ be a C^r-submanifold in X of codimension 1, i.e., $\dim \Sigma = \dim X - 1$. We say that Σ is the global transversal of the dynamical system φ if the following points hold.*

1. *For every $x \in X$ $\gamma_x \bar{\cap} \Sigma$, where γ_x is the trajectory of the dynamical system φ passing through the point x.*

2. *For arbitrary $x \in X$ there exists $t \in R$ such that $\varphi_t(x) \in \Sigma$. i.e. arbitrary trajectory γ_x of the dynamical system φ has at least one common point with the submanifold Σ.*

3. *If $x \in \Sigma$ then there exist numbers $t > 0$, $s < 0$ such that $\varphi_t(x) \in \Sigma$, $\varphi_s(x) \in \Sigma$.*

We say that the submanifold Σ is the local transversal of the dynamical system φ at the point $x \in X$ or also the local transversal of the trajectory γ_x at the point x if $x \in \Sigma$ if $\gamma_x \bar{\cap}_x \Sigma$. If φ is the flow of the vector field F then we say also that Σ is the local transversal of the vector field F at the point x.

3.23 Definition. *Let φ be a C^r-dynamical system on the smooth manifold X and Σ be the global transversal to φ. Then we define the mapping $f : \Sigma \to \Sigma$, $f(x) = \varphi_{\tau(x)}(x)$, where $\tau(x) = \min\{t \in R : t > 0, \varphi_t(x) \in \Sigma\}$. The mapping f is called the recurrent mapping derived from the dynamical system φ.*

3.24 Proposition. *Let X be a smooth manifold of dimension n and $f : X \to X$ be a C^r-diffeomorphism. Then there exists a C^r-manifold $\tilde{X}$ of dimension $n+1$, C^r-submanifold Σ in $\tilde{X}$ of dimension n and a continuous C^r-dynamical system φ on $\tilde{X}$ such that Σ is its global transversal and the* *recurrent mapping* $g : \Sigma \to \Sigma$ *derived from the dynamical system* φ *is the* C^r-*diffeomorphism* C^r-*conjugate with* f.

Proof

Let us define on the set $X_1 = \langle 0, 1 \rangle \times X$ the relation $\overset{f}{\sim}$ as follows: $(t_1, x_1) \overset{f}{\sim} (t_2, x_2)$ iff one of the following cases occurs:

1. $(t_1, x_1) = (t_2, x_2)$;
2. $t_1 = 1, t_2 = 0, x_2 = f(x_1)$;
3. $t_1 = 0, t_2 = 1, x_1 = f(x_2)$.

Obviously, the relation $\overset{f}{\sim}$ is the relation of equivalence on the set X_1. Let $\tilde{X} = X_1 / \overset{f}{\sim}$ be the quotient set, $[(t, x)]$ be the class of equivalence and $p : X_1 \to \tilde{X}$, $p((t, x)) = [(t, x)]$ be the natural projection. Let a relative topology be given on X_1 induced by the product topology on $R \times X$. Let us consider the factor topology on the set $\tilde{X}$. We shall prove that $\tilde{X}$ is a Hausdorff topological space. Let Δ_0 be the diagonal in $X_1 \times X_1$, $\Delta_1 = \{((0, x), (1, f^{-1}(x))) \in X_1 \times X_1\}$, $\Delta_2 = \{((1, x), (0, f(x)) \in X_1 \times X_1\}$, $\Delta = \Delta_0 \cup \Delta_1 \cup \Delta_2$, $\tilde{\Delta}$ is the diagonal in $\tilde{X} \times \tilde{X}$, $c(\Delta) = (X_1 \times X_1) \setminus \Delta$ and $c(\tilde{\Delta}) = (\tilde{X} \times \tilde{X}) \setminus \tilde{\Delta}$. Since X_1 is Hausdorff topological space, according to Theorem 1.7 the diagonal Δ_0 is a closed set in $X_1 \times X_1$. The sets $X^{(0)} = \{0\} \times X$, $X^{(1)} = \{1\} \times X$ are closed subsets in X_1. Let us consider these sets as the topological spaces with the relative topology induced by the topology on X_1. Then $X^{(0)}, X^{(1)}$ are Hausdorff topological spaces. The mappings $\varphi_1 : X^{(0)} \to X^{(1)}$, $\varphi_1((0, x)) = (1, f^{-1}(x))$, $\varphi_2 : X^{(1)} \to X^{(0)}$, $\varphi_2((1, x)) = (0, f(x))$, are obviously continuous, and since Δ_1 is the graph of the mapping φ_1 and Δ_2 is the graph of the mapping φ_2 the set Δ_1 is closed in $X_1 \times X_2$ and the set Δ_2 is closed in $X_1 \times X_1$. This means, that both sets are closed in $X_1 \times X_1$. We have proved that the set $c(\Delta)$ is open in $X_1 \times X_1$ and since $c(\Delta) = (p \times p)^{-1}(c(\tilde{\Delta}))$, where $(p \times p)(u, v) = (p(u), p(u))$, the set $c(\tilde{\Delta})$ is open in $\tilde{X} \times \tilde{X}$. This means that the diagonal $\tilde{\Delta}$ is closed in $\tilde{X} \times \tilde{X}$, and thus, according to Theorem 1.7 $\tilde{X}$ is a Hausdorff topological space. Now we shall introduce the C^∞-structure on $\tilde{X}$. Let $\{(U_i; \alpha_i)\}_{i \in I}$ be a C^∞-atlas on X and $U = (-1, 2)$, $\alpha : U \to R$, $\alpha(t) = t$. Let us define $\tilde{U}_i = p(U \times U_i) \subset \tilde{X}$, $\tilde{\alpha}_i : \tilde{U}_i \to R \times R^n$, $\tilde{\alpha}_i([(t, x)]) = (\alpha(t), \alpha_i(x))$. The system $\{(\tilde{U}_i, \tilde{\alpha}_i)\}_{i \in I}$ is obviously a C^∞-atlas on $\tilde{X}$ defining the C^∞-structure S on $\tilde{X}$, i.e., $(\tilde{X}, S)$ is a C^∞-manifold of dimension $n + 1$.

Let $\Sigma = \{[(1, x)] \in \tilde{X} : x \in X\}$ and $h : \Sigma \to X$, $h([(1, x)]) = x$, $g : \Sigma \to \Sigma$, $g([(1, x)]) = [(1, f(x))]$. The set Σ is obviously a C^r-submanifold in $\tilde{X}$. Let us define $U^{(i)} = \{[(1, x)] \in \tilde{X} : x \in U_i\}$, $\alpha^{(i)} : U^{(i)} \to R^n$, $\alpha^{(i)}([(1, x)]) = \alpha_i(x)$. The system $\{(U^{(i)}, \alpha^{(i)})\}_{i \in I}$ is a C^∞-atlas on Σ defining a C^∞-structure $\tilde{S}$ on Σ, and thus, $(\Sigma, \tilde{S})$ is a C^∞-manifold of dimension n. The mapping $h : \Sigma \to X$, $g : \Sigma \to \Sigma$ are

C^∞-diffeomorphisms, and thus, it holds that

$$h \circ f(x) = [(1, f(x))], g \circ h(x) = g([(1,x)]) = [(1,f(x))] \quad \text{for all } x \in X,$$

i.e., $h \circ f = g \circ h$. This means that the diffeomorphism g is C^∞-conjugate with the diffeomorphism f and h is the conjugating C^∞-diffeomorphism. Let us define the mapping $\varphi : \tilde{X} \times R \to \tilde{X}$, $\varphi([(s,x)], t) = (t+s-\{t+s\}, f^{\{t+s\}}(x))$, where $\{t+s\}$ is an integral part of the number $t+s$ and $f^0 \stackrel{\text{def}}{=} \text{id}$. Obviously, the mapping φ has all the properties of the continuous C^∞-dynamical system on X. If $[(s,x)] \in \tilde{X}$, then

$$\varphi([(s,x)], 1-s) = [(0,x)] = [(1, f^{-1}(x))] \in \Sigma,$$

i.e., Σ has the second property of global transversal. If $k \in Z$, then

$$\varphi([(1,x)], k) = [(0, f^{k+1}(x))] = [(1, f^k(x))] = g^k(x),$$

which means that Σ has the third property of global transversal. Now we shall prove that Σ has also the first property of global transversal. Let $[(1,x)] \in \Sigma$ and $\gamma_{[(1,x)]}$ be the trajectory of the dynamical system φ passing through the point $[(1,x)]$. Let $V = (-1,1), \beta : V \to R, \beta(t) = t$, (U, α) be an admissible chart on X such that $x \in U$. Then $(\tilde{U}, \tilde{\alpha})$, where $\tilde{U} = p(V \times U), \tilde{\alpha} = \beta \times \alpha$ is an admissible chart on $\tilde{X}$, $[(1,x)] \in \tilde{U}$ and $\tilde{\alpha}([(1,x)]) = (1, y)$, where $y = \alpha(x)$. The mapping $\Psi : \tilde{\alpha}(\tilde{U}) \times R \to R^{n+1}$, $\Psi((s,u), t) = \tilde{\alpha} \circ \varphi(\tilde{\alpha}^{-1}((s,u)), t)$ is a local representation of the dynamical system φ. For $t \in (-1,1)$ then $\Psi((1,y), t) = (t, f_\alpha(y))$, where $f_\alpha(y) = \alpha \circ f \circ \alpha^{-1}(y)$, and thus, $\dot{\Psi}_{(1,y)} = d_0 \Psi_{(1,y)}(1) = (1, 0, \ldots, 0) \in R^{n+1}$. Let $\tilde{\gamma} = \tilde{\varphi}(\gamma_{[(1,x)]} \cap \tilde{U})$ and $\tilde{\Sigma} = \tilde{\alpha}(\Sigma \cap \tilde{U}) = \{(1, v_1, v_2, \ldots, v_n) \in R^{n+1} : (v_1, v_2, \ldots, v_n) \in \alpha(U)\}$. Then $\hat{T}_{(1,y)}(\tilde{\gamma}) = \{(\lambda_1, 0, \ldots, 0) \in R^{n+1} : \lambda_1 \in R\}$, $\hat{T}_{(1,v)}(\tilde{\Sigma}) = \{(0, \lambda_2, \ldots, \lambda_{n+1}) \in R^{n+1} : (\lambda_2, \lambda_3, \ldots, \lambda_n) \in R^n\}$, and thus, $\hat{T}_{(1,y)}(\tilde{\Sigma}) + \hat{T}_{(1,y)}(\tilde{\gamma}) = \hat{T}_{(1,y)}(R^{n+1})$, i.e. $\tilde{\gamma} \overline{\cap}_{(1,y)} \tilde{\Sigma}$. Since the definition of transversality is independent of the system of coordinates, we have proved that $\gamma_{[(1,x)]} \overline{\cap}_{[(1,x)]} \Sigma$. We have proved that Σ is a global transversal of the dynamical system $\varphi, g : \Sigma \to \Sigma$ is a C^r-diffeomorphism C^r-conjugate with f and $\varphi_1 = g$. □

Various phenomena and processes that can be described with the use of differential equations very often depend upon some parameters. Therefore it is important to study the dependence of dynamical systems or vector fields on parameters. Now these mathematical objects will be defined in more detail.

3.25 Definition. *Let X and P be smooth manifolds, $\dim P = k$ and $1 \leq r \leq \infty$. The mapping $H : X \times P \to T(X)$ is called the parametrized*

C^r-vector field on X with the set of parameters P if it is of the class C^r and

$$\pi_X \circ H_\varepsilon = \operatorname{id}_X \qquad (3.10)$$

for arbitrary $\varepsilon \in P$, where $H_\varepsilon : X \to T(X)$, $H_\varepsilon(x) = H(x, \varepsilon)$, i.e., if for every $\varepsilon \in P$, H_ε is a vector field on X. The system $\{H_\varepsilon\}_{\varepsilon \in P}$ is called the k-parametric system of C^r-vector fields on X. The set of all the parametrized C^r-vector fields on X with the set of parameters P will be denoted as $V^r(P, X)$. The mapping $\varphi : X \times P \times R \to X$ of the class C^r is called the parametrized C^r-dynamical system on X with the set of parameters P if for every $\varepsilon \in P$ the mapping $\varphi_\varepsilon : X \times R \to X$, $\varphi_\varepsilon(x, t) = \varphi(x, \varepsilon, t)$, is the dynamical system on X. If $H \in V^r(P, X)$ and for every $\varepsilon \in P$, φ_ε is the flow of the vector field H_ε, then we say that φ is the parametrized C^r-flow of the parametrized vector field H. The system of differential equations $\dot{x} = f(x, \varepsilon)$, where $f \in C^r(U \times V, R^n)$, $U \in R^n$, $V \subset R^k$ are open sets, $\varepsilon \in V$ is called the k-parametric system of differential equations on U or also the parametrized differential equation on U. The set of all such equations is denoted as $(PDR)^r(V, U)$ or $(PDR)^r$.

3.26 Theorem. If P, X are smooth manifolds, X is compact, $1 \leq r \leq \infty$ and $H \in V^r(P, X)$, then there exists a unique parametrized C^r-flow of the parametrized vector field H.

Proof

According to Theorem 3.10 for every $\varepsilon \in P$ there exists a unique global C^r-flow φ_ε of the vector field H_ε. It suffices to prove that the mapping $\varphi : X \times P \times R \to X$, $\varphi(x, \varepsilon, t) = \varphi_\varepsilon(x, t)$ is of the class C^r. We define the mapping $G : X \times P \to T(X \times P) = T(X) \times T(P)$, $G(x, \varepsilon) = (H(x, \varepsilon), 0_\varepsilon)$, where 0_ε is the zero element of the tangent space $T_\varepsilon(P)$. Obviously, this mapping is a C^r-vector field on $X \times P$. From Theorem 3.10 it follows that for every $(x, \varepsilon) \in X \times P$ there exists a global C^r-flow $\Psi : U \times V \times I \to X \times P$ of the vector field G at the point (x, ε), where $U \times V$ is an open neighbourhood of the point (x, ε), $I \subset R$ being an interval. From the form of the mapping G, from Proposition 3.5 and from Theorem 1.41 it follows that $p_1 \circ \Psi = \varphi / U \times V \times I$, where $p_1 : X \times P \to X$ is the projection. This means that the mapping φ is of class C^r. $\square$

3.27 Definition. Let P, X be smooth manifolds, $\dim P = k$ and $f : X \times P \to X$ be a C^r-mapping such that for every $\varepsilon \in P$ is $f_\varepsilon : X \to X$, $f_\varepsilon(x) = f(x, \varepsilon)$ a diffeomorphism. Then the mapping f is called the parametrized C^r-diffeomorphism on X with the set of parameters P and the system $\{f_\varepsilon\}_{\varepsilon \in P}$ is called the k-parametric system of diffeomorphism on X. The set of all parametrized C^r-diffeomorphism on X with the set of parameters P is denoted as $\operatorname{Diff}^r(P, X)$.

3.28 Definition. *Let P, X be smooth manifolds, $\dim P = k$, $U \subset X$, $V \subset P$ be open sets and $f : U \times V \to X$ be a C^r-mapping such that for every $\varepsilon \in V$ is $f_\varepsilon : U \to X$, $f_\varepsilon(x) = f(x,\varepsilon)$, a C^r-diffeomorphism of the set U onto to the set $f_\varepsilon(U)$. Then we say that f is a local parametrized C^r-diffeomorphism. We say that the mapping $g : X \times P \to X$ is a local parametrized C^r-diffeomorphism at the point $(x,\varepsilon) \in X \times P$ if there exists an open neighbourhood $U \times V$ of the point (x,ε) such that $g/U \times V$ is a local parametrized C^r-difeomorphism.*

3.29 Theorem. *If X, P are smooth compact manifolds, and $1 \leq r < \infty$, then the space of parametrized vector fields $V^r(P,X)$ is a Banach space.*

Proof

Let us define:

$$(\lambda_1 G_1 + \lambda_2 G_2)(x,p) \stackrel{\text{def}}{=} \lambda_1 G_1(x,p) + \lambda_2 G_2(x,p)$$

for arbitrary $G_1, G_2 \in V^r(P,X), (x,p) \in X \times P$ and $\lambda_1, \lambda_2 \in R$. On $V^r(P,X)$ the metric d can be defined in the same way as in Definition 2.64 and since $r < \infty$, this metric is derived from the norm. From Proposition 2.65 it follows that $(V^r(P,X), d)$ is a Banach space. $\square$

3.2 Limit Properties of Dynamical Systems

In this section we shall introduce some limit properties of trajectories of dynamical systems on differentiable manifolds. However, first we shall outline how a metric on differential manifolds can be introduced.

If X is a smooth manifold of dimension n and $x \in X$, then the tangent space $T_x(X)$ is isomorphic with R^n and thus, there exists a scalar product on $T_x(X)$. Thus, in general on $T_x(X)$ there exists at least one positively definite bilinear form. Let us denote by S_x^2 the set of all positively definite bilinear forms on $T_x(X)$ and let $S^2(X) \stackrel{\text{def}}{=} \cup_{x \in X} S_x^2$.

3.30 Definition. *Let X be a smooth manifold and $x \in X$. The mapping $g : X \to S^2(X)$ is called the Riemannian C^∞-metrics on X if*

1. for every $x \in X$ is $\pi_2 \circ g(x) = x$, where $\pi_2 : S^2(X) \to X$, $\pi_2(s_x) = x$ for $s_x \in S_x^2$,

2. for arbitrary two vector fields $F, G \in V^\infty(X)$ the function $g_{F,G} : X \to R$, $g_{F,G}(x) = g(x)(F(x),G(x))$ is of the class C^∞.

The function g assigns to every point $x \in X$ a scalar product on $T_x(X)$, and thus, the function $\|\cdot\|_x^g : T_x(X) \to R$, $\|u\|_x^g = (g(x)(u,u))^{1/2}$ is the norm on $T_x(X)$. The number $\|u\|_x^g$ is the length of the vector $u \in T_x(X)$. From the second property of the Riemannian metric g it follows that if $F \in V^\infty(X)$, then $\|F\| : X \to R$, $\|F\|(x) = \|F(x)\|_x^g$ is a function of the class C^∞, i.e. the length of the vector $F(x)$ depends smoothly on X.

3.31 Theorem. *On every smooth manifold there exist Riemannian C^∞-metric.*

The proof of this theorem can be found in [43] and elsewhere.

Let $g : X \to S^2(X)$ be the Riemannian C^∞-metric on the smooth manifold X, $\gamma : \langle a, b \rangle \to X$ be a C^∞-curve and e_t be the unit vector in $T_t(R)$, where $t \in (a, b)$. Then $D_t\gamma : T_t(R) \to T_{\gamma(t)}(X)$ and the length of the vector $D_t\gamma(e_t)$ is $\|D_t\gamma(e_t)\|_{\gamma(t)}^g$.

3.32 Definition. *The number*

$$|\gamma|_g = \int_a^b \|D_t\gamma(e_t)\|_{\gamma(t)}^g$$

is called the length of the curve γ derived from the Riemannian metric g.

3.33 Definition. *We say that the C^0-curve $\gamma : \langle a, b \rangle \to X$ is piecewise of the class C^∞ if there exists a finite number of points $t_1, t_2, \ldots, t_k \in (a, b)$ such that the curves $\tilde{\gamma}_i = \gamma_i/(t, t_{i+1})$, $i = 0, 1, \ldots, k$, where $t_0 = a, t_{k+1} = b$ are of class C^∞. Such a curve is called the C^∞-polygon and its length derived from the Riemannian C^∞-metric g on X is defined as the number*

$$|\gamma|_g = \sum_{i=0}^k |\tilde{\gamma}_i|_g.$$

We say that the C^∞-polygon $\gamma :< a, b > \to X$ connects the points $x, y \in X$ if $\gamma(a) = x$ and $\gamma(b) = y$. The set of all the C^∞-polygons connecting the points x, y is denoted as $P^\infty(x, y)$.

3.34 Theorem. *Let X be a smooth manifold and $g : X \to S^2(X)$ be the Riemannian C^∞-metric on X. Then the function*

$$\varrho_g : X \times Y \to R, \quad \varrho_g(x, y) = \inf_{\gamma \in P^\infty(x, y)} |\gamma|_g$$

is the metric on X and the topology of the metric space (X, ϱ_g) is identical to the original topology on X.

Proof of this theorem can be found in [44]. The metric ϱ_g from Theorem 3.34 is called the metric derived from the Riemannian metric g.

In the rest of the present section we shall assume that on the smooth manifold X the metric $d = \varrho_g$ derived from the Riemannian metric g on X is given. All further considerations will concern the continuous dynamical system φ on X. However, it should be noted that the notions and results that will be presented in this section can also be reformulated for discrete dynamical systems.

3.35 Definition. *Let $y \in X$. The set of points $z \in X$ for which there exists a sequence of real numbers $\{t_n\}_{n=1}^\infty$, such that $\lim_{n \to \infty} t_n = \infty$ ($\lim_{n \to \infty} t_n = -\infty$), $\lim_{n \to \infty} \varphi(y, t_n) = z$ is called the ω-limit (α-limit) set of the point y and we denote it by $\omega(y)(\alpha(y))$. If $M \subset X$, then the set*

$$\omega(M) = \bigcup_{y \in M} \omega(y) \quad (\alpha(M) = \bigcup_{y \in M} \alpha(y))$$

is called the ω-limit (α-limit) set of the set M.

3.36 Lemma. *Let γ_x be a trajectory of the dynamical system φ passing through the point $x \in X$ and $y \in \gamma_x$. Then $\omega(y) = \omega(x)$ and $\alpha(y) = \alpha(x)$.*

Proof

If $u \in \omega(y)$ then there exists a sequence $\{t_n\}_{n=1}^\infty$ such that $\lim_{n \to \infty} t_n = \infty$, $u = \lim_{n \to \infty} \varphi(y, t_n)$. Since $y \in \gamma_x$, there exists $t \in R$ such that $y = \varphi(x, t)$. From the equality (3.5) it follows that

$$u = \lim_{n \to \infty} \varphi(\varphi(x, t), t_n) = \lim_{n \to \infty} \varphi(x, t_n + t),$$

i.e., $u \in \omega(x)$, and thus, $\omega(y) \subset \omega(x)$. For the same reasons $\omega(x) \subset \omega(y)$ and thus, we obtain that $\omega(x) = \omega(y)$. The proof for α-limit set is analogous. □

3.37 Lemma. *If $x \in X$, then*

$$\omega(x) = \bigcap_{\tau > 0} \overline{\omega_\tau(x)}, \quad \alpha(x) = \bigcap_{\tau < 0} \overline{\alpha_\tau(x)},$$

where $\omega_\tau(x) = \{\varphi(x, t) : \tau \leq t < \infty\}$ and $\alpha_\tau(x) = \{\varphi(x, t) : -\infty < t \leq \tau\}$.

Proof

Let $y \in \omega(x)$. Then there exists a non-decreasing sequence of positive numbers $\{t_n\}_{n=1}^\infty$ such that $\lim_{n \to \infty} t_n = \infty$ and $y = \lim_{n \to \infty} \varphi(x, t_n)$. It τ is an arbitrary positive number, then there exists $n \in Z$ such that $t_k \geq \tau$ for all $k \geq n$, and thus, $\varphi(x, t_k) \in \omega_\tau(x)$ for all $k \geq n$. This means that $y \in \overline{\omega_\tau(x)}$ for arbitrary $\tau > 0$, and thus, $y \in \bigcap_{\tau > 0} \overline{\omega_\tau(x)}$. If

$y \in \cap_{\tau>0}\overline{\omega_\tau(x)}$ and $s > 0$ is arbitrary then $y \in \overline{\omega_s(x)}$ and thus, there exists a sequence $\{t_n\}_{n=1}^\infty$, such that $t_n \geq s, n \in Z$, $\lim_{n\to\infty} t_n = \infty$ and $y = \lim_{n\to\infty} \varphi(x, t_n)$, i.e., $y \in \omega(x)$. The proof for the α-limit set $\alpha(x)$ is analogous. □

3.38 Definition. *Let $x \in X$. The trajectory γ_x of the dynamical system φ is called positively (negatively) stable in Lagrange's sense if there exists a compact set $K \subset X$ such that $\gamma_x^+ \subset K$ ($\gamma_x^- \subset K$). The trajectory γ_x is called stable in Langrange's sense if it is both positively and negatively stable in Langrange's sense.*

3.39 Theorem. *If the trajectory γ_x ($x \in X$) of the dynamical system φ is positively (negatively) stable in Langrange's sense, then $\omega(x)$ ($\alpha(x)$) is a non-empty, closed and connected set.*

Proof

From Lemma 3.37 and Theorem 1.10 it follows that $\omega(x) \neq \emptyset$. We shall proof the closeness of the set $\omega(x)$. Let $\{y_n\}_{n=1}^\infty$ be a sequence of points of $\omega(x)$ converging to $y \in X$ and let $\varepsilon > 0$. For every $n \in Z$ it is $y_n \in \omega(x)$, and thus, there exists $t_n \in R$ such that $\alpha(\varphi(x, t_n), y_n) < \varepsilon/2$. From the convergence of the sequence $\{y_n\}_{n=1}^\infty$ it follows that there exists $N \in R$ such that for all $n > N$ is $\alpha(y_n, y) < \varepsilon/2$. Thus, we obtain that $\alpha(\varphi(x, t_n), y) < d(\varphi(x, t_n), y) + d(y_n, y) < \varepsilon$ for all $n > N$ and this means that $y = \lim_{n\to\infty} \varphi(x, t_n)$, i.e., $y \in \omega(x)$. Now we shall prove that the set $\omega(x)$ is connected. The trajectory γ_s is positively stable in Langrange's sense and therefore there exists a compact set $K \subset X$ such that $\gamma_x^+ \subset K$. From Theorem 1.9 it follows that K is a closed set, and thus, $\omega_\tau(x) \subset K$ for all $\tau > 0$. From Lemma 2.38 it follows that $\omega(x) \subset K$. Let us assume that the set $\omega(x)$ is not connected. Then $\omega(x) = \omega_1 \cup \omega_2$, where ω_1, ω_2 are closed non-empty sets and $\omega_1 \cap \omega_2 = \emptyset$. Since K is a compact set and $\omega_1, \omega_2 \subset K$, according to Theorem 1.9 the sets ω_1, ω_2 are compact. According to Theorem 1.11 there exist open sets $U_1, U_2 \subset K$ such that $\omega_1 \subset U_1$, $\omega_2 \subset U_2$ and $U_1 \cap U_2 = \emptyset$. Since $\omega_1 \subset \omega(x)$, $\omega_2 \subset \omega(x)$, there exist the sequences $\{t_k\}_{k=1}^\infty$, $\{s_k\}_{k=1}^\infty$ such that $t_k < s_k$ for all $k \in N$, $\lim_{k\to\infty} t_k = \infty$, $\lim_{k\to\infty} s_k = \infty$, $\varphi(x, t_k) \in U_1$, $\varphi(x, s_k) \in U_2$, $k \in N$, $\lim_{k\to\infty} \varphi(x, t_k) = y \in \omega_1$, $\lim_{k\to\infty} \varphi(x, s_k) = z \in \omega_2$. Since $U_1 \cap U_2 = \emptyset$, there exists a sequence of numbers $\{r_k\}_{k=1}^\infty$ such that $t_k < r_k < s_k$ for all $k \in N$ and $\varphi(x, r_k) \in V = K \setminus (U_1 \cup U_2)$. The set V is compact, and thus there exists a subsequence of the sequence $\{\varphi(x, r_k)\}_{k=1}^\infty$ converging to a point $v \in V$. This means that $v \in \omega(x)$ and this is a contradiction. The proof for the set $\alpha(x)$ is analogous. □

3.40 Definition. *We say that the set $M \subset X$ is the locally invariant set of the dynamical system φ in a neighbourhood of the point $x_0 \in M$ if there* exists a neighbourhood U of the point x_0 such that for arbitrary $x \in M \cap U$ is $\varphi_t(x) \in M$ for all $t \in I$, where $I \subset R$ is an arbitrary interval such that $0 \in I$ and $\varphi_t(x) \in U$ for all $t \in I$. If $\varphi_t(M) \subset M$ for all $t \in (-\infty, \infty)$ then M is called the invariant set of the dynamical system φ. We say that the set $M \subset X$ is the local invariant set of a diffeomorphism or a homeomorphism $f : X \to X$ in a neighbourhood of the point $x_o \in M$ if there exists a neighbourhood U of the point x_0 such that for arbitrary $x \in M \cap U$ it is $f^n(x) \in M$ for all $n \in J$, where J is an arbitrary interval such that $0 \in J$ and $f^n(x) \in U$ for all $n \in J$. If $f^n(M) \subset M$ for all $n \in Z$ then we say that M is the invariant set of the diffeomorphism or homeomorphism f.

3.41 Theorem. *An ω-limit set $\omega(x)$ and an α-limit set $\alpha(x)$ of the dynamical system φ are invariant sets of this dynamical system.*

Proof

Let $y \in \omega(x)$. Then there exists a sequence of numbers $\{t_n\}_{n=1}^{\infty}$ such that $\lim_{n \to \infty} t_n = \infty$ and $y = \lim_{n \to \infty} \varphi(x, t_n)$. If $t \in R$ then $\varphi_t(y) = \varphi_t(\lim_{n \to \infty} \varphi(x, t_n)) = \lim_{n \to \infty} \varphi(\varphi(x, t_n), t) = \lim_{n \to \infty} \varphi(x, t + t_n)$, and thus, $\varphi_t(y) \in \omega(x)$. For $\alpha(x)$ this can be proved in an analogous way. $\square$

3.42 Theorem. *If $x \in X$ and γ_x is a periodic trajectory of the dynamical system φ passing through the point x, then $\omega(x) = \gamma_x$ and $\alpha(x) = \gamma_x$.*

Proof

Let $y \in \gamma_x$. Then $y = \varphi(x,t)$ for some $t \in R$. If $T > 0$ is the period of the trajectory γ_n, then $\varphi(x, t + nT) = \varphi(x,t)$ for all $n \in N$. The sequence $\{t_n\}_{n=1}^{\infty}$, where $t_n = t + nT$ is such that $\lim_{n \to \infty} t_n = \infty$ and $\lim_{n \to \infty} \varphi(x, t_n) = \varphi(x,t) = y$, and thus, $y \in \omega(x)$. Thus, the inclusion $\gamma_x \subset \omega(x)$ is proved. Before we prove an opposite inclusion, we shall prove that $\overline{\gamma}_x = \gamma_x$. Let $\{y_n\}_{n=1}^{\infty}$ be a sequence such that $y_n \in \gamma_x$ for all $n \in N$ and $\lim_{n \to \infty} y_n = y \in X$. Since $\gamma_x \subset \omega(x)$, then $y_n \in \omega(x)$ for all $n \in N$. Since $\gamma_x = \Phi(\langle 0, T \rangle)$, where $\Phi : \langle 0, T \rangle \to X$, $\Phi(t) = \varphi(x,t)$, according to Theorem 1.15, γ_x is a compact set. Therefore according to Theorem 3.39, $\omega(x)$ is a closed set. Since $y_n \in \omega(x)$ for all n, then $\lim_{n \to \infty} y_n = y \in \omega(x)$. From the definition of the set $\omega(x)$ we obtain that there exists a sequence of numbers $\{t_n\}_{n=1}^{\infty}$ such that $\lim_{n \to \infty} t_n = \infty$, $\lim_{n \to \infty} \varphi(y, t_n) = y$. Let $s_n = t_n \pmod{T}$. Then from the periodicity of the solution $\varphi_x(t)$ we get that $\varphi(x, t_n) = \varphi(x, s_n)$ for all $n \in N$, and thus, $y = \lim_{n \to \infty} \varphi(x, s_n)$. Since $s_n \in \langle 0, T \rangle$ for all $n \in N$, there exists a subsequence $\{s_{n_k}\}_{k=1}^{\infty}$ of the sequence $\{s_n\}_{n=1}^{\infty}$ such that $s = \lim_{k \to \infty} s_{n_k}$ exists and $s \in \langle 0, T \rangle$. We find that

$$y = \lim_{k \to \infty} \varphi(x, s_{n_k}) = \varphi(x, \lim_{n \to \infty} s_{n_k}) = \varphi(x, s),$$

and thus, $y \in \gamma_x$. The inclusion $\omega(x) \subset \gamma_x$ is a consequence of invariance and closeness of γ_x. The proof for an α-limit set is analogous. $\square$

3.43 Definition. *The set $M \subset X$ is called the minimal set of the dynamical system φ if it is a non-empty, closed and invariant set of the dynamical system φ not having a proper subset of the above properties.*

3.44 Theorem. *Let $K \subset X$ be a non-empty, closed, bounded set and an invariant set of the dynamical system φ on X. Then the set K contains at least one minimal set of the dynamical system φ.*

Proof

Let $P = \{L \subset K : L$ be non-empty, closed bounded and an invariant set of the dynamical system $\varphi\}$ and on P be given the relation R as follows: $L_1 R L_2$ iff $L_1 \supset L_2$. The set P with this relation is a partially ordered set. Since $K \in P$, then $P \neq \emptyset$. Let $\{L_k\}_{k \in I}$ be an ordered subset of the set P. Then $L = \cap_{k \in I} L_k$ is closed, bounded, non-empty, and an invariant set of the dynamical system φ (let us note that the intersection of invariant sets of the dynamical system φ is an invariant set of this dynamical system) and $L_k \leq L$ for all $k \in I$ (instead of R we write $\leq$). This means that L is the maximal element of the set $\{L_k\}_{k \in I}$. According to the Zorn Lemma, there exists a maximal element in P, obviously having all the properties of the minimal set of the dynamical system φ. □

3.45 Corollary. *If the semi-trajectory γ_x^+ (γ_x^-) of the dynamical system φ is stable in Langrange's sense, then $\omega(\gamma_x^+)$ ($\alpha(\gamma_x^-)$) contains at least one minimal set of the dynamical system φ.*

3.46 Proposition. *The minimal set of a dynamical system is a connected set.*

Proof

Let M be a minimal set of the dynamical system φ and $M = M_1 \cup M_2$, where M_1, M_2 are closed sets, where $M_1 \cap M_2 = \emptyset$. From the invariance of the set M and the continuity of the dynamical system φ it follows that M_1, M_2 are invariant sets of the dynamical system φ and this is a contradiction with the minimality of the set M. □

3.47 Proposition. *The singular point and periodic trajectory of a dynamical system are its minimal sets.*

Proof

Obviously, the singular point is the minimal set of a dynamical system. Let γ_x be a periodic trajectory of the dynamical system φ passing through the point x. According to Theorem 3.42 it is $\omega(x) = \gamma_x$ and according to Corollary 3.45 the set $\omega(x)$ contains at least one minimal set M. Since γ_x

does not contain any proper invariant subset of the dynamical system φ, $M = \gamma_x$. □

The set $\omega(\gamma_x)$ need not be minimal as can be seen from Fig. 13, where $\omega(\gamma_x) = \cup_{i=1}^{3}\{x_i\} \cup \gamma_i$, x_i is a singular point and γ_i is the trajectory for which $\omega(\gamma_i) = x_{i+1}$ and $\alpha(\gamma_i) = x_i$ ($i = 1, 2, 3, 4$), where $x_1 = x_5$.

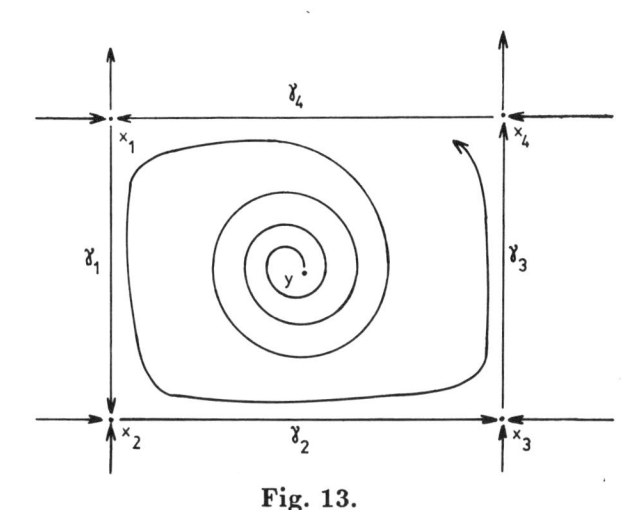

Fig. 13.

3.48 Schwartz Theorem. *Let X be a smooth manifold,* $\dim X = 2$ *and M be a non-empty compact and minimal set of a C^2-dynamical system φ on X. Then one of the following assertions holds:*
1. *M is a singular point*
2. *M is a periodic trajectory*
3. *$M = X$.*

The above theorem is a generalization of the Poincaré–Bendixon Theorem formulated for $X = R^2$ (see [58, 29]). The proof of the Schwarz Theorem can be found in [58, 136]. The minimal sets of the C^r-dynamical systems ($r \geq 2$) on two-dimensional manifolds are by the Schwarz Theorem topologically simple. If, however, $\dim X > 2$ then the minimal sets of dynamical systems on X can be very complicated. For the more detailed study of these problems, also covering continuous dynamical systems, the reader is referred to [23].

3.49 Definition. *The point $x_0 \in X$ is called the wandering point of the dynamical system φ if there exists a neighbourhood U of the point x_0 and a number $t_0 > 0$ such that*

$$\bigcup_{|t|>t_0} \varphi_t(U) \cap U = \emptyset.$$

The point $x_0 \in X$ is called the non-wandering point of the dynamical system φ if it is not wandering. If φ is the flow of the vector field F, then we also say that the point x_0 is the wandering respectively non-wandering point of the vector field F. The sets of non-wandering points of the dynamical system φ or the vector field F are denoted as $\Omega(\varphi)$ and $\Omega(F)$, respectively.

3.50 Proposition. *If γ_x is a bounded trajectory of the dynamical system φ passing through the point x, then $\omega(\gamma_x) \cup \alpha(\gamma_x) \subset \Omega(\varphi)$.*

Proof

If $y \in \omega(\gamma_x)$ and U is an arbitrary neighbourhood of the point y, then $U \cap \gamma_x \neq \emptyset$. If $z \in U \cap \gamma_x$, then according to Lemma 3.36 it is $\omega(z) = \omega(\gamma_x)$, and thus, $y \in \omega(z)$. From the definition of the set $\omega(z)$ it follows that there exists a sequence of positive numbers $\{t_n\}_{n=1}^{\infty}$ such that $\lim_{n \to \infty} t_n = \infty$ and $y = \lim_{n \to \infty} \varphi_{t_n}(z)$. Therefore there exists a number $N > 0$ such that for all $n > N$

$$\bigcup_{|t|>N} \varphi_t(U) \cap U \neq \emptyset.$$

For the set $\alpha(\gamma_x)$ the proof is analogous. □

From Theorem 3.39 and Proposition 3.50 we obtain the following proposition.

3.51 Proposition. *If the dynamical system φ has a trajectory positively or negatively stable in Langrange's sense, then the set $\Omega(\varphi)$ is non-empty.*

3.52 Proposition. *The set $\Omega(\varphi)$ is a closed invariant set of the dynamical system φ.*

Proof

It is sufficient to prove that the set $B = X \setminus \Omega(\varphi)$ is open and it is the invariant set of the dynamical system φ. If $x_0 \in B$ then there exists a neighbourhood U of the point x_0 and a number $t_0 > 0$ such that $\cup_{|t|>t_0} \varphi_t(U) \cap U = \emptyset$, i.e., $U \subset B$, and thus, B is an open set. Let $s \in R$. Since the mapping $\varphi_s : X \to X$ is a homeomorphism, then $\tilde{U} = \varphi_s(U)$ is a neighbourhood of the point $x_1 = \varphi_s(x_0)$ and from the equality (3.5) we obtain that

$$\bigcup_{|t|>t_0} \varphi_t(\tilde{U}) \cap \tilde{U} = \bigcup_{|t|>t_0} \varphi_t \circ \varphi_s(U) \cap \varphi_s(U) = \bigcup_{|t|>t_0} \varphi_s(\varphi_t(U) \cap U)$$

$$= \varphi_s(\bigcup_{|t|>t_0} \varphi_t(U) \cap U) = \emptyset,$$

i.e., $x_1 \in B$, and thus, $\varphi_s(B) \subset B$. □

3.53 Theorem on Closing of Trajectories. *Let X be a smooth, compact manifold, $F \in V^1(X)$ and $x \in \Omega(F)$. Then for an arbitrary neighbourhood W of the vector field F in $V^1(X)$ there exists a vector field $G \in W$ having a periodic trajectory passing through the point x.*

Proof of this theorem can be found in [129] and it is one of the most complicated in the theory of dynamical systems. It is still not known whether the theorem is also valid for the set $V^r(X)$, where $r > 1$.

3.54 Definition. *The point $x_0 \in X$ is called the wandering point of the diffeomorphism $f \in \text{Diff}^r(X)$, if there exists a neighbourhood U of the point x_0 and a number $n_0 > 0$ such that $\cup_{|n|>n_0} f^n(U) \cap U = \emptyset$. The point x_0 is called the non-wandering point of the diffeomorphism f if it is not its wandering point. The set of all non-wandering points of the diffeomorphism f is denoted as $\Omega(f)$.*

3.55 Remark. For the set $\text{Diff}^1(X)$ a theorem analogous to Theorem 3.53 holds. The interesting point about it is that for an arbitrary natural number r the above theorem was proved only in case of the set $\text{Diff}(S^1)$, its proof can be found in [119]. An introduction to this topic is presented in [131].

3.56 Definition. *Let X be a smooth manifold, d be a metric on X, φ be a continuous (discrete) dynamical system on X and $M \subset X$ be its invariant set. The set M is called orbitally stable if for arbitrary $\varepsilon > 0$ there exists $\delta > 0$ such that if*

$$x_0 \in U(M, \delta) = \{x \in X : d(M, x) = \inf_{y \in M} d(y, x) < \delta\},$$

then $\varphi_t(x_0) \in U(M, \varepsilon)$ for all $t \geq 0$, $t \in R$ ($t \in Z$). If moreover $\lim_{n \to \infty} d(M, \varphi_t(x_0)) = 0$ then the set M is called asymptotically orbitally stable. An invariant set which is not orbitally stable is called orbitally unstable. The set

$$O(M) = \{x \in X : \lim_{t \to \infty} d(M, \varphi_t(x_0)) = 0\}$$

is called the region of attractivity of the set M. In the case that M is a trajectory, we say that this trajectory is orbitally stable (unstable) and asymptotically orbitally stable, respectively. The periodic trajectory γ, which is asymptotically stable is called the stable limit cycle. If $\lim_{n \to -\infty} d(\gamma, \varphi_t(x)) = 0$ for all $x \in U(\gamma, \delta)$, then the periodic trajectory γ is called the unstable limit cycle. If M is a singular point then we also say that this point is orbitally stable (unstable) and asymptotically orbitally stable, respectively. When the type of stability concerned is known then the term "orbital" will be omitted. If φ is a flow of the vector field F, then all the notions defined for φ will also be related to the vector field F.

For further studies of the stability or orbital stability of solutions or of the trajectories of differential equations the reader is referred to [30].

3.3 Examples of Vector Fields

3.57 Example: *Gradient vector field.* Let X be a smooth manifold on which the Riemannian C^∞-metric $g : X \to S^2(X)$ is defined. Thus for every $x \in X$ the scalar product $(u, v)_x = g(x)(u, v)$ is defined on $T_x(X)$. Let a C^{r+1}-function $g : X \to R$ be given, where $1 \leq r \leq \infty$. The tangent space $T_x(X)$ is isomorphic with R^n and thus, according to Theorem 1.25 there exists a unique point $F(x) \in T_x(X)$ such that $(F(x), v) = D_x f(v)$ for all $v \in T_x(X)$. Since such an element $F(x)$ exists for every $x \in X$ and g is the Riemannian C^∞-metric, the mapping $F : X \to T(X)$, $x \mapsto F(x)$ is a C^r-vector field on X denoted as ∇f. The vector field $-\nabla f$ is called the gradient vector field on X defined by the function f. The flow of the gradient vector field is called the gradient flow or the gradient dynamical system on X. From the definition of gradient vector field it follows that $-\nabla f(x) = 0$ iff $D_x f = 0$ i.e. x is a singular point of the vector field $-\nabla f$ iff it is a critical point of the function f. If $X = R^n$, then $-\nabla f(x) = (x, -\operatorname{grad} f(x))$. The gradient vector fields have such special properties due to which the gradient dynamical systems have quite a simple structure of trajectories. Let φ be a flow of the gradient vector field $-\nabla f$ on R^n. Since $d_x f(v) = (\operatorname{grad} f(x), v)$ for arbitrary $x, v \in R^n$, then

$$\frac{d}{dt} f(\varphi_t(x)) = d_{\varphi_t(x)} f \left(\frac{d}{dt} \varphi_t(x) \right) = -\|\operatorname{grad} f(\varphi_t(x))\|^2.$$

From this equality it follows that if x is not a singular point, then the function $t \mapsto f(\varphi_t(x))$ is descreasing, and thus, the gradient dynamical system cannot have periodic trajectories. If the reader is interested in more detail about gradient vector fields he is referred to [68].

3.58 Example: *Hamiltonian vector field.* Let a C^{r+1}-function $H : R^{2n} \to R$ be given, where $1 \leq r \leq \infty$. The vector field F_H on R^{2n} whose main part of local representation with respect to the chart $(R^{2n}, \operatorname{id})$ is

$$f_H(x, y) = \Big(\frac{\partial H(x, y)}{\partial y_1}, \frac{\partial H(x, y)}{\partial y_2}, \ldots, \frac{\partial H(x, y)}{\partial y_n}, -\frac{\partial H(x, y)}{\partial x_1}, \\ -\frac{\partial H(x, y)}{\partial x_2}, \ldots, -\frac{\partial H(x, y)}{\partial x_n} \Big),$$

$x = (x_1, x_2, \ldots, x_n)$, $y = (y_1, y_2, \ldots, y_n) \in R^n$ is called the Hamiltonian vector field on $R^2 n$. The function H is called the Hamiltonian function. The vector field F_H defines the so-called Hamiltonian system of differential equations:

$$\frac{dx_i}{dt} = \frac{\partial H(x,y)}{\partial y_i}, \frac{dy_i}{dt} = -\frac{\partial H(x,y)}{\partial x_i}, \quad i = 1, 2, \ldots, n.$$

Hamiltonian vector fields are studied in connection with the so-called Hamilton mechanics, in which the Hamiltonian function has the meaning of energy. Let us introduce at least one important property of Hamiltonian vector fields: the Hamiltonian function H is constant on trajectories of the vector field F_H. In fact, if φ is the flow of the vector field F_H then from the Hamiltonian system of differential equations we obtain that

$$\frac{d}{dt} H(\varphi_t(z)) = 0$$

for arbitrary $z \in R^{2n}$. In [68] the Hamiltonian vector field is defined independently of coordinates. This definition is possible, in general, on special so-called simplectic manifolds of dimension $2n$ (see, e.g. [1, 2, 43]).

3.59 Example: *Second-order differential equation.* The vector field F on R^{2n} having its main part of local representation with respect to the chart (R^{2n}, id) of the form $f_h(x, y) = (y, g(x, y))$, where $g \in C^r(R^{2n}, R^n)$, $1 \leq r \leq \infty$ defines the system of differential equations $\dot{x} = y$, $\dot{y} = g(x, y)$, where $x, y \in R^n$, that can be written in the from of the second-order differential equation $\ddot{x} = g(x, \dot{x})$. This differential equation can represent a mechanical system with n degrees of freedom. If the mapping g does not depend on y, i.e. $g = g(x)$ and there exists a C^{r+1}-function $U : R^n \to R$ such that $g(x) = -\text{grad}\, U(x)$ then the trajectories of the vector field F can represent, for instance, a motion in a force field and then the value $U(x(t))$ ($x(t) = \varphi_t(x_0)$, where φ is the flow of the vector field F) is the potential energy, $T(x(t)) = 1/2 \|\dot{x}(t)\|^2$ is the kinetic energy and $E(x(t)) = T(x(t)) + U(x(t))$ is the total energy of the system at the point $x(t)$. The energy conservation law can be formulated like this: $E(x(t)) = E(x_0)$ for all $t \in R$. Indeed,

$$\frac{dE(x(t))}{dt} = (\ddot{x}(t), \dot{x}(t)) + (\text{grad}\, U(x(t)), \dot{x}(t)) = 0 \quad \text{for all} \quad t \in R.$$

The second-order differential equation can be defined independently of coordinates. Let us assume that there is given an n-dimensional manifold X. According to Proposition 2.26 the tangent bundle $T(X)$ is a smooth $2n$-dimensional manifold. Let $\pi_X : T(X) \to X$ be the natural projection. The second-order differential equation on X of the class C^r is defined as a C^r-vector field F on the tangent bundle $T(X)$ fulfilling the condition $D\pi_X \circ F = \text{id}_{T(X)}$. Let (U, α) be an admissible chart on X, $(\pi_X^{-1}(U), T_\alpha)$ be the chart on $T(X)$ derived from the chart (U, α) and (V, T_α^2) be an admissible chart on $T^2(X) = T(T(X))$ derived from the map $(\pi_X^{-1}(U), T_\alpha)$. Let π_α, F_α be local representations of the mappings π and F, respectively, with respect to the above charts. Then $\pi_\alpha(y, v) = y$ and there exist C^r-mappings g, h such that $F_\alpha(y, v) = (y, v, h(y, v), g(y, v))$, where (y, v) belongs to the domain of the mappings π_α, F_α. If $(D\pi)_\alpha$ is a local representation of the mapping $D\pi$, then according to Proposition 2.30 the mapping $(D\pi)_\alpha$ is of the form: $(D\pi)_\alpha(y, v, w_1, w_2) = (\pi_\alpha(y, v), d_{(y,v)}\pi_\alpha(w_1, w_2)) = (y_1, w_1)$. From this equality and from the condition defining the second-order differential equation it follows that $(D\pi)_\alpha \circ F_\alpha(y, v) = (y, h(y, v)) = (y, v)$ for all (y, v). This means that the mapping F_α is of the form $F_\alpha(y, v) = (y, v, v, g(y, v))$ and defines the second-order differential equation $\ddot{y} = g(y, \dot{y})$.

3.4 Generic Properties of Parameter-Dependent Matrices

This part prepares the reader for the study of generic bifurcations of the parametrized vector fields. There the generic classification of parameter-dependent matrices will be presented with respect to the signs of the real parts of their eigenvalues and according to their position with respect to the unit circle. From the following paragraphs it will easily be seen how this classification is important for the study of generic properties of trajectories both of the continuous and discrete parametrized dynamical systems in neighbourhoods of their singular points and periodic trajectories.

Now we shall introduce an example to outline the problems presented above. Let us consider the set of matrices $M(2)$ with Euclidean metric $\varrho(A, B)$ and the space $C(I, M(2))$, where $I = \langle -1, 1 \rangle$ with the metric $d(F, G) = \max_{t \in I} \varrho(F(t), G(t))$. The matrix

$$A = \begin{bmatrix} 0 & 0 \\ 0 & 1 \end{bmatrix}$$

has the zero eigenvalue of multiplicity 1. Let us consider the mapping $B : I \to M(2)$,

$$B(t) = \begin{bmatrix} t & 0 \\ 0 & 1 \end{bmatrix}.$$

The matrix $B(t)$ has the eigenvalue $\lambda(t) = 0$ just for $t = 0$, whereby $\lambda(t) > 0$ for $t > 0$, $\lambda(t) < 0$ for $t < 0$ and $\varphi(B(t), A) \leq |t|$ for $t \in I$. Therefore for arbitrary $\varepsilon > 0$ there exists $t_0 \in I$, $t_0 \neq 0$ such that $\varrho(B(t_0), A) < \varepsilon$ which means that in an arbitrary neighbourhood of the matrix A there exists a matrix that does not have a zero eigenvalue. Since the function $\det : M(2) \to R$, $A \mapsto \det A$ is continuous, there exists a small neighbourhood U of the mapping B in $C(I, M(2))$ such that it holds that: For arbitrary $\tilde{B} \in U$ there exists a continuous function $\tilde{\lambda} : I \to R$ such that for arbitrary $t \in I$, $\tilde{\lambda}(t)$ is an eigenvalue of the matrix $\tilde{B}(t)$, whereby $\tilde{\lambda}(1) > 0$, $\tilde{\lambda}(-1) < 0$. Therefore there exists $s \in I$ such that $\tilde{\lambda}(s) = 0$. This means that in this case the zero eigenvalue for some values of the parameter cannot be avoided by a small perturbation of the mapping B. The matrix $B(t)$ has the zero eigenvalue for one value of the parameter, only. We shall demonstrate that the above properties of the mapping B are generic.

As for the methodology, this paragraph is related to [26, 109]. With respect to the considerable extent and diversity of the material needed it is not possible to present here proofs of the two following lemmas.

3.60 Lemma. *Let $K \subset R^n$ be a semi-algebraic set defined by the polynomials $P_1, P_2, \ldots, P_k, Q_1, Q_2, \ldots, Q_m \in R[X_1, X_2, \ldots, X_n]$, i.e. $K = \{x \in R^n : P_i(x) = 0, i = 1, 2, \ldots, k, Q_j(x) \geq 0, j = 1, 2, \ldots, n\}$ and $K = K_1 \cup K_2 \cdots \cup K_s$ is its ordered Whitney stratification. Then*

$$\operatorname{codim} K_1 \geq \operatorname{rank} \left(\frac{\partial P_i(x)}{\partial x_j} \right) \quad \text{for all} \quad x \in K_1.$$

The proof of this lemma can be found in Whitney's paper [161]. Let us denote by $S_n(A)$ the set of all the matrices $M(n)$ that are similar to the matrix $A \in M(n)$.

3.61 Lemma. *If $A \in M(n)$ then the set $S_n(A)$ is an immersive submanifold in $M(n)$ and $\operatorname{codim} S_n(A) \geq n$.*

An outline of the present lemmas is presented in [26]. For the completness of this proof several results of the theory of Lie groups and matrix theory are needed (see [155, Proposition 2], [5, part 2], [39, VIII, 2, Theorem 2]).

Let $\tilde{N}(k)$ be a semi-algebraic set in $M(n) \times R^{2k}$ of the form $\tilde{N}(k) = \{(A, \alpha_1, \alpha_2, \ldots, \alpha_{2k}) \in M(n) \times R^{2k} : P_1^{(j)}(\alpha_{1+s}, \alpha_{2+s}) = P_2^{(j)}(\alpha_{1+s}, \alpha_{2+s}) = 0, j = 0, 1, \ldots, m_s - 1, Q_r(\alpha_1, \alpha_2, \ldots, \alpha_{2k}) \geq 0, r = 1, 2, \ldots, p; s = 0, 1, \ldots, 2k - 2\}$, where

$$P(y) = P_1(y_1, y_2) + iP(y_1, y_2) = y^n + a_1 y^{n-1} + \cdots + a_n$$

is the characteristic polynomial of the matrix A, $y = y_1 + iy_2$,

$$P_1^{(j)}(y_1, y_2) + P_2^{(j)}(y_1, y_2) = \frac{d^j P(y)}{dy^j},$$

$j = 1, 2, \ldots, m_{s-1}$, $P_1^{(0)} = P_1$, $P_2^{(0)} = P_2$ and Q_r, $r = 1, 2, \ldots, s$ are some polynomials. The equalities from the definition of the set $\tilde{N}(k)$ defined by the polynomials $P_1, P_2, P_1^{(j)}, P_2^{(j)}$, $j = 1, 2, \ldots, m_{s-1}$ reflect the fact that $\lambda_{1+s} = \alpha_{1+s} + i\alpha_{2+s}$, $s = 0, 1, \ldots, 2k - 2$ are eigenvalues of the matrix A of multiplicity m_s. Let us note that the space of matrices $M(n)$ can be identified with the space R^{n^2}, and thus, the set $\tilde{N}(k)$ is semi-algebraic in the sense of Definition 2.94.

Let us define the mapping $p_{2k} : M(n) \times R^{2k} \to R^n \times R^{2k}$, $p_{2k}(A, \alpha) = (f(A), \alpha)$, where $f(A) = (a_1, a_2, \ldots, a_n)$, $P(y) = y^n + a_1 y^{n-1} + \cdots + a_n$ is the characteristic polynomial of the matrix A, i.e., $p_{2k} = f \times \mathrm{id}$. According to Theorem 2.109 the set $W(k) = p_{2k}(\tilde{N}(k))$ is a semi-algebraic set. According to Theorem 2.107 there exist Whitney stratifications $\tilde{N}(k) = \bigcup_{j=1}^{r,k} N_j(k)$, $W(k) = \bigcup_{j=1}^{s,k} W_j(k)$. Let us denote $N(k) = \pi_{2k}(\tilde{N}(k))$ and $\tilde{W}(k) = \tau_{2k}(W(k))$, where $\pi_{2k} : M(n) \times R^{2k} \to M(n)$, $\tau_{2k} : R^n \times R^{2k} \to R^n$ are projections. The sets $\tilde{N}(k)$, $N(k)$, $W(k)$ and $\tilde{W}(k)$ will be taken for topological spaces with the relative topology induced by the topologies on $M(n) \times R^{2k}$, $M(n)$, $R^n \times R^{2k}$ and R^n, respectively. Let us denote $\tilde{p}_{2k} = p_{2k}/\tilde{N}(k) : \tilde{N}(k) \to W(k)$.

3.62 Lemma. *For every* $x = (a_1, a_2, \ldots, a_n, \alpha) \in R^{n+k}$ *there exists a finite number of immersive* C^∞*-submanifolds* $L_1(x), L_2(x), \ldots, L_m(x)$ *in* $M(n) \times R^{2k}$ *such that*

$$p_{2k}^{-1}(x) = \bigcup_{j=1}^{m} L_j(x)$$

and $\operatorname{codim} L_j(x) \geqq n$ *for* $j = 1, 2, \ldots, m$.

Proof

For $\tilde{x} = (a_1, a_2, \ldots, a_n) \in R^n$, $f^{-1}(\tilde{x})$ is the set of all matrices from $M(n)$ having their characteristic polynomial of the form $P(y) = y^n + a_1 y^{n-1} + \cdots + a_n$. There exists a finite number of the matrices $A_1, A_2, \ldots, A_m \in M(n)$ having the same eigenvalues of the same multiplicity but differing by the number of blocks of the Jordan form corresponding to individual roots of the polynomial $P(y)$ (see Theorem 1.18) and it holds that:

$$f^{-1}(\tilde{x}) = \bigcup_{j=1}^{m} S_n(A_j).$$

According to Lemma 3.61 the sets $S_n(A_j)$, $j = 1, 2, \ldots, n$ are immersive C^∞-submanifolds in $M(n)$ and $\operatorname{codim} S_n(A_j) = n^2 - \dim S_n(A_j) \geqq n$. Therefore the sets $L_j(x) = S_n(A_j) \times R^{2k}$ are immersive C^∞-submanifolds in $M(n) \times R^{2k}$, where $\operatorname{codim} L_j(x) \geqq n$ and $p_{2k}^{-1}(\tilde{x}) = \cup_{j=1}^{m} L_j(x)$ for every $x = (\tilde{x}, \alpha) \in R^{n+k}$. □

3.63 Lemma. *The mapping $\tilde{p}_{2k} : \tilde{N}(k) \to W(k)$ is open, i.e. it maps open sets onto open ones.*

Proof

Since $N(k) = \pi_{2k}(\tilde{N}(k))$, it is sufficient to prove that the mapping $f/N(k) : N(k) \to \tilde{W}(k)$ is open. It is to be proved that if $A \in N(k)$, U being an open neighbourhood of the matrix A in $N(k)$, then for an arbitrary point $\tilde{u} = (\tilde{a}_1, \tilde{a}_2, \ldots, \tilde{a}_n) \in R^n$ close enough to the point $u = f(A) = (a_1, a_2, \ldots, a_n)$ (in the topology on $\tilde{W}(k)$) there exists a matrix $B \in U$ such that $f(B) = \tilde{u}$. Since $f(TAT^{-1}) = f(A)$ for the arbitrary matrix $T \in GL(n)$ it is sufficient to suppose that the matrix A is in the Jordan canonical form. The matrix B in the Jordan canonical form with the characteristic polynomial $\tilde{P}(y) = y^n + \tilde{a}_1 y^{n-1} \cdots + \tilde{a}_n$ is the matrix to be found. From the continuous dependence of eigenvalues of the matrix on its coefficients it follows that if the point $\tilde{u} = (\tilde{a}_1, a_2, \ldots, \tilde{a}_n)$ is close enough to the point $u = (a_1, a_2, \ldots, a_n)$ then $B \in U$. □

3.64 Lemma. *If $\dim W_1(k) \leqq n - q$ then $\operatorname{codim} N_1(k) \geqq 2k + q$.*

Proof

Let i be such that $\tilde{p}_{2k}(N_1(k)) \cap W_i(k) \neq \emptyset$ and $\tilde{p}_{2k}(N_1(k)) \cap W_j(k) = \emptyset$ for all $j < i$. From the boundary conditions of the Whitney stratification it follows that the set $\cup_{j=1}^{i} W_j(k)$ is open in $W(k)$. Since

$$N_1(k) \cap \tilde{p}_{2k}^{-1}(W_i(k)) = \tilde{p}_{2k}^{-1}\left(p_{2k}(N_1(k)) \cap \bigcup_{j=1}^{i} W_j(k)\right)$$

$$= N_1(k) \cap \tilde{p}_{2k}^{-1}\left(\bigcup_{j=1}^{i} W_j(k)\right),$$

then the set $M_0 \stackrel{\text{def}}{=} \tilde{p}_{2k}^{-1}(W_i(k))$ is open in $N_1(k)$. From Lemma 3.63 it follows that the set $\tilde{p}_{2k}(M_0)$ is open in $W_i(k)$. Therefore according to the Sard Theorem there exists a point $\tilde{A} \in M_0$ that is a regular point of the mapping $\tilde{p}_{2k}$. Thus, there exists an open neighbourhood of the point $\tilde{A}$ in $N_1(k)$ such that the mapping $g_k = \tilde{p}_{2k}/U : U \to W_i(k)$ is a C^∞-submersion.

Therefore according to Theorem 2.48 $g_k^{-1}(\tilde{p}_{2k}(\tilde{A}))$ is a C^∞-submanifold in $N_1(k)$ and it holds that:

$$\text{codim}\, g_k^{-1}(\tilde{p}_{2k}(\tilde{A})) = \dim N_1(k) - \dim g_k^{-1}(\tilde{p}_{2k}(\tilde{A})) = \dim W_i(k)$$
$$\leq \dim W_1(k) \leq n - q.$$

From Lemma 3.62 we obtain that $\dim g_k^{-1}(\tilde{p}_{2k}(\tilde{A})) \leq n^2 - n$, and thus from the above inequality the inequality $\dim N_1(k) \leq \dim g_k^{-1}(\tilde{p}_{2k}(\tilde{A})) + n - q \leq n^2 - q$ follows. Therefore $\text{codim}\, N_1(k) \geq n^2 + 2k - (n^2 - q) = 2k + q$. $\square$

3.65 Lemma. *Let $K \subset R^m$ be an open set in R^m with compact closure, $B_0 \in C^r(\overline{K}, M(n))$, $1 \leq r < \infty$, $\varepsilon_0 \in K$ and the matrix $B_0(\varepsilon_0)$ has the complex eigenvalues $\lambda_j^0 = \lambda_{j1}^0 + i\lambda_{j2}^0$, $j = 1, 2, \ldots, k$ and the real eigenvalues $\lambda_{2k+1}^0, \lambda_{2k+2}^0, \ldots, \lambda_n^0$, all of multiplicity 1. Then there exists an open neighbourhood $U(\varepsilon_0) \subset K$ of the point ε_0, an open neighbourhood $V(b_0)$ of the mapping B_0 in $C^r(\overline{K}, M(n))$, C^r-mappings $C : V(B_0) \times U(\varepsilon_0) \to GL(n)$, $\lambda_{js} : V(B_0) \times U(\varepsilon_0) \to R$, $j = 1, 2, \ldots, k$; $s = 1, 2$, $\lambda_p : V(B_0) \times U(\varepsilon_0) \to R$, $p = 2k+1, 2k+2, \ldots, n$ such that for arbitrary $(B, \varepsilon) \in V(B_0)$ it holds that*

1. $\lambda_j(B, \varepsilon) = \lambda_{j1}(B, \varepsilon) + i\lambda_{j2}(B, \varepsilon)$, $j = 1, 2, \ldots, k$, $\lambda_p(B, \varepsilon)$, $p = 2k+1, 2k+2, \ldots, n$ *are the eigenvalues of the matrix $B(\varepsilon)$, where $\lambda_j(B_0, \varepsilon_0) = \lambda_j^0$, $j = 1, 2, \ldots, k, 2k+1, 2k+2, \ldots, n$;*

2. $C(B, \varepsilon)B(\varepsilon)(C(B, \varepsilon))^{-1} = \text{diag}\{D_1, D_2, \ldots, D_k, \lambda_{2k+1}, \lambda_{2k+2}, \ldots, \lambda_n\} \stackrel{\text{def}}{=} D(\varepsilon)$,

$$D_j = D_j(B, \varepsilon) = \begin{bmatrix} \lambda_{j1}(B, \varepsilon) & \lambda_{j2}(B, \varepsilon) \\ -\lambda_{j2}(B, \varepsilon) & \lambda_{j1}(B, \varepsilon) \end{bmatrix},$$

$j = 1, 2, \ldots, k$, $\lambda_s = \lambda_s(B, \varepsilon)$, $s = 2k+1, 2k+2, \ldots, n$. *If the mapping B_0 is analytical then for arbitrary $B \in V(B_0)$ all the mappings $C(B, \varepsilon), \lambda_{js}(B, \varepsilon), \lambda_p(B, \varepsilon)$ are analytical in the variable ε.*

Proof

Without loss of generality we can suppose that the matrix $B_0(\varepsilon_0)$ is in the Jordan canonical form. The mappings $C, \lambda_{js}, \lambda_p$ can be found as the solution of the system of equations as follows: $CB(\varepsilon) - D(\varepsilon)C = 0$, $\|c_j\|^2 - 1 = 0$, $j = 1, 2, \ldots, n$, where $c_j \in R^n$ is the j-th row of the matrix C, $D(\lambda) = \text{diag}\{D_1, D_2, \ldots, D_k, \lambda_{2k+1}, \lambda_{2k+2}, \ldots, \lambda_n\}, \lambda = (\lambda_{11}, \lambda_{12}, \ldots, \lambda_{k1}, \lambda_{k2}, \lambda_{2k+1}, \ldots, \lambda_n)$,

$$D_j = \begin{bmatrix} \lambda_{j1} & \lambda_{j2} \\ -\lambda_{j2} & \lambda_{j1} \end{bmatrix},$$

$j = 1, 2, \ldots, k$. Let us define the functions $F_{ij} : C^r(\overline{K}, M(n)) \times R^m \times R^{n^2} \times R^n \to R$, $i, j = 1, 2, \ldots, n$ so that the value $F_{ij}(B, \varepsilon, c_1, c_2, \ldots, c_n, \lambda)$

is the ij-th element of the matrix $CB(\lambda) - D(\lambda)C$, where c_j is the j-th row of the matrix C and let $F_p : C^r(\overline{K}, M(n)) \times R^m \times R^{n^2} \times R^n \to R$, $F_p(B, \varepsilon, c_1, c_2, \ldots, c_n, \lambda) = \|c_p\|^2 - 1$, $p = 1, 2, \ldots, n$. Let $F = (F_{11}, F_{12}, \ldots, F_{nn}, F_1, F_2, \ldots, F_n)$ and $F_{(B_0, \varepsilon_0)} : R^{n^2} \times R^n \to R^{n^2} \times R^n$, $F_{(B_0, \varepsilon_0)}(c_1, c_2, \ldots, c_n, \lambda) = F(B_0, \varepsilon_0, c_1, c_2, \ldots, c_n, \lambda)$. Since the matrix $B(\varepsilon_0)$ according to the assumption is in the Jordan canonical form, obviously, $F(B_0(\varepsilon_0), e, \lambda^0) = 0$, where $e = (e_1, e_2, \ldots, e_n)$, $e_j (j = 1, 2, \ldots, n)$ is the vector of R^n, whose j-th coordinate equals 1, the rest of the coordinates being equal to 0, $\lambda^0 = (\lambda_{11}^0, \lambda_{12}^0, \ldots, \lambda_{k1}^0, \lambda_{k2}^0, \lambda_{2k+1}^0, \lambda_{2k+2}^0, \ldots, \lambda_n^0)$. Since rank $[d_{(e, \lambda^0)} F_{(B_0, \varepsilon_0)}] = n^2 + n$ (proof is left to the reader), from the Implicit Function Theorem it follows that there exists a neighbourhood $U(\varepsilon_0) \subset K$ of the point ε_0, the neighbourhood $V(B_0)$ of the mapping B_0 and C^r-mappings $C : V(B_0) \times U(\varepsilon_0) \to GL(n)$, $\lambda : V(B_0) \times U(\varepsilon_0) \to R^n$ such that $C(B_0, \varepsilon_0) = I_n$, $\lambda(B_0, \varepsilon_0) = I_n$, $\lambda(B_0, \varepsilon_0) = \lambda^0$ and $F(B(\varepsilon), \varepsilon, c_1(B, \varepsilon), \ldots, c_n(B, \varepsilon), \lambda(B, \varepsilon)) = 0$ for all $(B, \varepsilon) \in V(B_0) \times U(\varepsilon_0)$, where $c_j(B, \varepsilon)$ $(j = 1, 2, \ldots, n)$ is the j-th row of the matrix $C(B, \varepsilon)$. This means that the mappings C, λ represent solutions of the above system of equations. The assertion on analyticity of these mappings in the variable ε follows from the analytical version of the Implicit Function Theorem. □

3.66 Remark. If the multiplicity of one of the eigenvalues of the matrix $B_0(\varepsilon_0)$ is greater than 1 then the assertion of Lemma 3.65 need not be true. The matrix

$$B_0(\varepsilon_1, \varepsilon_2) = \begin{bmatrix} \varepsilon_1 & \varepsilon_2 \\ \varepsilon_2 & -\varepsilon_1 \end{bmatrix}$$

has the eigenvalues $\lambda = (\varepsilon_1^2 + \varepsilon_2^2)^{1/2}, \lambda_2 = -(\varepsilon_1^2 + \varepsilon_2^2)^{1/2}$ that are not differentiable at the point $\varepsilon_0 = (0, 0)$.

Let us define the sets of matrices as follows:

$I_n^k = \{A \in M(n) : A \text{ has a zero eigenvalue of multiplicity} \geq k\}$, $k \geq 1$,

$J_n^k = \{A \in M(n) : A \text{ has a pure imaginary eigenvalue of multiplicity} \geq k\}$, $k \geq 1$,

$K_n^1 = \{A \in M(n) : A \text{ has a zero eigenvalue of multiplicity 1 and at least one more pure imaginary eigenvalue of multiplicity} \geq 1\}$,

$L_n^1 = \{A \in M(n) : A \text{ has at least two different pure imaginary eigenvalues of multiplicity} \geq 1, \text{ their complex conjugates not included}\}$,

$K_n^2 = \{A \in M(n) : A \text{ has at least one pure imaginary eigenvalue, its real part being of multiplicity} \geq 2 \text{ and at least one more pure imaginary eigenvalue of multiplicity} \geq 1\}$,

$L_n = \{A \in M(n) : A \text{ has at least three pure imaginary eigenvalues of multiplicity} \geq 1, \text{ their complex conjugates not included}\}$.

Chapter 3

Now to each of the above sets, an algebraic or semi-algebraic set will be defined in one of the spaces $M(n) \times R^{2p}$, $p = 1, 2, 3$. Let us define the sets as follows:

$\tilde{I}_n^1 = \{(A, \lambda_1, \lambda_2) \in M(n) \times R^2 : P_1(\lambda_1, \lambda_2) = P_2(\lambda_1, \lambda_2) = 0, \lambda_1 = \lambda_2 = 0\}$,

$\tilde{I}_n^k = \{(A, \lambda_1, \lambda_2) \in M(n) \times R^2 : P_1(\lambda_1, \lambda_2) = P_2(\lambda_1, \lambda_2) = 0, P_1^{(j)}(\lambda_1, \lambda_2) = P_2^{(j)}(\lambda_1, \lambda_2) = 0, j = 1, 2, \ldots, k-1, \lambda_1 = \lambda_2 = 0\}$, $k > 1$,

$\tilde{J}_n^1 = \{(A, \lambda_1, \lambda_2) \in M(n) \times R^2 : P_1(\lambda_1, \lambda_2) = P_2(\lambda_1, \lambda_2) = 0, \lambda_1 = 0, \lambda_2 \neq 0\}$,

$\tilde{J}_n^k = \{(A, \lambda_1, \lambda_2) \in M(n) \times R^2 : P_1(\lambda_1, \lambda_2) = P_2(\lambda_1, \lambda_2) = 0, \lambda_1 = 0, \lambda_2 \neq 0, P_1^{(j)}(\lambda_1, \lambda_2) = P_2^{(j)}(\lambda_1, \lambda_2) = 0, j = 1, 2, \ldots, k-1\}$, $k > 1$,

$\tilde{K}_n^1 = \{(A, \lambda_1, \lambda_2, \mu_1, \mu_2) \in M(n) \times R^4 : P(\lambda_1, \lambda_2) = P_2(\lambda_1, \lambda_2) = 0, P_1(\mu_1, \mu_2) = P_2(\mu_1, \mu_2) = 0, \lambda_1 = \lambda_2 = 0, \mu_1 = 0, \mu_2 \neq 0\}$,

$\tilde{L}_n^1 = \{(A, \lambda_1, \lambda_2, \mu_1, \mu_2) \in M(n) \times R^4 : P(\lambda_1, \lambda_2) = P_2(\lambda_1, \lambda_2) = 0, P_1(\mu_1, \mu_2) = P_2(\mu_1, \mu_2) = 0, \lambda_1 = 0, \mu_1 = 0, \lambda_2 - \mu_2 \neq 0, \lambda_2 \neq 0, \mu_2 \neq 0\}$,

$\tilde{K}_2^n = \{(A, \lambda_1, \lambda_2, \mu_1, \mu_2) \in M(n) \times R^4 : P(\lambda_1, \lambda_2) = P_2(\lambda_1, \lambda_2) = 0, P_1(\mu_1, \mu_2) = P_2(\mu_1, \mu_2) = 0, P_1^{(1)}(\lambda_1, \lambda_2) = P_2^{(1)}(\lambda_1, \lambda_2) = 0, \lambda_1 = 0, \mu_1 = 0, \mu_2 \neq 0\}$,

$\tilde{L}_n = \{(A, \lambda_1, \lambda_2, \mu_1, \mu_2, \nu_1, \nu_2) \in M(n) \times R^6 : P(\lambda_1, \lambda_2) = P_2(\lambda_1, \lambda_2) = 0, P_1(\mu_1, \mu_2) = P_2(\mu_1, \mu_2) = 0, P_1(\nu_1, \nu_2) = P_2(\nu_1, \nu_2) = 0, \lambda_1 = 0, \mu_1 = 0, \nu_1 = 0, \lambda_2 \neq 0, \mu_2 \neq 0, \nu_2 \neq 0, \lambda_2 - \mu_2 \neq 0, \lambda_2 - \nu_2 \neq 0, \mu_2 - \nu_2 \neq 0\}$.

According to Theorem 2.106 or 2.107 there exist ordered Whitney stratifications

$\tilde{I}_n^k = \cup_{j=1}^{m_k} \hat{I}_j^k$, $\tilde{J}_n^k = \cup_{j=1}^{n_k} \hat{J}_j^k$, $k = 1, 2, 3$, $\tilde{K}_n^p = \cup_{j=1}^{q_p} \hat{K}_j^p$, $p = 1, 2$,

$\tilde{M}_n^1 = \cup_{j=1}^{s_1} \hat{L}_j^1$, $\tilde{L}_n = \cup_{j=1}^{s} \hat{L}_j$.

Let us introduce the notations as follows $V(\tilde{I}_n^k) = p_2(\tilde{I}_n^k) \subset R^{n+2}$, $V(\tilde{J}_n^k) = p_2(\tilde{J}_n^k) \subset R^{n+2}$, $k = 1, 2, 3$, $V(\tilde{K}_n^p) = p_4(\tilde{K}_n^p) \subset R^{n+4}$, $p = 1, 2$, $V(\tilde{L}_n^1) = p_4(\tilde{L}_n^1) \subset R^{n+4}$ and $V(\tilde{L}_n) = p_6(\tilde{L}_n) \subset R^{n+6}$. According to Theorem 2.109 the above sets are algebraic or semi-algebraic, and thus, according to Theorem 2.107 there exist their Whitney stratifications $V(\tilde{I}_n^k) = \cup_{j=1}^{m'_k} V_j(\tilde{I}_n^k)$, $V(\tilde{J}_n^k) = \cup_{j=1}^{n'_k} V_j(\tilde{J}_n^k)$, $k = 1, 2, 3$, $V(\tilde{K}_n^p) = \cup_{j=1}^{q'_p} V_j(\tilde{K}_n^p)$, $p = 1, 2$, $V(\tilde{L}_n^1) = \cup_{j=1}^{s'_1} V_j(\tilde{L}_n^1)$, $V(\tilde{L}_n) = \cup_{j=1}^{s'} V_j(\tilde{L}_n)$.

3.67 Lemma.
1. $\dim V_1(\tilde{I}_n^k) \leq n - k$, $k = 1, 2, 3$;
2. $\dim V_1(\tilde{J}_n^k) \leq n - k$, $k = 1, 2, 3$;
3. $\dim V_1(\tilde{K}_n^p) \leq n - p - 1$, $p = 1, 2,$;
4. $\dim V_1(\tilde{L}_n^1) \leq n - 1$;
5. $\dim V_1(\tilde{L}_n) \leq n - 3$.

In the proof of Lemma 3.67 the following notations will be used

$$dP_i(\tilde{\lambda}) = \left(\frac{\partial P_i(\tilde{\lambda})}{\partial a_1}, \ldots, \frac{\partial P_i(\tilde{\lambda})}{\partial a_n}, \frac{\partial P_i(\tilde{\lambda})}{\partial y_1}, \frac{\partial P_i(\tilde{\lambda})}{\partial y_2}\right),$$

$$dP_i^{(j)}(\tilde{\lambda}) = \left(\frac{\partial P_i^{(j)}(\tilde{\lambda})}{\partial a_1}, \ldots, \frac{\partial P_i^{(j)}(\tilde{\lambda})}{\partial a_n}, \frac{\partial P_i^{(j)}(\tilde{\lambda})}{\partial y_1}, \frac{\partial P_i^{(j)}(\tilde{\lambda})}{\partial y_2}\right),$$

$$dQ_i(\tilde{\lambda}) = \left(\frac{\partial Q_i(\tilde{\lambda})}{\partial a_1}, \ldots, \frac{\partial Q_i(\tilde{\lambda})}{\partial a_n}, \frac{\partial Q_i(\tilde{\lambda})}{\partial y_1}, \frac{\partial Q_i(\tilde{\lambda})}{\partial y_2}\right),$$

$i = 1, 2$; $j = 1, 2, \ldots, k - 1$, $\tilde{\lambda} = (\lambda_1, \lambda_2)$, $Q_1(y_1, y_2) = y_1$, $Q_2(y_1, y_2) = y_2$. When it is obvious at which point the partial derivatives are calculated we shall use just dP_i, $dP_i^{(j)}$ and dQ_i, respectively. Let the following system of vectors be given

$$v_1 = (v_{1m}, v_{1m-1}, \ldots, v_{1s}, \ldots, v_{1k}, \ldots, v_{11}),$$

$$v_2 = (v_{2m}, v_{2m-1}, \ldots, v_{2s}, \ldots, v_{2k}, \ldots, v_{21}),$$

$$\ldots$$

$$v_k = (v_{km}, v_{km-1}, \ldots, v_{ks}, \ldots, v_{kk}), \ldots, v_{k1}), \quad k \leq s \leq m.$$

Let us denote as $D[v_1, v_2, \ldots, v_k]$ the matrix from $M(k, s)$ whose i-th row ($1 \leq i \leq k$) is represented by the vector $(v_{is}, v_{is-1}, \ldots, v_{i1})$.

The polynomials P_1, P_2, $P_1^{(j)}$, $P_2^{(j)}$, $j = 1, 2, 3$, are of the form

$$P_1(y_1, y_2) = \operatorname{Re}(y^n) + a_1(\operatorname{Re}(y^{n-1})) + \cdots + a_{n-3}y_1(y_1^2 - 3y_2^2)$$
$$+ a_{n-2}(y_1^2 - y_2^2) + a_{n-1}y_1 + a_n,$$

$$P_2(y_1, y_2) = \operatorname{Im}(y^n) + a_1(\operatorname{Im}(y^{n-1})) + \cdots + a_{n-3}y_2(3y_1^2 - y_2^2)$$
$$+ a_{n-2}(2y_1, y_2) + a_{n-1}y_2,$$

$$P_1^{(1)}(y_1, y_2) = n(\operatorname{Re}(y^{n-1})) + (n-1)(\operatorname{Re}(y^{n-2})) + \cdots + 2a_{n-2}y_1 + a_{n-1},$$

$$P_2^{(1)}(y_1, y_2) = n(\operatorname{Im}(y^{n-1})) + (n-1)(\operatorname{Im}(y^{n-2})) + \cdots + 2a_{n-2}y_2,$$

$$dP_1^{(2)}(y_1, y_2) = n(n-1)(\operatorname{Re}(y^{n-2})) + \cdots + 6a_{n-3}y_1 + 2a_{n-2},$$

$$dP_2^{(2)}(y_1, y_2) = n(n-1)(\operatorname{Im}(y^{n-2})) + \cdots + 6a_{n-3}y_2.$$

Chapter 3

Proof of Lemma 3.67

We shall prove proposition 1. Let $k = 1$. Let us assume that $x = (a_1, a_2, \ldots, a_n, \tilde{\lambda}) \in V_1(\tilde{I}_n^1)$, $\tilde{\lambda} = (\lambda_1, \lambda_2)$ and let us denote $D_s(x) = D_s[dP_1(\tilde{\lambda}), dQ_1(\tilde{\lambda}), dQ_2(\tilde{\lambda})]$. Since $\det D_3(x) = 1$ for all $x \in V_1(\tilde{I}_n^1)$, from Lemma 3.60 it follows that $\operatorname{codim} V_1(\tilde{I}_n^1) \geq 3$, and thus $\dim V_1(\tilde{I}_n^1) = n + 2 - \operatorname{codim} V_1(\tilde{I}_n^1) \leq n - 1$. Let $k = 2$. Let us assume that $x = (a_1, a_2, \ldots, a_n, \tilde{\lambda}) \in V_1(\tilde{I}_n^2)$, $\tilde{\lambda} = (\lambda_1, \lambda_2)$ and let us denote $D_s(x) = D_s[dP_1(\tilde{\lambda}), dP_1^{(1)}(\tilde{\lambda}), dQ_1(\tilde{\lambda}), dQ_2(\tilde{\lambda})]$. Since $\det D_4(x) = -1$ for all $x \in V(\tilde{I}_n^2)$, from Lemma 3.60 it follows immediately that $\dim V_1(\tilde{I}_n^2) \leq n - 2$. Let $k = 3$, $x = (a_1, a_2, \ldots, a_n, \lambda_1, \lambda_2) \in V_1(\tilde{I}_n^3)$ and $D_s(x) = D_s[dP_1, dP_1^{(1)}, dP_1^{(2)}, dQ_1, dQ_2]$. Since $\det D_5(x) = -2$, then $\dim V_1(\tilde{I}_n^3) = n + 2 - \operatorname{codim} V_1(\tilde{I}_n^3) \leq n - 3$. Now let us prove assertion 2. Let $k = 1$, $x = (a_1, a_2, \ldots, a_n, \tilde{\lambda}) \in V_1(\tilde{J}_n^1)$ and $D_s(x) = D_s[dP_1, dP_2, dQ_1]$. The matrix derived from matrix $D_4(x)$ by omitting the second column has its determinant equal to

$$\lambda_2 \frac{\partial P_1(\tilde{\lambda})}{\partial y_2}.$$

Since the set $V_1(\tilde{J}_n^1)$ is open in $V(\tilde{J}_n^1)$ and $\lambda_2 \neq 0$, it is sufficient to prove that the set

$$H = \left\{ x = (a_1, a_2, \ldots, a_n, 0, \lambda_2) \in V_1(\tilde{J}_n^1) : \frac{\partial P_1(0, \lambda_2)}{\partial y_2} \neq 0 \right\}$$

is dense in $V(\tilde{J}_n^1)$. Let $x = (a_1, a_2, \ldots, a_n, 0, \lambda_2) \in V(\tilde{J}_n^1) \setminus H$. Let us define the polynomial $P^\varepsilon(y) = P(y) + \varepsilon \varphi(y)$, where $\varphi(y) = (y - i\lambda_2)(y + i\lambda_2) = y_1^2 - y_2^2 + \lambda_2^2 + 2iy_1 y_2$, ε being a real number. Then $P^\varepsilon(y) = y^n + a_1(\varepsilon)y^{n-1} + \cdots + a_n(\varepsilon) = P_1^\varepsilon(y_1, y_2) + iP_2^\varepsilon(y_1, y_2)$, where $P_1^\varepsilon(y_1, y_2) = P_1(y_1, y_2) + \varepsilon(y_1^2 - y_2^2 + \lambda_2^2)$, and thus,

$$\frac{\partial P_1^\varepsilon(0, \lambda_2)}{\partial y_2} = -2\lambda_2 \varepsilon \neq 0 \quad \text{for} \quad \varepsilon \neq 0.$$

Obviously, $x(\varepsilon) = (a_1(\varepsilon), a_2(\varepsilon), \ldots, a_n(\varepsilon), 0, \lambda_2) \in V(\tilde{J}_n^1)$ and for an arbitrary neighbourhood U of the point x in $V(\tilde{J}_n^1)$ there exists $\varepsilon \neq 0$ such that $x(\varepsilon) \in U \cap H$. Let $k = 2$, $x = (a_1, a_2, \ldots, a_n, \lambda_1, \lambda_2) \in V_1(\tilde{J}_n^2)$ and $D_s(x) = D_s[dP_1, dP_1^{(1)}, dP_2^{(1)}, dQ_1]$. Since

$$\det D_4(x) = \frac{\partial P_2^{(1)}(\lambda_1, \lambda_2)}{\partial y_2},$$

it is sufficient to prove that the set

$$H_1 = \left\{ x = (a_1, a_2, \ldots, a_n, 0, \lambda_2) \in V(\tilde{J}_n^2) : \frac{\partial P_2^{(1)}(0, \lambda_2)}{\partial y_2} \neq 0 \right\}$$

is dense in $V(\tilde{J}_n^2)$. Let $x = (a_1, a_2, \ldots, a_n, 0, \lambda_2) \in V(\tilde{J}_n^2) \setminus H_1$. Let us define the polynomial $P^\varepsilon(y) = P(y) + \varepsilon\varphi(y)$, where $\varphi(y) = (y^2 + \lambda_2^2)^2$, ε being a real number. If

$$\frac{\partial P^\varepsilon(y)}{\partial y} = P_{\varepsilon 1}^{(1)}(y_1, y_2) + i P_{\varepsilon 2}^{(1)}(y_1, y_2),$$

then $P_{\varepsilon 2}^{(1)}(y_1, y_2) = P_2^{(1)}(y_1, y_2) = P_2^{(1)}(y_1, y_2) + 4\varepsilon[(y_1^2 - y_2^2 + \lambda_2^2)y_2 + 2y_1^2 y_2]$, and thus,

$$\frac{\partial P_2^{(1)}(0, \lambda_2)}{\partial y_2} = -8\lambda_2^2 \varepsilon \neq 0$$

for $\varepsilon \neq 0$. Obviously, $x(\varepsilon) = (a_1(\varepsilon), a_2(\varepsilon), \ldots, a_n(\varepsilon), 0, \lambda_2) \in V(\tilde{J}_n^2)$ and for an arbitrary neighbourhood U of the point x there exists $\varepsilon \neq 0$ such that $x(\varepsilon) \in U \cap H_1$. Let $k = 3$ and $D_s(x) = D_s[dP_1, dP_1^{(1)}, dP_1^{(2)}, dQ_2]$. Since $\det D_5(x) = -2$ for all $x \in V_1(\tilde{J}_n^3)$, then $\dim V_1(\tilde{J}_n^3) \leq n - 3$. Now we shall prove assertion 3. Let $p = 1$, $x = (a_1, a_2, \ldots, a_n, \lambda_1, \lambda_2, \mu_1, \mu_2) \in V_1(\tilde{K}_n^1)$ and

$$D_s(x) = D_s[dP_1(\tilde{\lambda}), dP_1(\tilde{\mu}), dP_2(\tilde{\mu}), dQ_1(\tilde{\lambda}), dQ_2(\tilde{\lambda}), dQ_1(\tilde{\mu})],$$

$\tilde{\lambda} = (\lambda_1, \lambda_2)$, $\tilde{\mu} = (\mu_1, \mu_2)$. The matrix derived from the matrix $D_7(x)$ by omitting the second column has its determinant equal to

$$-\mu_2^2 \frac{\partial P_2(0, \mu_2)}{\partial y_2},$$

thus, it is sufficient to prove that

$$G_1 = \left\{ x = (a_1, a_2, \ldots, a_n, 0, 0, 0, \mu_2) \in V(\tilde{K}_n^1) : \frac{\partial P_2(0, \mu_2)}{\partial y_2} \neq 0 \right\}$$

is a dense set in $V(\tilde{K}_n^1)$. Let $x = (a_1, a_2, \ldots, a_n, 0, 0, 0, \mu_2) \in V(\tilde{K}_n^1)$. Let us define the polynomial $P^\varepsilon(y) = P(y) + \varepsilon\varphi(y) = y^n + a_1(\varepsilon)y^{n-1} + \cdots + a_n(\varepsilon) = P_1^\varepsilon(y_1, y_2) + iP_2^\varepsilon(y_1, y_2)$, where $\varphi(y) = y(y^2 + \mu_2^2)$, ε being a real number. Then $P_2^\varepsilon(y_1, y_2) = P_2(y_1, y_2) + \varepsilon[y_2(y_1^2 - y_2^2 + \mu_2^2) - 2y_1^2 y_2]$ and thus

$$\frac{\partial P_2(0, \mu_2)}{\partial y_2} = -2\mu_2^2 \varepsilon \neq 0$$

for $\varepsilon \neq 0$. Obviously, $x(\varepsilon) = (a_1(\varepsilon), a_2(\varepsilon), \ldots, a_n(\varepsilon), 0, 0, 0, \mu_2) \in V(\tilde{K}_n^1)$ and for an arbitrary neighbourhood U of the point x there exists $\varepsilon \neq 0$ such that $x(\varepsilon) \in U \cap G_1$. Now let $p = 2$, $x = (a_1, a_2, \ldots, a_n, \tilde{\lambda}, \tilde{\mu}) \in V(\tilde{K}_n^2)$, $\tilde{\lambda} = (\lambda_1, \lambda_2)$, $\tilde{\mu} = (\mu_1, \mu_2)$ and $D_s(x) = D_s[dP_1(\tilde{\lambda}), dP_1^{(1)}(\tilde{\lambda}), dP_1(\tilde{\mu}), dP_2(\tilde{\mu}), dQ_1(\tilde{\lambda}), dQ_2(\tilde{\lambda}), dQ_1(\tilde{\mu})]$. The matrix

obtained from the matrix $D_8(x)$ by omitting its second column has its determinant equal to
$$-\mu_2^3 \frac{\partial P_1(0, \mu_2)}{\partial y_2},$$
and thus, it is sufficient to prove that the set
$$G_2 = \left\{ x = (a_1, a_2, \ldots, a_n, 0, 0, 0, \mu_2) \in V(\tilde{K}_n^2) : \frac{\partial P_1(0, \mu_2)}{\partial y_2} \neq 0 \right\}$$
is dense in $V(\tilde{K}_n^2)$. Let us define the polynomial $P^\varepsilon(y) = P(y) + \varepsilon\varphi(y) = y^n + a_1(\varepsilon)y^{n-1} + \cdots + a_n(\varepsilon) = P_1^\varepsilon(y_1, y_2) + iP_2^\varepsilon(y_1, y_2)$, where ε is a real number and $\varphi(y) = y^2(y^2 + \mu_2^2)$. Then $P_1^\varepsilon(y_1, y_2) = P_1(y_1, y_2) + \varepsilon[(y_1^2 - y_2^2)(y_1^2 - y_2^2 + \mu_2^2) - 4y_1^2 y_2^2]$, and thus
$$\frac{\partial P_1^\varepsilon(0, \mu_2)}{\partial y_2} = 4\varepsilon\mu_2^3 \neq 0 \quad \text{for} \quad \varepsilon \neq 0.$$

Since $x(\varepsilon) = (a_1(\varepsilon), a_2(\varepsilon), \ldots, a_n(\varepsilon), 0, 0, 0, \mu_2) \in G_2$ for all $\varepsilon \neq 0$, the density of the set G_2 is proved. Now we shall prove assertion 4. Let us assume that $x = (a_1, a_2, \ldots, a_n, \tilde{\lambda}, \tilde{\mu}) \in V(\tilde{L}_n^1)$, $\tilde{\lambda} = (\lambda_1, \lambda_2)$, $\tilde{\mu} = (\mu_1, \mu_2)$ and let $D_5(x) = D_5[dP_1(\tilde{\lambda}), dP_1(\tilde{\mu}), dP_2(\tilde{\mu}), dQ_1(\tilde{\lambda}), dQ_1(\tilde{\mu})]$. Then
$$\det D_5(x) = \mu_2 \frac{\partial P_1(\tilde{\lambda})}{\partial y_2} \cdot \frac{\partial P_1(\tilde{\mu})}{\partial y_2}.$$

Let us assume that $\det D_5(x) = 0$ and let us define the polynomial $P^\varepsilon(y) = P(y) + \varepsilon\varphi(y) = y^n + a_1(\varepsilon)y^{n-1} + \cdots + a_n(\varepsilon) = P_1^\varepsilon(y_1, y_2) + iP_2^\varepsilon(y_1, y_2)$, where ε is a real number and $\varphi(y) = (y^2 + \lambda_2^2)(y^2 + \mu_2^2) = \varphi_1(y_1, y_2) + i\varphi_2(y_1, y_2)$, $\varphi_1(y_1, y_2) = (y_1^2 - y_2^2 + \lambda_2^2)(y_1^2 - y_2^2 + \mu_2^2) - 4y_1^2 y_2^2$, $\varphi_2(y_1, y_2) = 4y_1 y_2[2(y_1^2 + y_2^2) + \lambda_2^2 + \mu_2^2]$. Obviously, $x(\varepsilon) = (a_1(\varepsilon), a_2(\varepsilon), \ldots, a_n(\varepsilon), 0, \lambda_2, 0, \mu_2) \in V(\tilde{L}_n^1)$. Since $\det D_5(x) = 0$, then $d(\varepsilon) = \det D_5[dP_1^\varepsilon(\tilde{\lambda}), dP_1^\varepsilon(\tilde{\mu}), dP_2^\varepsilon(\tilde{\mu}), dQ_1(\tilde{\lambda}), dQ_1(\tilde{\mu})] = \varepsilon\alpha_1 + \varepsilon^2\alpha_2$, where the numbers α_1, α_2 are independent of ε and
$$\alpha_2 = \frac{\partial \varphi_1(0, \lambda_2)}{\partial y_2} \cdot \frac{\partial \varphi_1(0, \mu_2)}{\partial y_2} \mu_2 = 4\lambda_2 \mu_2^2 (\lambda_2^2 - \mu_2^2).$$

Moreover, we can assume that $\lambda_2 + \mu_2 \neq 0$ since in the opposite case the point $x(\varepsilon)$ is situated at the first stratum of a semi-algebraic set which is a subset of the set $V_1(\tilde{L}_n^1)$, its codimension not being greater than $\operatorname{codim} V_1(\tilde{L}_n^1)$. Then we obtain that $\alpha_2 \neq 0$. If $\alpha_1 \neq 0$ then for small enough $\varepsilon \neq 0$ is $d(\varepsilon) \neq 0$. If $\alpha_1 = 0$, then $d(\varepsilon) = \varepsilon^2 \alpha_2 \neq 0$ for $\varepsilon \neq 0$, and thus, assertion 4 is proved. The only thing left to be proved now is assertion 5. Let $x = (a_1, a_2, \ldots, a_n, \tilde{\lambda}, \tilde{\mu}, \tilde{\nu}) \in V(\tilde{L}_n)$, $\tilde{\lambda} = (\lambda_1, \lambda_2)$, $\tilde{\mu} = (\mu_1, \mu_2)$, $\tilde{\nu} = (\nu_1, \nu_2)$ and let $D_{10}(x) = D_{10}[dP_1(\tilde{\lambda}), dP_2(\tilde{\lambda}), dP_1(\tilde{\mu}), dP_2(\tilde{\mu}), dP_1(\tilde{\nu}),$

$dP_2(\tilde{v})$, $dQ_1(\tilde{\lambda})$, $dQ_1(\tilde{\mu})$, $dQ_1(\tilde{v})$]. Let $S(x) \in M(9)$ be the matrix that will be obtained from the matrix $D_{10}(x)$ by omitting its second column. Then

$$g(x) = \det S(x) = \frac{\partial P_1(0, \lambda_2)}{\partial y_2} \cdot \frac{\partial P_1(0, \mu_2)}{\partial y_2} \lambda_2\mu_2(\mu_2^2 - \lambda_2^2) + \frac{\partial P_2(o, \lambda_2)}{\partial y_2} \cdot$$
$$\left[\frac{\partial P_1(0, \mu_2)}{\partial y_2}\mu_2 v_2(v_2^2 - \mu_2^2) + \frac{\partial P_2(0, \mu_2)}{\partial y_2}\lambda_2 v_2(\lambda_2^2 - v_2^2)\right].$$

Let us assume that $g(x) = 0$ and let us define the polynomial $P^\varepsilon(y) = P(y) + \varepsilon\varphi(y) = P_1^\varepsilon(y_1, y_2) + iP_2^\varepsilon(y_1, y_2) = \Psi_1(y_1, y_2) + i\Psi_2(y_1, y_2)$, where

$$\Psi_1(y_1, y_2) = (y_1^2 - y_2^2 + v_2^2)\varphi_1(y_1, y_2) - 2y_1 y_2\varphi_2(y_1, y_2),$$
$$\Psi_2(y_1, y_2) = 2y_1 y_2\varphi_1(y_1, y_2) + (y_1^2 - y_2^2 + v_2^2)\varphi_2(y_1, y_2),$$

where φ_1, φ_2 are the polynomials defined in the proof of assertion 4. Let $D_{10} = D_{10}[dP_1^\varepsilon(\tilde{\lambda}), dP_2^\varepsilon(\tilde{\lambda}), dP_1^\varepsilon(\tilde{\mu}), dP_2^\varepsilon(\tilde{\mu}), dP_1^\varepsilon(\tilde{v}), dP_2^\varepsilon(\tilde{v}), dQ_1(\tilde{\lambda}), dQ_1(\tilde{\mu}), dQ_1(\tilde{v})]$ and let $T(\varepsilon) \in M(9)$ be the matrix obtained from the matrix D_{10} by omitting its second column. Since $g(x) = 0$, then $\det T(\varepsilon) = \varepsilon\alpha_1 + \varepsilon^2\alpha_2$, where the numbers α_1, α_2 are independent of ε and

$$\alpha_2 = \frac{\partial \Psi_1(0, \lambda_2)}{\partial y_2} \cdot \frac{\partial \Psi_1(0, \mu_2)}{\partial y_2}\lambda_2\mu_2(\mu_2^2 - \lambda_2^2) = -4\lambda_2^2\mu_2^2(\mu_2^2 - \lambda_2^2)^2.$$

For the same reasons as in the proof of assertion 4 we can assume that $\mu_2 + \lambda_2 \neq 0$. Since $x \in V(\tilde{L}_n)$, then $\lambda_2 \neq 0, \mu_2 \neq 0, \lambda_2 \neq \mu_2$, and thus, we obtain at $\alpha_2 \neq 0$. If $\alpha_1 \neq 0$ then for small enough $\varepsilon \neq 0$ is $\det T(\varepsilon) \neq 0$. If $\alpha_1 = 0$, then $\det T(\varepsilon) = \varepsilon^2\alpha_2 \neq 0$ for $\varepsilon \neq 0$. □

3.68 Lemma.
1. codim $\hat{I}_1^k \geq 2 + k$, $k = 1, 2, 3$;
2. codim $\hat{J}_1^k \geq 2 + k$, $k = 1, 2, 3$;
3. codim $\hat{K}_1^p \geq 5 + p$, $p = 1, 2,$;
4. codim $\hat{L}_1^1 \geq 6$;
5. codim $\hat{L}_1 \geq 9$.

Lemma 3.68 is a direct consequence of Lemma 3.64 and Lemma 3.67. Since $I_n^k = \pi_2(\tilde{I}_n^k), J_n^k = \pi_2(\tilde{J}_n^k)$, $k = 1, 2, 3, K_n^p = \pi_4(\tilde{K}_n^p)$, $p = 1, 2$, $L_n^1 = \pi_4(\tilde{L}_n^1)$, $L_n = \pi_6(\tilde{L}_n)$, where $\pi_{2k} : M(n) \times R^{2k} \to M(n)$, $k = 1, 2, 3$ are projections, according to Consequence 2.110 these sets are semi-algebraic sets in $M(n)$ and according to Theorem 2.107 there exist their ordered Whitney stratifications $I_n^k = \cup_{j=1}^{m_k} I_{nj}^k$, $J_n^k = \cup_{j=1}^{n_k} J_{nj}^k$, $k = 1, 2, 3$, $K_n^p = \cup_{j=1}^{q_p} K_{nj}^p$, $p = 1, 2, L_n^1 = \cup_{j=1}^{s_1} L_{nj}^1$, $L_n = \cup_{j=1}^{s} L_{nj}$. From Lemma 3.68 the following lemma results.

3.69 Lemma.
1. codim $I_{n1}^k \geq k$, $k = 1, 2, 3$;
2. codim $J_{n1}^k \geq k$, $k = 1, 2, 3$;
3. codim $K_{n1}^p \geq 1 + p$, $p = 1, 2$;
4. codim $L_{n1}^1 \geq 2$;
5. codim $L_{n1} \geq 3$.

3.70 Lemma.
1. codim $I_{n1}^k = k$, $k = 1, 2, ;$
2. codim $J_{n1}^1 = 1$;
3. codim $K_{n1}^1 = 2$;
4. codim $L_{n1}^1 = 2$.

To prove Lemma 3.70 two more lemmas will be needed and before formulating them let us define some mappings. Let $K \subset R^p$ be a p-dimensional compact interval and $B^r = C^r(K, M(n))$. For an arbitrary mapping $A \in B^r$ let $G_A = A/\text{int } K$ and let us define the mapping $F : B^r \to C^r(\text{int } K, M(n))$, $F(A) = G_A$. The evaluation mapping to the mapping F is defined as follows: $\text{ev}_F : B^r \times (\text{int } K) \to M(n)$, $\text{ev}_F(A, x) = A(x)$.

3.71 Lemma.
1. *The evaluation mapping ev_F is of the class C^r;*
2. *For arbitrary $A, B \in B^r$, where $1 \leq r \leq \infty$ and $x, v \in \text{int } K$, it holds that*

$$d_{(A,x)}\text{ev}_F((B, v)) = d_x A(v) + B(x). \tag{3.11}$$

Proof

(For $r = 1$). If ε is a number small enough then with the use of the Taylor Theorem for the mappings A and B we obtain that $\text{ev}_F((A, x) + \varepsilon(B, v)) = \text{ev}_F((A + \varepsilon B, x + \varepsilon v)) = A(x + \varepsilon v) + \varepsilon B(x + \varepsilon v) = A(x) + \varepsilon d_x A(v) + \varepsilon B(x) + o(|\varepsilon|)$ and thus,

$$d_{(A,x)}\text{ev}_F((B, v)) = \lim_{\varepsilon \to 0} \varepsilon^{-1}(\text{ev}_F((A, x) + \varepsilon(B, v)) - \text{ev}_F((A, x)))$$
$$= d_x A(v) + B(x).$$

Therefore from Theorem 1.27 it follows that the mapping ev_F is of class C^1 and the Frechet derivative of this mapping at (A, x) is given by the formula (3.11). □

The proof of Lemma 3.71 for $r > 1$ can be done by induction and is left to the reader.

Vector Fields and Dynamical Systems

3.72 Lemma. *Let $1 \leq r \leq \infty$ and Z be an arbitrary C^r-submanifold in $M(n)$. Then $\text{ev}_F \overline{\cap} Z$.*

Proof

From the definition of transversality it follows immediately that it is sufficient to prove the surjectivity of the mapping $D_{(A,x)}\text{ev}_F$ for arbitrary $(A,x) \in B^r \times (\text{int } K)$. From Proposition 2.24 it follows that this mapping is surjective iff is the mapping $d_{(A,x)}\text{ev}_F$ is surjective. According to Lemma 3.71 it is $d_{(A,x)}\text{ev}_F(B,v) = d_x A(v) + B(x)$. Obviously for arbitrary $C \in M(n)$ there exists a mapping $B \in B^r$ such that $B(x) = C$, and thus, from the above formula we obtain that $d_{(A,x)}\text{ev}_F(0,v) = B(x) = C$ (0 is the zero element in B^r). □

3.73 Lemma. *Let $K \subset R^p$ be a compact p-dimensional interval, Z be a C^r-submanifold in $M(n)$, $\text{codim } Z = q$ and $r > \max(0, p - q)$. Then the following assertions hold*

1. If $W \subset \text{int } K$ is a compact set and Z is closed in $M(n)$ then the set $H_{W,Z} = \{A \in C^r(K, M(n)) : G_A \overline{\cap}_W Z\}$ is open and dense in $C^r(K, M(n))$,

2. The set $H_Z = \{A \in C^r(K, M(n)) : G_A \overline{\cap} Z\}$ is massive (and thus, dense) in $C^r(K, M(n))$, where $G_A = A/\text{int } K$.

Proof

Lemma 3.72 and the assumptions of the lemma guarantee that all the assumptions of the Abraham Transversality Theorem are fulfilled, from which both the assertions of the above lemma follow. □

Proof of Lemma 3.70

It is sufficient to prove the inequalities reversed to the inequalities 1. for $k = 1, 2, 3$ for $k = 1, 3$ and 4 from Lemma 3.69. In the proof we shall use the notation $P_B(\lambda) = \lambda^n + \alpha_1(B)\lambda^{n-1} + \cdots + \alpha_n(B)$ for the characteristic polynomial of the matrix $B \in M(n)$. We shall prove assertion 1. Let $k = 1$. Let us assume that $\text{codim } I_{n1}^1 > 1$. Let us consider the mapping $A : K \to M(n)$, $A(t) = \text{diag}\{t, 1, 1, \ldots, 1\}$, where $K = \langle -1, 1 \rangle$ and let for $B \in C^r(K, M(n))$ be $G_B = B/\text{int } K$. Obviously, $G_A(0) = A(0) \in I_{n1}^1$. From Lemma 3.73 it follows that for an arbitrary neighbourhood $U(A)$ of the mapping A in $C^r(K, M(n))$ there exists a mapping $\tilde{A} \in U(A)$ such that $G_{\tilde{A}} \overline{\cap} I_{n1}^1$. From Proposition 2.75 it follows that $G_{\tilde{A}}^{-1}(I_{n1}^1) = \emptyset$. This, however, would mean that the matrix $A(t)$ has no zero eigenvalue for any $t \in (-1, 1)$ which for a small enough neighbourhood $U(A)$ is obviously impossible. Let $k = 2$. Let us assume that $\text{codim } I_{n1}^2 > 2$ and let us consider the mapping $A : K \to M(n)$,

99

$$A(t,s) = \text{diag}\left\{\begin{bmatrix} 0 & 1 \\ s & t \end{bmatrix}, 1, 1, \ldots, 1\right\},$$

where $K = \langle -1, 1\rangle \times \langle -1, 1\rangle$ and let $G_B = B/\text{int } K$. Obviously, $G_A(0,0) = A(0,0) \in I_{n1}^2$. From Lemma 3.73 and from Proposition 2.75 it follows that for an arbitrary neighbourhood $U(A)$ of the mapping A in $C^r(K, M(n))$ there exists a mapping $\tilde{A} \in U(A)$ such that $G_{\tilde{A}}^{-1}(I_{n1}^2) = \emptyset$. In this case, however, the contradiction is not as clear as for $k = 1$. It can be verified very easily that $\alpha_{n-1}(A(t,s)) = f_1(t,s) \stackrel{\text{def}}{=} (-1)^{n+1}(t - (n-2)s)$ and $\alpha_n(A(t,s)) = f_2(t,s) \stackrel{\text{def}}{=} (-1)^{n+1}s$. Let $\alpha_{n-1}(\tilde{A}(t,s)) = g_1(t,s), \alpha_n(\tilde{A}(t,s)) = g_2(t,s)$ and let us consider the mappings $f = (f_1, f_2), g = (g_1, g_2)$. Let $F : C^r(K, R^2) \times K \to R^2, F(h,t,s) = h(t,s)$ and let $F_h : K \to R^2, F_h(t,s) = F(h,t,s)$ for $h \in C^r(K, R^2)$. Then, obviously, $F(f, 0, 0) = f(0, 0) = (0, 0)$, the mapping F is of the class C^r and $d_{(0,0)}F_f : R^2 \to R^2$ is a linear isomorphism. From the Implicit Function Theorem it follows that there exists a neighbourhood $V(f)$ of the mapping f in $C^r(K, R^2)$, a neighbourhood W of the point $(0,0)$ in R^2 and a C^r-mapping $w : V(f) \to W$, such that $w(f) = (0,0), F(h, w(h)) = (0,0)$ for all $h \in V(f)$. If $U(A)$ is a small enough neighbourhood of the mapping A, then $g \in V(f)$, and thus, $F(g, w(g)) = 0$. This means that $g_1(t,s) = 0, g_2(t,s) = 0$ for some $(t,s) \in \text{int } K$ and thus, the matrix $A(t,s)$ has a double zero eigenvalue. This, however contradicts the fact that $G_{\tilde{A}}^{-1}(I_{n1}^2 = \emptyset$. The proof of assertion 2 is the same as that of assertion 1 for $k = 1$ with the difference resting in the fact that the mapping is defined as follows:

$$A(t) = \text{diag}\left\{\begin{bmatrix} t & 1 \\ -1 & t \end{bmatrix}, 1, 1, \ldots, 1\right\}.$$

Now we shall prove assertion 3. Let us assume that codim $K_{n1}^1 > 2$. Let us consider the mapping

$$A : K \to M(n), \quad A(t,s) = \text{diag}\left\{\begin{bmatrix} s & 1 \\ -1 & s \end{bmatrix}, t, 4, 5, \ldots, n\right\}$$

and let $G_B = B/\text{int } K$, where $K = \langle -1, 1\rangle \times \langle -1, 1\rangle$. Obviously, $G_A(0,0) = A(0,0) \in K_{n1}^1$. From Lemma 3.73 and Proposition 2.75 it follows that for arbitrary neighbourhood $U(A)$ of the mapping A there exists a mapping $\tilde{A} \in U(A)$ such that $G_{\tilde{A}}^{-1}(K_{n1}^1) = \emptyset$. In this case the contradiction is still not clear enough. Moreover, the set K_n^1 cannot be explicitly defined through the coefficients of characteristic polynomials as it was possible in case of the set I_n^2. In this case, however, all the eigenvalues $\lambda_1^0 = i, \lambda_2^0 = -i, \lambda_3^0 = 0, \lambda_j^0 = j, j = 4, 5, \ldots, n$, of the matrix $A(0,0)$ are of multiplicity 1. According to Lemma 3.65 there exists an open neighbourhood $V(A)$ of the mapping A in $C^r(K, M(n))$, an open neighbourhood $U \subset \text{int } K$ of the point $(0,0)$ in R^2, a C^r-mapping $(r \geq 1)$ $C : V(A) \times U \to GL(n)$ and a C^r-mappings $\lambda_{1k} : V(A) \times U \to R$, $k = 1, 2, \lambda_j : V(A) \times U \to R$, $j = 3, 4, \ldots, n$ such that for $(B, t, s) \in V(A) \times U$, $\lambda_1(B, t, s) = \lambda_{11}(B, t, s) + i\lambda_{12}(B, t, s)$, $\lambda_2(B, t, s) = \overline{\lambda_1(B, t, s)}$, $\lambda_j(B, t, s)$, $j = 3, 4, \ldots, n$, are the eigenvalues of the matrix $B(t, s)$ and it holds that $C(B, t, s) B(t, s)(C(B, t, s))^{-1} = \text{diag}\{D(B, t, s), \lambda_3(B, t, s), \lambda_4(B, t, s), \ldots, \lambda_n(B, t, s)\}$, where $D(B, t, s) = (d_{ij})$, $d_{11} = d_{22} = \lambda_{11}(B, t, s)$, $d_{12} = -d_{21} = \lambda_{12}(B, t, s)$. Let us define the mapping $F : V(A) \times U \to R^2$, $F(B, t, s) = (\lambda_{11}(B, t, s), \lambda_3(B, t, s))$ and the mapping $F_A : U \to R^2$, $F_A(t, s) = F(A, t, s)$. Obviously, F is of the class C^r, $F(A, 0, 0) = (0, 0)$ and the mapping $d_{(0,0)} F_A : R^2 \to R^2$ is an isomorphism. From the Implicit Function Theorem it follows that there exists a neighbourhood $V_1(A)$ of the mapping A, a neighbourhood W of the point $(0, 0)$ and a C^r-mapping $\sigma : V_1(A) \to W$ such that $\sigma(A) = (0, 0)$, $F(B, \sigma(B)) = (0, 0)$ for all $B \in V_1(A)$. If the neighbourhood $U(A)$ is so small that $U(A) \subset V_1(A)$, then $F(\tilde{A}, \sigma(\tilde{A})) = (0, 0)$, and thus, there exists a point $(t_0, s_0) \in \text{int } K$ such that $\tilde{A}(t_0, s_0)$ has the zero eigenvalue and a pure imaginary eigenvalue. This, however, is in contradiction with the fact that $G_{\tilde{A}}^{-1}(K_{n1}^1) = \emptyset$. Proposition 4 is still to be proved. Let us assume that $\text{codim } L_{n1}^1 > 2$. Let us consider the mapping $A : K \to M(n)$, $A(t, s) = \text{diag}\{C(s), D(s), 5, 6, \ldots, n\}$, where $K = \langle -1, 1\rangle \times \langle -1, 1\rangle$, $C(s) = (c_{ij}(s)) \in M(2)$, $c_{11}(s) = c_{22}(s) = s$, $c_{12}(s) = -c_{21}(s) = 1$, $D(t) = (d_{ij}(t)) \in M(2)$, $d_{11}(t) = d_{22}(t) = t$, $d_{12}(t) = -d_{21}(t) = 2$. Let the mapping G_A be defined in the same way as in the preceding parts of the proof. Obviously, $G_A(0, 0) \in L_{n1}^1$. For an arbitrary neighbourhood $U(A)$ of the mapping A there exists a mapping $\tilde{A} \in U(A)$ such that $G_{\tilde{A}} \cap L_{n1}^1$, and thus, $G_{\tilde{A}}^{-1}(L_{n1}^1) = \emptyset$. According to Lemma 3.65 there exists a neighbourhood $V(A)$ of the mapping A, a neighbourhood $U \subset \text{int } K$ of the point $(0, 0)$ and a C^r-mapping $H : V(A) \times U \to GL(n)$, such that for $(B, t, s) \in V(A) \times U$ it is $H(B, t, s) B(t, s)(H(B, t, s))^{-1} = \text{diag}\{D_1, D_2, \lambda_5, \lambda_6, \ldots, \lambda_n\}$, where $D_k = (d_{ij}^k) \in M(2)$, $d_{11}^k = d_{22}^k = \lambda_{k1}(B, t, s)$, $d_{12}^k = -d_{21}^k = \lambda_{k2}(B, t, s)$, $k = 1, 2, \lambda_j = \lambda_j(B, t, s)$, $j = 5, 6, \ldots, n$. Let us define the mapping $F : V(A) \times U \to R^2$, $F(B, t, s) = (\lambda_{11}(B, t, s), \lambda_{21}(B, t, s))$. Obviously F is of class C^r, $F(A, 0, 0) = (0, 0)$ and the mapping $d_{(0,0)} F_A$ is an isomorphism, $F_A : U \to R^2$, $F(t, s) = F(A, t, s)$. Analogously to the proof of assertion 3, from the Implicit Function Theorem the existence of a point $(t_0, s_0) \in \text{int } K$ follows, such that the matrix $\tilde{A}(t_0, s_0)$ has two different pure imaginary eigenvalues, excluding their complex conjugates. This is in contradiction with $G_{\tilde{A}}^{-1}(L_{n1}^1) = \emptyset$. □

Let us denote the subset of matrices of $M(n)$ having the zero eigenvalue of multiplicity k, m pure imaginary eigenvalues different from each other of multiplicity $j_1, j_2, \ldots, j_m$ with no other eigenvalues with zero real part except eigenvalues that are their complex conjugates as

$\sum_n(0[k], I[j_1, j_2, \ldots, j_m])$. Let $\sum_n(0[k])$ be the set of matrices of $M(n)$ with zero eigenvalue of multiplicity k and having no other eigenvalue with zero real part and let us denote as $\sum_n(I[j_1, j_2, \ldots, j_m])$ the set of matrices of $M(n)$ having n pure imaginary eigenvalues different from each other of multiplicity $j_1, j_2, \ldots, j_m$ and having no other eigenvalues with zero real part except their complex conjugate eigenvalues.

3.74 Theorem. *Let $1 \leq r \leq \infty$ and X be a C^∞-manifold of dimension k or $X \subset R^k$ be a k-dimensional interval. Then there exists a massive subset N^r in $C^r(X, M(n))$ such that the following assertions hold.*

1. If $k = 1$ and $A \in N^r$, then the sets $A^{-1}(\sum_n(0, [1])), A^{-1}(\sum_n I[1]))$ consist of isolated points;

2. If $k = 2$ and $A \in N^r$, then the sets $A^{-1}(\sum_n(0[1])), A^{-1}(\sum_n(I[1]))$ are one-dimensional C^r-submanifolds in X and the sets $A^{-1}(\sum_n(0[2]))$, $A^{-1}(\sum_n(0[1], I[1])), A^{-1}(\sum_n(I[1, 1]))$ consist of isolated points.

All other sets $A^{-1}(\sum_n(0[k])), A^{-1}(\sum_n(I[j_1, j_2, \ldots, j_m])), A^{-1}(\sum_n(0[k])), A^{-1}(\sum_n(0[k], I[j_1, j_2, \ldots, j_m]))$ except the above sets in individual cases are empty. If X is a compact manifold or a compact interval then the set $N^r(X)$ is also an open set.

Proof

From Theorem 2.113 it follows that the sets $T(I_n^k) = \{A \in C^r(X, M(n)) : A \overline{\pitchfork} I_n^k\}, T(J_n^k) = \{A \in C^r(X, M(n)) : A \overline{\pitchfork} J_n^k\}, k = 1, 2, 3, T(K_n^p) = \{A \in C^r(X, M(n)) : A \overline{\pitchfork} K_n^p\}, p = 1, 2, T(L_n^1) = \{A \in C^r(X, M(n)) : A \overline{\pitchfork} L_n^1\}, T(L_n) = \{A \in C^r(X, M(n)) : A \overline{\pitchfork} L_n\}$ are massive in $C^r(X, M(n))$ (see the notations introduced before Lemma 3.69) and thus, also the set

$$N^r(X) = (\cap_{k=1}^3 (T(I_n^k) \cap T(J_n^k))) \cap (\cap_{p=1}^2 T(K_n^p)) \cap T(L_n^1) \cap T(L_n)$$

is also massive in $C^r(X, M(n))$. If X is a compact manifold or a compact interval then according to the same theorem the set $N^r(X)$ is open, as well. From Proposition 2.76, from Lemma 3.69 and Lemma 3.70 all other properties of the set $N^r(X)$ presented in the theorem follow. □

Let us denote by $\Omega_n(1[k], -1[p], S[j_1, j_2, \ldots, j_m])$ the set of matrices of $M(n)$ having the eigenvalue 1 of multiplicity k, eigenvalue -1 of multiplicity p, and having m other distinct eigenvalues on the unit circle S^1, of multiplicity $j_1, j_2, \ldots, j_m$ respectively. None of the latter set of eigenvalues is equal to the complex conjugate of any other, and there are no other eigenvalues on S^1 except the m complex conjugates of these. In the same way in which we defined the sets $\sum_n 0[k])$ and $\sum_n(I[j_1, j_2, \ldots, j_m])$ we can also define the sets $\Omega_n(1[k]), \Omega_n(-1[k])$ and $\Omega_n(S[j_1, j_2, \ldots, j_m])$.

3.75 Theorem. *Let X be a smooth, compact manifold and $1 \leq r \leq \infty$. Then for an arbitrary natural number k there exists an open dense subset D_{nk}^r in $C^r(X, M(n))$ such that the following assertions hold.*

1. If $\dim X = 1$ and $A \in D_{nk}^r(X)$, then the sets $A^{-1}(\Omega_n(1[1]))$, $A^{-1}(\Omega_n(-1[1]))$, $A^{-1}(\Omega_n(S[1]))$ consist of isolated points and if the matrix $A(\mu)$ has eigenvalue $\lambda \in S^{-1}$, $\lambda \neq \pm 1$ for some $\mu \in X$, then $\lambda^k \neq \pm 1$.

2. If $\dim X = 2$, then the sets $A^{-1}(\Omega_n(1[1]))$, $A^{-1}(\Omega_n(-1[1]))$, $A^{-1}(\Omega_n(S[1]))$ are one-dimensional C^r-submanifolds in X and there exists a set $X_0(A) \subset X$ consisting of isolated points such that for $\mu \in X_0(A)$ the matrix $A(\mu)$ has at least two eigenvalues, let us denote them by λ, and v, respectively, on S^1, where one of the following cases occurs:

1. $\lambda = +1$ of multiplicity 2,
2. $\lambda = -1$ of multiplicity 2,
3. $\lambda \in S^1$ of multiplicity 2, $\lambda \neq \pm 1$,
4. $\lambda = +1, \lambda = -1$ both of multiplicity 1,
5. $\lambda \in S^1$ of multiplicity 1, $\lambda \neq \pm 1$, $\lambda^m = 1$ for some $m \in N$,
6. $\lambda = +1, v \in S^1$ both of multiplicity 1, $v \neq \pm 1, v^k \neq \pm 1$,
7. $\lambda = -1, v \in S^1$ both of multiplicity 1, $v \neq \pm 1, v^k \neq \pm 1$,
8. $\lambda \in S^1, v \in S^1$ both of multiplicity 1, $\lambda \neq \pm 1$, $v \neq \pm 1$, $\lambda^k \neq \pm 1$, $v^k \neq \pm 1$, $\mathrm{Re}\,\lambda \neq \mathrm{Re}\,v$.

Except for the above eigenvalues in no of the cases 1–8 does the matrix $A(\mu)$ have any other eigenvalues on S^1 except of their complex conjugates. If

$$\mu \in X \setminus (A^{-1}(\Omega_n(1[1])) \cup A^{-1}(\Omega_n(-1[1])) \cup A^{-1}(\Omega_n(S[1])) \cup X_0(A)),$$

then the matrix $A(\mu)$ has no eigenvalues on S^1.

The set $D_n^r(X)$ of all the mappings for which assertion 1 is valid, if $\dim X = 1$ and assertion 2, if $\dim X = 2$, namely for an arbitrary natural number k, is massive in $C^r(X, M(n))$ (X need not be compact).

Proof of assertion 1 can be found in [25] while the proof of assertion 2 is presented in [109].

3.5 Linear Dynamical Systems and Some Notions From the Theory of Non-linear Dynamical Systems

Let the following linear differential equation on n-dimensional Euclidean space E_n be given:

$$\dot{x} = Ax, \qquad (3.12)$$

where $A \in L(E_n)$. From Theorem 1.42 it follows that the above differential equation is complete.

3.76 Theorem. *If $A \in L(E_n)$ then the following assertions hold.*

1. The mapping $U : R \to L(E_n)$, $U(t) = e^{tA}$ is an operator solution of the differential equation (3.12) fulfilling the condition $U(0) = \mathrm{id}$.

2. For arbitrary $x_0 \in E_n$ is $x : R \to E_n$, $x(t) = U(t)x_0$ the solution of the differential equation (3.12) fulfilling the condition $x(0) = x_0$.

3. The mapping $\varphi : E_n \times R \to E_n, \varphi(x,t) = U(t)x$, is a C^∞-dynamical system on E_n.

Proof

Let $k \in N$,

$$S_k : L(E_n) \to L(E_n), \quad S_k(B) = \mathrm{id} + \sum_{i=1}^{k} \frac{1}{i!} B^i,$$

$I = \langle -T, T \rangle, T > 0$ and $\{U_j\}_{j=1}^\infty$, where $U_j : I \to L(E_n), U_j(t) = S_j(tA)$. For every $j \in N$ the mapping U_j is of class C^∞. The sequence $\{U_j\}_{j=1}^\infty$ converges in the metric space $C(I, L(E_n))$ to the mapping U. For every $k \in N$ and $t \in I$, $\dot{U}_k(t) = AU_{k-1}(t)$, and thus, for arbitrary $t, \tau \in I$ is

$$\int_\tau^t AU_{k-1}(s)\mathrm{d}s = U_k(t) - U_k(\tau).$$

Since $\lim_{k \to \infty} U_k = U$ and T is an arbitrary positive number, then

$$\int_\tau^t Ae^{sA}\mathrm{d}s = e^{tA} - e^{\tau A}$$

for arbitrary $t, \tau \in R$. By differentiating this equality by t we obtain that

$$\dot{U}(t) = \frac{\mathrm{d}}{\mathrm{d}t}(e^{tA}) = Ae^{tA} = AU(t),$$

where obviously $U(0) = \mathrm{id}$, i.e. assertion 1 holds. Since $x(t) = U(t)x_0$, assertion 2 results from Proposition 1. Obviously, the mapping φ is of class C^∞ and $\varphi_0(x) = U(0)x = x$ for every $x \in E_n$. Since $e^{(t+s)A} = e^{tA} \circ e^{sA}$ for all $t, s \in R$ (for the proof of this equality see, e.g. [68]), and thus, $\varphi_{t+s} = \varphi_t \circ \varphi_s$. □

The dynamical system φ from Theorem 3.72 on a smooth manifold E_n defines a smooth vector field called the *linear vector field* defined by the mapping A denoted as A_v or equally as the mapping A. The set of all linear vector fields on E_n is denoted as $V_L(E_n)$. If $A_v, B_v \in V_L(E_n), \lambda \in R$, then we define $A_v + B_v \stackrel{\text{def}}{=} (A+B)_v$ and $\lambda A_v \stackrel{\text{def}}{=} (\lambda A)_v$. The set $V_L(E_n)$ with these operations and with the norm $\|A_v\| = \|A\|$ is a Banach space. The dynamical system φ is called the linear dynamical system on E_n defined by the mapping A or the linear flow of the vector field A_v.

3.77 Definition. *Let E_n, F_n be n-dimensional Euclidean spaces, φ being a linear flow of the vector field $A \in V_L(E_n)$, and Ψ be a linear flow of the vector field $B \in V_L(F_n)$. We say that φ is linearly equivalent to Ψ (we write $\varphi \sim_L \Psi$), if there exists a linear mapping $T \in GL(E_n, F_n)$ and a number $a > 0$ such that $T(\varphi(x,t)) = \varphi(T(x), a \cdot t)$ for all $(x,t) \in E_n \times R$.*

The relation $\sim_L$ is obviously the equivalence relation on the set of all the linear dynamical systems whose state spaces are n-dimensional euclidean spaces. With the mark γ_x^φ the trajectory of the dynamical system φ passing through the point x is denoted. If φ is a flow of the field F then this trajectory is also denoted as γ_x^F. If it is clear which dynamical system or vector field is concerned, the upper index will be omitted.

If the linear flows φ and Ψ of the vector fields $A \in V_L(E_n)$ or $B \in V_L(F_n)$, respectively, are linearly equivalent and $x \in E_n$ then, obviously, the mapping T from Definition 3.77 maps the trajectory γ_x^φ onto the trajectory $\gamma_{T(x)}^\Psi$.

3.78 Proposition. *Let φ and Ψ be linear flows of vector fields $A \in V_L(E_n)$ and $B \in V(E_n)$, respectively. Then $\varphi \sim_L \Psi$ iff there exists a number $a > 0$ and a mapping $T \in GL(E_n)$ such that $A = a(T^{-1}BT)$.*

Proof

If $\varphi \sim_L \Psi$, then there exists a number $a > 0$ and a mapping $T \in GL(E_n)$ such that for arbitrary $(x,t) \in E_n \times R$ is $\varphi_t(x) = T^{-1}\Psi_{a \cdot t}(Tx)$, and thus,

$$A\varphi_t(x) = \frac{d\varphi_t(x)}{dt} = T^{-1}\frac{d\Psi_{a \cdot t}(Tx)}{dt} = a \cdot T^{-1}B\Psi_{a \cdot t}(Tx)$$
$$= a \cdot T^{-1}BT\varphi_t(x),$$

i.e. $A\varphi_t = a(T^{-1}BT)\varphi_t$. Since $\varphi_0 = \text{id}$, then $A = a(T^{-1}BT)$. Now let us assume that there exists a number $a > 0$ and a mapping $T \in GL(E_n)$, such that $A = a(T^{-1}BT)$. Then for arbitrary $(x,t) \in E_n \times R$ it is

$$\varphi_t(x) = e^{tA}x = e^{a \cdot t(T^{-1}BT)}x = T^{-1}e^{a \cdot tB}Tx = T^{-1}\Psi_{a \cdot t}(Tx),$$

i.e. $\varphi \sim_L \Psi$. □

3.79 Definition. *If E_n is an n-dimensional Euclidean space, $A \in V_L(E_n)$ and $\lambda_1, \lambda_2, \ldots, \lambda_n$ are eigenvalues of the mapping A. The singular point $x_0 = 0$ of the vector field A is called hyperbolic if $\text{Re}\,\lambda_i \neq 0$ for $i = 1, 2, \ldots, n$. In the case $n = 2$ this hyperbolic singular point is called the node (focus) if eigenvalues of the mapping A are real (complex) and have equal signs of their real parts. If there exist $i, j \in \{1, 2, \ldots, n\}$, such that, $\text{Re}\,\lambda_i < 0$ and $\text{Re}\,\lambda_j > 0$ then the hyperbolic singular point x_0 is called* the saddle. *If the singular point x_0 is hyperbolic then the vector field A is called hyperbolic and its flow is called also hyperbolic. The set of all linear hyperbolic vector fields on E_n is denoted as $HV_L(E_n)$.*

3.80 Lemma. *The set $HV_L(R^n)$ is open and dense in $V_L(R^n)$.*

Proof

Let a basis $\mathscr{B}$ in R^n be given and $HM(n)$ be the set of the matrices $A \in M(n)$ such that the mapping $\tilde{A} \in L(R^n)$ whose representation $[\tilde{A}]$ with respect to the basis $\mathscr{B}$ is equal to A, defines a hyperbolic vector field $\tilde{A}_v \in HV_L(R^n)$. It is sufficient to prove that the set $HM(n)$ is dense in $M(n)$. Let $A \in M(n) \setminus HM(n)$, $I = \langle -2, 2 \rangle$, $J = \langle -1, 1 \rangle$ and $F \in C^r(I, M(n))$ ($r \geq 1$) is a mapping such that $F(0) = A$. From Theorem 3.74 it follows that in an arbitrary neighbourhood of the mapping F there exists a mapping $G \in C^r(I, M(n))$, such that the set $\{\mu \in J: \text{matrix } G(\mu) \text{ has an eigenvalue with zero real part}\}$ consists of a finite number of points. This, however, means that for arbitrary $\varepsilon > 0$ there exists $\mu_\varepsilon \in J$ such that $\|A - G(\mu_\varepsilon)\| < \varepsilon$, where $G(\mu_\varepsilon) \in HM(n)$. Thus, the density of the set $HM(n)$ is proved. Since the function $\det : M(n) \to R$, $A \mapsto \det A$ is continuous and for arbitrary $A_0 \in HM(n)$ $\det A_0 \neq 0$, for every $A_0 \in HM(n)$ there exists a neighbourhood $U(A_0)$ of the matrix A_0 such that for all $A \in U(A_0)$ $\det A \neq 0$, i.e., the matrix A has no zero eigenvalue. It is, however, still to be proved that if the neighbourhood $U(A_0)$ is small enough then no matrix from $U(A_0)$ has imaginary eigenvalues. Let us assume that such a neighbourhood of matrix A_0 does not exist. Then there exists a sequence of matrices $\{A_k\}_{k=1}^\infty$ converging to the matrix A_0 such that $A_k \notin HM(n)$ for all $k \in N$, where each of these matrices has a pure imaginary eigenvalue $\lambda_k = ib_k$, $b_k \neq 0$. From the convergence of this sequence it follows that there exists a number $K > 0$ such that $\|A_k\| \leq K$ for all $k \in N$. According to Theorem 1.23 for every λ_k there exist vectors $x_k, y_k \in R^n$ such that $A_k x_k = -b_k y_k$, $A_k y_k = b_k x_k$ and $\|x_k\|^2 + \|y_k\|^2 = 1$. The vector $z_k = x_k + iy_k$ is obviously an eigenvector of the matrix A_k coresponding to the eigenvalue λ_k, and thus, for every $k \in N$ it is

$$|b_k| = |\lambda_k| = \|\lambda_k z_k\| = \|A_k z_k\| \leq \max_{\|z\|=1} \|A_k z\| = \|A_k\| \leq K.$$

Let $P = \{w \in R : |w| \leq K\}$ and $Q = \{(u, v) \in R^n \times R^n : \|u\|^2 + \|v\|^2 = 1\}$. Since the set $P \times Q \times Q$ is compact, then from Theorem 1.8 it follows that there exists a subsequence $\{k_j\}_{j=1}^\infty$ in N, such that, the sequence $\{(b_{k_j}, x_{k_j}, y_{k_j})\}_{j=1}^\infty$ converges to $(b, x, y) \in P \times Q \times Q$. Therefore we obtain that $A_0 x = -by$, $A_0 y = bx$, and thus, $A_0 z = \lambda z$, where $z = x + iy$, $\lambda = ib$. This means that λ is either zero or a pure imaginary eigenvalue of the matrix A_0, which is in contradiction with the assumption. $\square$

3.81 Theorem. *Let E_n be an n-dimensional Euclidean space. Then the set $HV_L(E_n)$ is open and dense in $V_L(E_n)$.*

Proof

Let $[T]$ be the matrix representation of the mappings $T \in L(E_n)$ with respect to a basis $\mathscr{B}$ in E_n. According to Theorem 1.24 λ is an eigenvalue of the mapping T iff it is an eigenvalue of the matrix $[T]$. This means that $A_v \in HV_L(E_n)$ iff $\tilde{A}_v \in HV_L(R^n)$, where $\tilde{A}_v$ is the vector field on R^n defined by the matrix $[A]$. Since $\|A_v\| = \|A\|$ is the norm on $V_l(E_n)$, then the set $HV_L(E_n)$ is open and dense in $V_L(E_n)$ iff the set $HV_L(R^n)$ is open and dense in $V_L(R^n)$. Therefore the assertion is a consequence of Lemma 3.80.

The assertion of Theorem 3.81 in fact, means that the hyperbolicity of linear vector fields on E_n is a generic property in $V_L(E_n)$.

Now we shall study asymptotic properties of trajectories of dynamical systems in neighbourhoods of their hyperbolic singular points.

3.82 Lemma. *Let E_n be an n-dimensional Euclidean space, $A \in L(E_n)$ and there exist the numbers α, β such that*

$$\alpha < \operatorname{Re}\lambda < \beta \tag{3.13}$$

for arbitrary eigenvalue λ of the mapping A. Then there exists a basis $\mathscr{B}$ in E_n such that

$$\alpha \|x\|_{\mathscr{B}}^2 \leq (Ax, x) \leq \beta \|x\|_{\mathscr{B}}^2 \tag{3.14}$$

for all $x \in E_n$, where $(.,.)_{\mathscr{B}}$ is the representation of the scalar product on E_n with respect to the basis $\mathscr{B}$ and $\|x\|_{\mathscr{B}} = (x,x)_{\mathscr{B}}^{1/2}$.

Proof

From the definition of the representation of the scalar product with respect to the basis $\mathscr{B}$ it follows that it suffices to prove the lemma for $E_n = R^n$. In this case a matrix from $M(n)$ corresponds to the mapping A. We shall prove that there exists a regular matrix $S \in M(n)$ and a matrix $B \in M(n)$ such that $A = SBS^{-1}$ and

$$\alpha \|y\|^2 \leq (B, y) \leq \beta \|y\|^2 \tag{3.15}$$

for all $y \in R^n$, where $(.,.)$ is the scalar product on R^n defined by the basis of unit vectors $e_1, e_2, \ldots, e_n$ on R^n. Let us assume at first that such matrices S and B do exist. Then the vectors $Se_1, Se_2, \ldots, Se_n$ form a new basis $\mathscr{B}$ on R^n and if $(.,.)_{\mathscr{B}}$ is the representation of scalar product $(.,.)$

with respect to the basis $\mathscr{B}$, then $(x,y)_{\mathscr{B}} = (S^{-1}x, S^{-1}y)$ for all $x, y \in R^n$.
We shall show that the inequality (3.14) follows from the inequality (3.15).
For arbitrary $x \in R^n$ there exists $y \in R^n$ such that $x = Sy$ and under the assumption of validity of (3.15) we find that

$$\alpha\|x\|_{\mathscr{B}}^2 = \alpha(S^{-1}x, S^{-1}x) = \alpha\|y\|^2 \leq (By,y) = (SBS^{-1}x,x)_{\mathscr{B}} = (Ax,x)_{\mathscr{B}}$$
$$\leq \beta\|y\|^2 = \beta(Sy, Sy)_{\mathscr{B}} = \beta\|x\|_{\mathscr{B}}^2.$$

Since $(By,y) = (BS^{-1}x, S^{-1}x) = (SBS^{-1}x,x)_{\mathscr{B}} = (Ax,x)_{\mathscr{B}}$, we find that the inequality (3.14) is fulfilled. Now we shall prove the existence of the matrices S and B. According to Theorem 1.18 there exists a regular matrix $R \in M(n)$ such that $A = R\tilde{A}R^{-1}$, where $\tilde{A} = \text{diag}\{A_1, A_2, \ldots, A_m\}$, where the matrix A_j ($1 \leq j \leq m$) has one of the following form:

1. $A_j = \lambda I_{n(j)}$;
2. $A_j = \lambda I_{n(j)} + N$,

where λ is an eigenvalue of the matrix A, $I_{n(j)} \in M(n(j))$ is the unit matrix, $N \in M(n(j))$ is a matrix whose all elements above its main diagonal are equal to 1 the other being zero;

3. $A_j = \text{diag}\{D, D, \ldots, D\} \in M(n(j))$;
4. $A_j = \text{diag}\{D, D, \ldots, D\} + Q \in M(n(j))$,

where $D = (d_{ij}) \in M(2), d_{11} = d_{22} = a, d_{12} = -d_{21} = -b, \lambda = a + ib$ is an eigenvalue of the matrix A, $b \neq 0$, Q is a block matrix having all elements above its main diagonal equal to the unit matrix $I_2 \in M(2)$, its other elements being equal to zero matrices from $M(2)$.

Every linear subspace $L_j \subset R^n$ corresponding to the block A_j is an invariant subspace for A, i.e. $AL_j \subset L_j$, and thus, it suffices to assume that the matrix A has one of the forms 1—4. Let us assume that the matrix A has the form 1, i.e., $A = \lambda I_n$, where λ is a real number. Then, obviously $(Ay,y) = \lambda\|y\|^2$ for all $y \in R^n$, and thus, the equality (3.15) is fulfilled. Let the matrix A have the form 2, i.e., $A = \lambda I_n + N$, where λ is a real number. Let the number $\varepsilon \neq 0$ be given and let $R_\varepsilon = (r_{ij}) \in M(n)$, where $r_{ij} = \varepsilon^{n-j}$ for $i = 1, 2, \ldots, j$ and $r_{ij} = 0$ for $i = j+1, j+2, \ldots, n$ ($j = 1, 2, \ldots, n$). Obviously, the matrix R_ε is regular and it holds that $R_\varepsilon N R_\varepsilon^{-1} = \varepsilon N$, $A_\varepsilon = R_\varepsilon A R_\varepsilon^{-1} = \lambda I_n + \varepsilon N$. With the use of the inequality (3.15) for the case of the matrix $B = \lambda I_n$ that has already been proved we find that if the number ε is positive and small enough, then

$$\alpha\|y\|^2 \leq (\lambda - \varepsilon)\|y\|^2 \leq (A_\varepsilon y, y) \leq (\lambda + \varepsilon)\|y\|^2 \leq \beta\|y\|^2$$

for all $y \in R^n$. Thus, it suffices to choose $B = A_\varepsilon$. Now, let us assume that the matrix A is of the form 3, i.e., $A = \text{diag}\{D, D, \ldots, D\}$, where D is of the same form as in the preceding part of the proof and $\lambda = a + ib$. Then, obviously, $(Ay, y) = (\text{Re }\lambda)\|y\|^2$ for all $y \in R^n$, and thus, the inequality

(3.15) holds. Let $A = \text{diag}\{D, D, \ldots, D\} + Q$. Let $\tilde{R}_\varepsilon = (\tilde{r}_{ij}) \in M(2n)$ be the block matrix whose elements $\tilde{r}_{ij}$ are matrices from $M(2)$ defined as follows: $\tilde{r}_{ij} = \varepsilon^{n-1}I_2$ for $i = 1, 2, \ldots, j$ and $\tilde{r}_{ij} = 0 \in M(2)$ for $i = j+1, j+2, \ldots, n$ ($j = 1, 2, \ldots, n$). Obviously, the matrix $\tilde{R}_\varepsilon$ is regular and it holds that $\tilde{R}_\varepsilon Q \tilde{R}_\varepsilon^{-1} = \varepsilon Q$, $\tilde{R}_\varepsilon \text{diag}\{D, D, \ldots, D\} \tilde{R}_\varepsilon^{-1} = \text{diag}\{D, D, \ldots, D\}$, and thus, $\tilde{A}_\varepsilon = \tilde{R}_\varepsilon A \tilde{R}_\varepsilon^{-1} = \text{diag}\{D, D, \ldots, D\} + \varepsilon Q$. Using the inequality (3.15) for the matrix in the form 3, that has already been proved we obtain, in an analogous way to case 2 that for ε positive and small enough the inequality $\alpha \|y\|^2 \leq (\tilde{A}_\varepsilon y, y) \leq \beta \|y\|^2$ for all $y \in R^n$ is valid. Thus, it is sufficient to choose $B = \tilde{A}_\varepsilon$. □

3.83 Lemma. *Let φ be a linear dynamical system on an n-dimensional Euclidean space E_n defined by the mapping $A \in L(E_n)$ (e.g. $\varphi_t(x) = e^{tA}x$) and let for an arbitrary eigenvalue λ of the mapping A the inequality (3.13) hold. Then there exists such a basis $\mathscr{B}$ on E_n that*

$$e^{\alpha t}\|x\|_{\mathscr{B}} \leq \|\varphi_t(x)\|_{\mathscr{B}} \leq e^{\beta t}\|x\|_{\mathscr{B}}, t \geq 0, \quad (3.16)$$

$$e^{\beta t}\|x\|_{\mathscr{B}} \leq \|\varphi_t(x)\|_{\mathscr{B}} \leq e^{\alpha t}\|x\|_{\mathscr{B}}, t \leq 0, \quad (3.17)$$

where $\|\cdot\|_{\mathscr{B}}$ is the norm on E_n derived from the scalar product $(.,.)_{\mathscr{B}}$ on E_n defined by the basis $\mathscr{B}$.

Proof

Let $\mathscr{B}$ be a basis on E_n for which the inequality (3.14) holds. Let $t \geq 0, x \in E_n$ and $x(t) = \varphi_t(x)$. Since

$$\frac{d}{dt}(\|x(t)\|_{\mathscr{B}}^2) = \frac{d}{dt}((x(t), x(t))_{\mathscr{B}} = 2(x(t), \dot{x}(t))_{\mathscr{B}}, \text{ so}$$

$$\frac{d}{dt}(\|x(t)\|_{\mathscr{B}}) = \|\dot{x}(t)\|_{\mathscr{B}}^{-1}(Ax(t), x(t))_{\mathscr{B}}$$

and from the inequality (3.14) the inequality

$$\alpha \leq \|x(t)\|_{\mathscr{B}}^{-1} \frac{d}{dt}(\|x(t)\|_{\mathscr{B}}) \leq \beta \quad (3.18)$$

follows. By integrating this inequality from 0 to $t \geq 0$ we get the inequality $\alpha t \leq \ln\|x(t)\|_{\mathscr{B}} \leq \beta t$. Since $x(t) = \varphi_t(x)$, this inequality is equivalent to the inequality (3.16). By multiplying the inequality (3.18) by the number -1 and integrating from $t \leq 0$ to 0 the inequality equivalent with the inequality (3.17) is obtained. □

The next theorem follows from Lemma 3.83.

3.84 Theorem. *Let E_n be an n-dimensional Euclidean space, φ being a linear dynamical system on E_n defined by the mapping $A \in L(E_n)$. Then there exists a basis $\mathscr{B}$ on E_n such that*

  1. if $\operatorname{Re}\lambda < -a < 0$ for arbitrary eigenvalue λ of the mapping A, then

$$\|\varphi_t(x)\|_{\mathscr{B}} \leq e^{-at}\|x\|_{\mathscr{B}} \qquad (3.19)$$

for all $x \in E_n$ and $t \geq 0$;

  2. if $\operatorname{Re}\lambda > a > 0$ for an arbitrary eigenvalue λ of the mapping A, then

$$e_{at}\|x\|_{\mathscr{B}} \leq \|\varphi_t(x)\|_{\mathscr{B}} \qquad (3.20)$$

for all $x \in E_n, t \geq 0$, where $\|\cdot\|_{\mathscr{B}}$ is a norm on E_n derived from the scalar product $(.,)_{\mathscr{B}}$ defined by the basis $\mathscr{B}$.

3.85 Corollary. *If φ is a linear dynamical system on E_n defined by the mapping $A \in L(E_n)$ and $\operatorname{Re}\lambda < 0$ ($\operatorname{Re}\lambda > 0$) for an arbitrary eigenvalue λ of the mapping A, then for an arbitrary $x \in E_n$ is $\lim_{t\to\infty}\varphi_t(x) = 0$ ($\lim_{t\to-\infty}\varphi_t(x) = 0$).*

3.86 Definition. *If $\operatorname{Re}\lambda < 0$ ($\operatorname{Re}\lambda > 0$) for an arbitrary eigenvalue λ of the mapping $A \in L(E_n)$, then the singular point $0 \in E_n$ of the dynamical system $\varphi(x,t) = e^{tA}x$ is called the sink (source). We also say that the dynamical system φ is contractive (expansive).*

3.87 Theorem. *Let E_n be an n-dimensional Euclidean space and $\varphi(x,t) = e^{tA}x$ be a linear dynamical system on E_n. Then there exist such linear subspaces in E^s, E^c, E^u of the space E_n, that the following assertions hold.*

  1. E^s, E^c, E^u are invariant sets of the dynamical system.

  2. The dynamical system $\varphi/E^s \times R$ ($\varphi/E^u \times R$) is contractive (expansive).

  3. The space E_n is decomposed into the direct sum of the subspaces E^s, E^c and E^u, i.e., $E_n = E^s \oplus E^c \oplus E^u$.

  4. If the dynamical system φ is hyperbolic, then $E^c = \{0\}$, i.e., $E_n = E^s \oplus E^u$.

Proof

It is sufficient to prove the theorem for the case $E_n = R^n$ in which the mapping A can be identified with a matrix of $M(n)$. From the definition of the exponent of the mapping A it follows that every invariant subspace of the mapping A is also an invariant subspace of the mapping e^{tA} for arbitrary $t \in R$. On the basis of Theorem 1.18, without loss of generality it can be assumed that on R^n such a basis $\mathscr{B}$ is given with respect to which the matrix A is of the Jordan canonical form $\tilde{A} = \text{diag}\{B, C, D\}$, where $B \in M(p)$, $C \in M(q)$ and $D \in M(r)(p+q+r=n)$ are block diagonal matrices whose diagonal blocks are the Jordan blocks corresponding to the eigenvalues of the matrix A with negative, zero and positive real parts, respectively. The subspaces $E^s = \{(u,0,0) \in R^n : u \in R^p\}$, $E^c = \{(0,v,0) \in R^n : v \in R^q\}$ and $E^u = \{(0,0,w) \in R^n : w \in R^r\}$ are obviously invariant sets of the mapping A, and thus, also for the mapping $\varphi_t : R^n \to R^n$, $\varphi_t(x) = e^{tA}x$ for all $t \in R$. From Theorem 3.84 it follows that $\lim_{t \to \infty} \varphi_t(x) = 0$ ($\lim_{t \to -\infty} \varphi_t(x) = 0$) for arbitrary $x \in E^s$ ($x \in E^u$) this means that the dynamical system $\varphi/E^s \times R$ ($\varphi/E^u \times R$) is contractive (expansive). If φ is a hyperbolic dynamical system, then obviously $E^c = \{0\}$, i.e. $R^n = E^s \oplus E^u$.

3.88 Definition. *The subspaces E^s, E^c, E^u possessing the properties 1–4 from Theorem 3.87, are called respectively the stable, centre and unstable subspace of the linear dynamical system φ.*

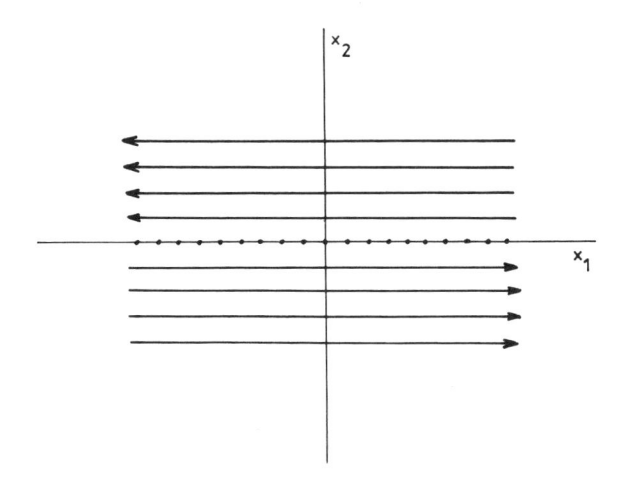

Fig. 14. $\lambda = 0$.

3.89 Example. *Linear dynamical systems in the plane.* From Theorem 3.18 and Theorem 3.76 it follows that an arbitrary linear dynamical system in the plane is linearly equivalent to the dynamical system $\varphi(x,t) = (\varphi_1(x,t)), \varphi_2(x,t) = e^{tA}x$, where the matrix $A \in M(2)$ is of one of the forms as follows:

$$1. \begin{bmatrix} \lambda & 1 \\ 0 & \lambda \end{bmatrix}; \quad 2. \begin{bmatrix} \lambda_1 & 0 \\ 0 & \lambda_2 \end{bmatrix}; \quad 3. \begin{bmatrix} a & b \\ -b & a \end{bmatrix}.$$

A detailed analysis of all the three cases can be found, e.g. in [58, 29, 86, 114]. For cases 1 and 2 we shall introduce their corresponding flows and figures of their trajectories. Case 3 will be studied in more detail.

In case 1 coordinates of the flow are of the form $\varphi_1(x,t) = e^{\lambda t}x_1 + te^{\lambda t}x_2$, $\varphi_2(x,t) = e^{\lambda t}x_2$. For $\lambda = 0$ (non-hyperbolic flow) the structure of trajectories is illustrated in Fig. 14 and for $\lambda \neq 0$ in Fig. 15, 16. In case 2 $\varphi_1(x,t) = e^{\lambda_1 t}x_1, \varphi_2(x,t) = e^{\lambda_2 t}x_2$. If $\lambda_1 = \lambda_2 = 0$, then every point of R^2 is a singular point of the dynamical system φ. For $\lambda_1 = 0, \lambda_2 \neq 0$ the structure of trajectories is shown in Figs. 17, 18, for $\lambda_1 \neq \lambda_2, \lambda_1\lambda_2 > 0$ ($\lambda_1 = \lambda_2, \lambda_1\lambda_2 > 0$) in Figs. 19, 20 (21, 22) and for $\lambda_1\lambda_2 < 0$ in Fig. 23.

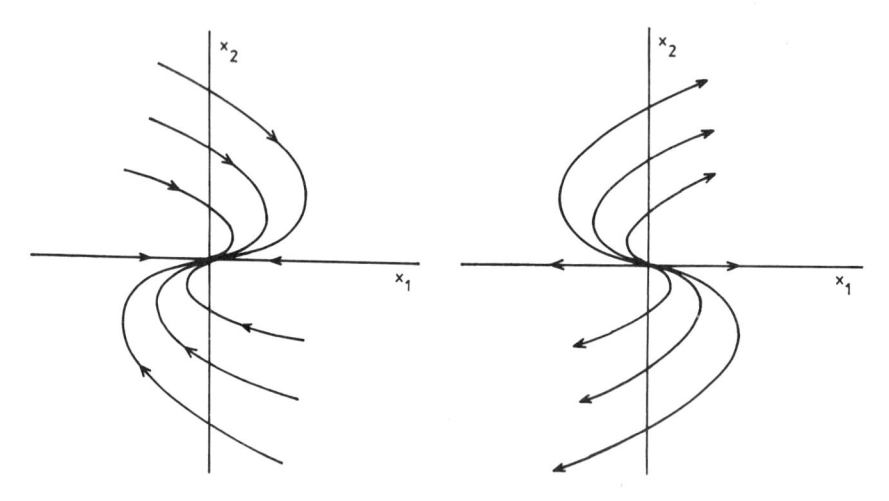

Fig. 15. $\lambda < 0$ (stable node). **Fig. 16.** $\lambda > 0$ (unstable node).

Now we shall consider the case 3. The matrix A can be written in the form $A = A_0 + B$, where

$$A_0 = \begin{bmatrix} a & 0 \\ 0 & a \end{bmatrix}, \quad B = \begin{bmatrix} 0 & b \\ -b & 0 \end{bmatrix}, \quad B^2 = \begin{bmatrix} -b^2 & 0 \\ 0 & -b^2 \end{bmatrix}, \quad B^3 = \begin{bmatrix} 0 & -b^3 \\ b^3 & 0 \end{bmatrix},$$

$$B^4 = \begin{bmatrix} b^4 & 0 \\ 0 & b^4 \end{bmatrix}, \quad B^5 = \begin{bmatrix} 0 & b^5 \\ -b^5 & 0 \end{bmatrix} \quad \text{etc.}$$

From the form of powers of the matrix B it follows that

$$e^{tB} = \begin{bmatrix} \cos tb & \sin tb \\ -\sin tb & \cos tb \end{bmatrix},$$

Vector Fields and Dynamical Systems

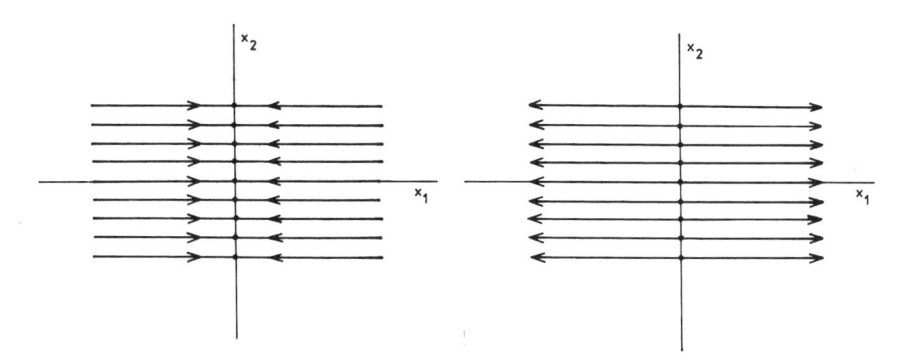

Fig. 17. $\lambda_2 < 0$.

Fig. 18. $\lambda_2 > 0$

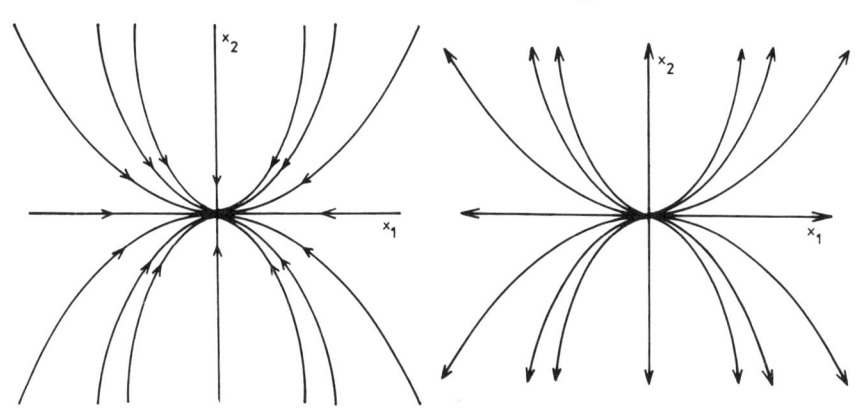

Fig. 19. $\lambda_1 < 0, \lambda_2 < 0, |\lambda_1| < |\lambda_2|$.

Fig. 20. $\lambda_1 > 0, \lambda_2 > 0, |\lambda_1| < |\lambda_2|$.

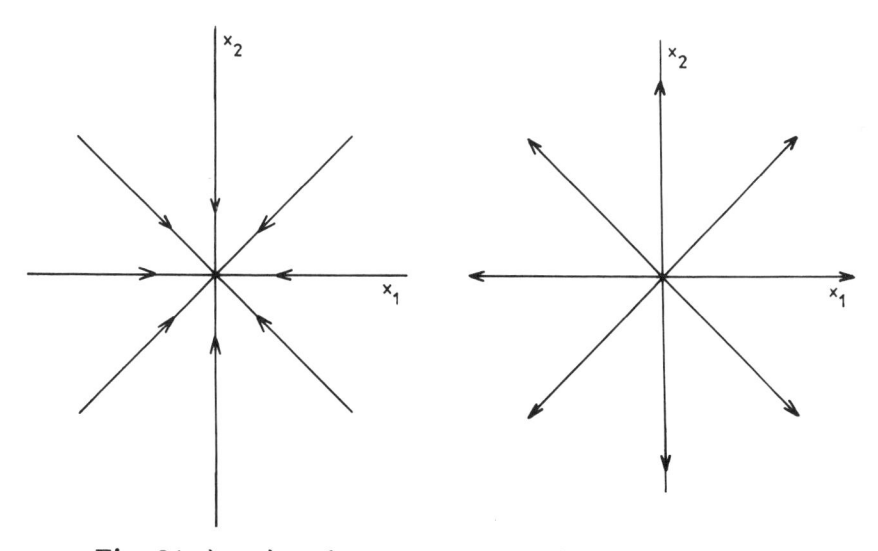

Fig. 21. $\lambda_1 = \lambda_2 < 0$.

Fig. 22. $\lambda_1 = \lambda_2 > 0$.

Chapter 3

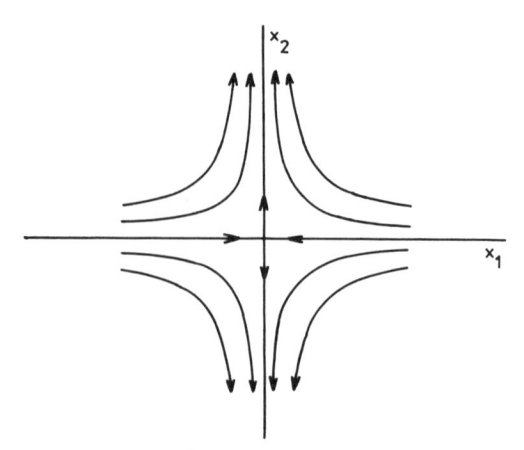

Fig. 23. Saddle.

and thus, it is

$$y_1(t) = \varphi_1(x,t) = e^{ta}(x_1 \cos tb + x_2 \sin tb),$$
$$y_2(t) = \varphi_2(x,t) = e^{ta}(-x_1 \sin tb + x_2 \cos tb). \tag{3.21}$$

Let us introduce on R^2 polar coordinates $x_1 = r \cdot \cos \vartheta$, $x_2 = r \cdot \sin \vartheta$, $r \geq 0$. Then there exist the functions $r = r(t)$, $\vartheta = \vartheta(t)$, such that $y_1(t) = r(t) \cos \vartheta(t)$ and $y_2(t) = r(t) \cdot \sin \vartheta(t)$. From (3.21) we get

$$\dot{r} \cdot \vartheta - r \cdot (\sin \vartheta)\dot{\vartheta} = ar \cdot \cos \vartheta + br \cdot \sin \vartheta,$$

$$\dot{r} \cdot \vartheta + r \cdot (\cos \vartheta)\dot{\vartheta} = -br \cdot \cos \vartheta + ar \cdot \sin \vartheta.$$

From these equations it follows that (r, ϑ) is the solution of the linear autonomous system

$$\dot{r} = ar, \ \dot{\vartheta} = -b. \tag{3.22}$$

The dynamical system $\varphi(x,t) = e^{tA}x$ thus in polar coordinates is of the form $(r(t), \vartheta(t)) = \Psi(r, \vartheta, t) = (e^{at}r, \vartheta - bt)$. The function $r = r(t)$ determines the distance of the vector $(y_1(t), y_2(t))$ from the point $(0,0)$. This distance can also be expressed as a function of ϑ (change of parametrization), i.e. $r = r(\vartheta)$. From the system (3.22) we obtain the differential equation for $r = r(\vartheta)$:

$$\frac{dr}{d\vartheta} = \frac{\dot{r}(t)}{\dot{\vartheta}(t)} = -\frac{a}{b}r. \tag{3.23}$$

The solution $r = r(\vartheta)$ of the differential equation (3.23) fulfilling the condition $r(0) = r_0$ is of the form $r(\vartheta) = r_0 \cdot \exp(-a/b\vartheta)$. From this expression for the distance of two points lying on trajectories from the point $(0,0)$ and from (3.22) we can arrive at the conclusion as follows. If $a = 0$, then for arbitrary $r_0 \in R$ we have $r(\vartheta) = r_0$ for all ϑ which means that for arbitrary point $x \in R^2$ the trajectory γ_x of the dynamical system φ passing through the point x is represented by the circle having its centre at the point $(0,0)$ and with the radius r_0. Since $r(0) = r(2\pi)$, then γ_x is in fact a 2π-periodic trajectory. In this case the singular point $(0,0)$ is called the centre (Fig. 24). If $a < 0, b < 0$, then

$$\lim_{\vartheta \to \infty} r(\vartheta) = 0 \quad \text{and} \quad \lim_{t \to \infty} \varphi_t(x) = (0,0)$$

for arbitrary $x \in R^2$ (Fig. 25). In this case the singular point $(0,0)$ is called the stable focus (non-degenerate). If $a > 0, b > 0$ then $\lim_{\vartheta \to -\infty} r(\vartheta) = \infty$ and $\lim_{t \to -\infty} \varphi_t(x) = (0,0)$ for arbitrary $x \in R^2$ (Fig. 26). In this case the singular point $(0,0)$ is called the unstable focus (non-degenerate).

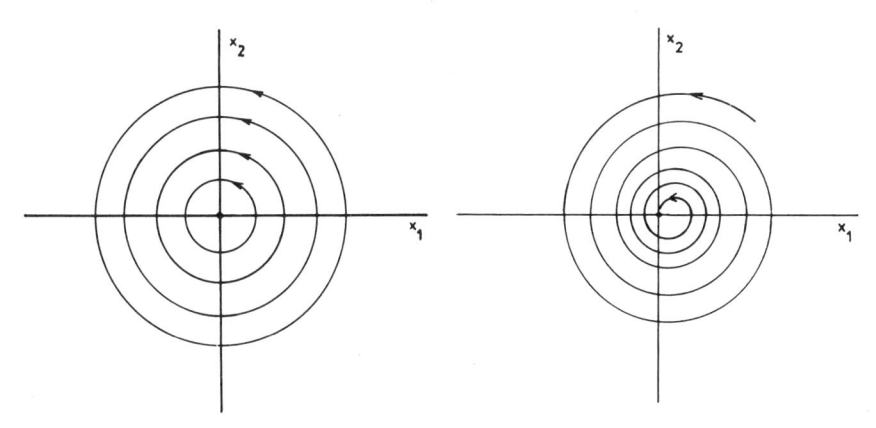

Fig. 24. $a = 0, b \neq 0$ (centre).   **Fig. 25.** $a < 0, b < 0$ (stable focus).

3.90 Example. *Linear dynamical systems on R^3.* Every linear dynamical system on R^3 is linearly equivalent to the dynamical system $\varphi(x,t) = e^{tA}x$, where $A \in M(3)$ is of one of the following forms:

1.
$$\begin{bmatrix} \lambda & 1 & 0 \\ 0 & \lambda & 1 \\ 0 & 0 & \lambda \end{bmatrix}$$

2. $A = \text{diag}\{\lambda, B\}$, where $B \in M(2)$ is of one of the forms 1–3 from Example 3.89. We shall not study all the above cases in detail. From the individual cases of planar dynamical systems discussed in Example 3.89

Chapter 3

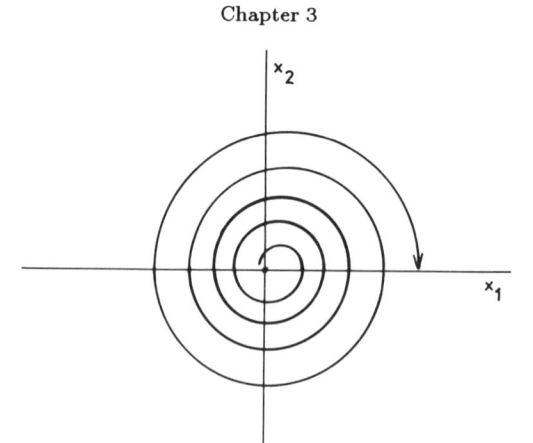

Fig. 26. $a > 0, b > 0$ (unstable focus).

the image of trajectories for some dynamical systems on R^3 can also be constructed. If, for instance, $\varphi(x,t) = e^{tA}x$, where $A = \text{diag}\{\lambda, B\}, \lambda > 0$ and the matrix $B \in M(2)$ has the eigenvalues λ_1, λ_2, where $\lambda_1 < 0, \lambda_2 < 0, \lambda_1 \neq \lambda_2$, then φ has the structure of its trajectories as illustrated in Fig. 27.

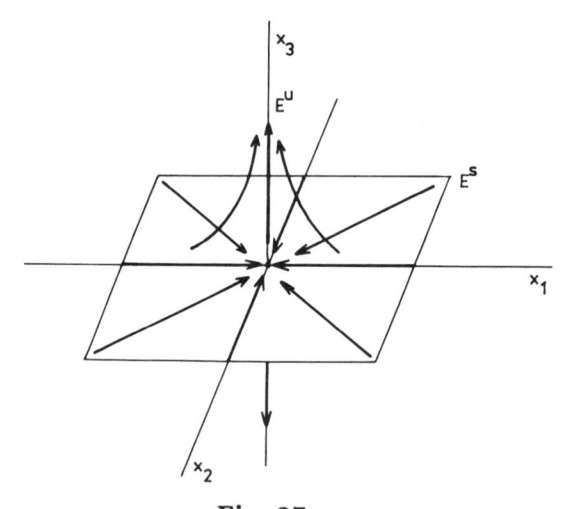

Fig. 27.

Since all similar matrices have the same eigenvalues, from Proposition 3.78 it follows that if the flows of the linear vector fields A_v, B_v are linearly equivalent, then the eigenvalues of the mapping A are positive multiples of eigenvalues of the mapping B. This, in turn, means that the set of the classes of linearly equivalent dynamical systems is uncountable, which from the viewpoint of classification is absolutely unsatisfactory. Therefore this definition must be modified. We shall now introduce the definition of equivalence related also to non-linear dynamical systems and complies with the definition of C^r-conjugacy of C^r-diffeomorphisms.

3.91 Definition. *Let X, Y be smooth manifolds and $\varphi : X \times R \to X$, $\Psi : Y \times R \to Y$ be C^r-dynamical systems. We say that these dynamical systems are C^k-equivalent (orbitally topologically equivalent, if $k = 0$), where $0 \leq k \leq \infty$ if there exists a C^k-diffeomorphism (homeomorphism) $h : X \to Y$ such that for arbitrary $x \in X$ it is $h(\gamma_x^\varphi) = \gamma_{h(x)}^\Psi$, where if $y, z \in \gamma_x^\varphi$, $y < z$, then $h(y) < h(z)$ ($<$ is the orientation of trajectories), i.e., h maps trajectories of the dynamical system φ onto trajectories of the dynamical system Ψ, preserving the orientation of trajectories. We say that the vector fields $F \in V^r(X), G \in V^r(Y)$ are globally C^k-equivalent (orbitally topologically equivalent, if $k = 0$), if their flows are orbitally C^k-equivalent (orbitally topologically equivalent). The diffeomorphism (homeomorphism) h is called the conjugating diffeomorphism (homeomorphism). We also say that the dynamical systems φ and Ψ are globally orbitally C^k-equivalent. If $X = Y$ and φ, Ψ are the dynamical systems generated by solutions of the differential equations $\dot{x} = f(x)$, and $\dot{x} = g(x)$, respectively, in R^n, then we also say that these differential equations are globally orbitally C^k-equivalent.*

The relation of orbital C^k-equivalence on the set of dynamical systems or on the set of vector fields is obviously the relation of equivalence.

With respect to complexity of the problem of global classification of non-linear dynamical systems according to the above equivalence it is advantageous to define the orbital C^k-equivalence locally.

3.92 Definition. *Let X, Y be smooth manifolds, $\varphi : X \times R \to X$, $\Psi : X \times R \to Y$ be C^r-dynamical systems, $x_0 \in X$, $y_0 \in Y$ and $0 \leq k \leq \infty$. We say that the couple (φ, x_0) is orbitally C^k-equivalent (orbitally topologically equivalent, if $k = 0$) to the couple (Ψ, y_0) (we write $(\varphi, x_0) \sim^k (\Psi, y_0)$), if there exists such an open neighbourhood U of the point x_0, an open neighbourhood V of the point y_0 and a C^k-diffeomorphism (homeomorphism) $h : U \to V$ such that maps the components of connectivity of intersections of trajectories of the dynamical system φ with the set U onto components of connectivity of intersections of trajectories of the dynamical system Ψ with the set V where the orientation of the trajectories is preserved.*

Let us note that from the preceding definition it does not follow that $h(\gamma_x^\varphi \cap U) = \gamma_{h(x)}^\Psi \cap V$. In fact, if $\lim_{t \to \infty} \varphi_t(x) = \lim_{t \to -\infty} \varphi_t(x)$, where x is not a singular point of the dynamical system φ (i.e., γ_x^φ is the so-called homoclinic trajectory; see Fig. 53), then $\gamma_x^\varphi \cap U$ can be an unconnected set having two (it can also be more) components of connectivity γ_1, γ_2 and the mapping h maps γ_1, γ_2 onto two different trajectories of the dynamical systems Ψ.

3.93 Definition. *We say that the vector fields $F, G \in V^r(X)$ are germ equivalent at the point $x_0 \in X$ (we write $F \sim_{x_0} G$), if there exists such a neighbourhood U of the point x_0 that $F/U = G/U$. Two differential equations or two systems of differential equations defined on an open set $V \subset R^n$ are called germ equivalent at the point $y_0 \in V$, if the vector fields on V generated by them are germ equivalent at the point y_0. The class $[F]_{x_0} = \{G \in V^r(X) : F \sim_{x_0} G\}$ is called the germ of the vector field F at the point x_0. The set of all such germs is denoted as $V_{x_0}^r(X)$. The class $[f]_{y_0}$ of differential equations on U germ equivalent at the point y_0 with the differential equation $\dot{x} = f(x)$, where $f \in C^r(U, R^n)$, is called the germ of this differential equation. The set of all such germs is denoted as $(DR)_{y_0}^r(U)$.*

3.94 Definition. *Let X, Y be smooth manifolds, $0 \leqq r \leqq \infty$, $0 \leqq k \leqq \infty$, $x_0 \in X$ and $y_0 \in Y$. We say that the germs $[F]_{x_0}, [G]_{y_0}$ are orbitally C^k-equivalent (orbitally topologically equivalent, if $k = 0$) and we write $[F]_{x_0} \sim^k [G]_{y_0}$, if for arbitrary vector fields $\tilde{F} \in [F]_{x_0}, \tilde{G} \in [G]_{y_0}$ is the couple (φ, x_0) orbitally C^k-equivalent (orbitally topologically equivalent, if $k = 0$) to the couple (Ψ, y_0), where φ, Ψ are C^r-flows of the vector field $\tilde{F}$ and $\tilde{G}$, respectively. We say that the vector fields $F \in V^r(X), G \in V^r(Y)$ are orbitally C^k-equivalent (orbitally topologically equivalent, if $k = 0$) at the point $(x_0, y_0) \in X \times Y$ and we write $F \sim^k_{(x_0, y_0)} G$, if $[F]_{x_0} \sim^k [G]_{y_0}$. We also say that the vector fields F and G are orbitally C^k-equivalent at the point (x_0, y_0). If $X = Y$ and $x_0 = y_0$ then we also say that F and G are C^k-equivalent at the point x_0 and we write $F \sim^k_{x_0} G$. All the notions defined up to now for vector fields and their germs can be also defined in a natural way for differential equations and their germs through vector fields generated by these differential equations.*

A notion that is very important from the viewpoint of physical realization of the models described by dynamical systems is the notion of the so-called structural stability that was introduced for the first time by A. A. Andronov and L. S. Pontrjagin in [6]. In the following definition this notion was generalized to an arbitrary topological space, enabling us to speak of a structural stability on the sets of different special classes of dynamical systems or vector fields and differential equations with respect to various types of equivalence.

3.95 Definition. *Let T be a topological space on which the relation E is defined. The element $x_0 \in T$ is called structurally stable in T (with respect* to the equivalence E) or also E-structurally stable in T if there exists its neighbourhood U such that every element $x \in U$ is equivalent with x_0, i.e. xEx_0.

If $T = V^r(X)$ with the Whitney C^r-topology and the equivalence on T is a global (local) C^k-equivalence, we get the notion of structural stability also called the global (local) structural stability in $V^r(X)$.

If $T = V_L(E_n)$ with the topology of Banach space then we can speak about structural stability in $V_L(E_n)$ with respect to orbital C^k- equivalence. Now we shall show that linear hyperbolic vector fields on R^n are structurally stable in $V_L(R^n)$ with respect to C^0-equivalence. For this, several lemmas will be needed.

3.96 Lemma. *Let all eigenvalues of the mapping $A \in L(R^n)$ have negative real parts, $x \in R^n \setminus \{0\}$ and $u : R \to R^n, u(t) = e^{tA}x$. Then there exists a basis $\mathscr{B}$ on R^n such that the mapping $\varrho : R \to R, \varrho(t) = ln\|u(t)\|_{\mathscr{B}}$, is a C^∞-diffeomorphism.*

Proof

Let $\alpha < \text{Re}\,\lambda < \beta < 0$ for arbitrary eigenvalue λ of the mapping A. According to Lemma 3.82 on R^n there exists a basis $\mathscr{B}$ such that if $\alpha\|u(t)\|_{\mathscr{B}}^2 \leq (Au(t), u(t))_{\mathscr{B}} \leq \beta\|u(t)\|_{\mathscr{B}}^2$ for all $t \in R$, and thus,

$$\frac{d}{dt}(\|u(t)\|_{\mathscr{B}}) = \|u(t)\|_{\mathscr{B}}^{-1}(Au(t), u(t))_{\mathscr{B}} \leq \beta\|u(t)\|_{\mathscr{B}} < 0$$

for all $t \in R$. This means that the mapping ϱ is injective. According to Lemma 3.83 it is

$$e^{\alpha t}\|x\|_{\mathscr{B}} \leq \|u(t)\|_{\mathscr{B}} \leq e^{\beta t}\|x\|_{\mathscr{B}}$$

for all $t \in R$, and thus,

$$\lim_{t \to \infty} \|u(t)\|_{\mathscr{B}} = 0 \quad \text{and} \quad \lim_{t \to -\infty} \|u(t)\|_{\mathscr{B}} = \infty.$$

Thus, the mapping ϱ is also surjective and since it is of class C^∞, then it is a C^∞-diffeomorphism (smoothness of ϱ^{-1} follows from Theorem 1.37). □

3.97 Lemma. *Let all eigenvalues of the mapping $A \in L(R^n)$ have negative real parts and $\varphi : R^n \times R \to R^n$, $\varphi(x, t) = e^{tA}x$. Then there exists a basis $\mathscr{B}$ on R^n, such that the mapping $\overline{\varphi} = \varphi/S_{\mathscr{B}} \times R$ is a C^∞-diffeomorphism of the set $S_{\mathscr{B}} \times R$ onto the set $R^n \setminus \{0\}$, where $S_{\mathscr{B}} = \{x \in R^n : \|x\|_{\mathscr{B}} = 1\}$.*

Proof

Let $x \in R^n \setminus \{0\}$. According to Lemma 3.96 on R^n there exists a basis $\mathscr{B}$ such that the mapping $\varrho : R \to R, \varrho(t) = ln\|\varphi_t(x)\|_{\mathscr{B}}$ is a C^∞-diffeomorphism. Therefore there exists $\tau \in R$, such that $\|\varphi_\tau(x)\|_{\mathscr{B}} = 1$. If

$y = \varphi_\tau(x)$, then, obviously, $\overline{\varphi}(y, -\tau) = x$ which means, that the mapping $\overline{\varphi}$ is surjective. From injectivity of the mapping ϱ the injectivity of the mapping $\overline{\varphi}$ follows. It also follows from the fact that the trajectory coming from the point x is unique and it crosses the sphere $S_\mathscr{B}$ in a single point. From Theorem 1.48 it follows that the mapping $\overline{\varphi}$ is of the class C^∞ and from the Inverse Mapping Theorem it also follows that the inverse mapping $\overline{\varphi}^{-1}$ is of the class C^∞. □

3.98 Lemma. *If the signs of real parts of all eigenvalues of the mappings $A, \tilde{A} \in L(R^n)$ are equal, then the dynamical systems $\varphi : R^n \times R \to R^n$, $\varphi(x,t) = e^{tA}x$, $\Psi : R^n \times R \to R^n$, $\Psi(x,t) = e^{t\tilde{A}}x$ are globally orbitally topologically equivalent.*

Proof

Let us suppose that the signs of real parts of all eigenvalues of the mappings $A, \tilde{A}$ are negative (for positive real parts the proof is analogous). According to Lemma 3.97 there exist bases $\mathscr{B}$ and $\mathscr{B}'$ on R^n such that the mapping $\overline{\varphi} = \varphi/S_\mathscr{B} \times R$ and $\overline{\Psi} = \Psi/S_{\mathscr{B}'} \times R$ are C^∞-diffeomorphisms on $R^n \setminus \{0\}$, where $S_\mathscr{B} = \{x \in R^n : \|x\|_\mathscr{B} = 1\}$, $S_{\mathscr{B}'} = \{x \in R^n : \|x\|_{\mathscr{B}'} = 1\}$. Let us define the mapping $h : R^n \to R^n$ as follows: $h(x) = 0$ for $x = 0$ and $h(x) = \overline{\Psi} \circ (\overline{h} \times \mathrm{id}_R) \circ \overline{\varphi}^{-1}(x)$ for $x \neq 0$, where $\overline{h} : S_\mathscr{B} \to S_{\mathscr{B}'}$, $\overline{h}(y) = \|y\|_{\mathscr{B}'}^{-1} y$. Obviously, the mapping h is of the class C^∞ on $R^n \setminus \{0\}$ and it is bijective. We shall prove that h is continuous at the point 0. Let V be a neighbourhood of the point 0. It will suffice to find a neighbourhood U of the point 0, such that $h(U) \subset V$. From Theorem 3.84 it follows that there exists $\tau > 0$ such that for all $t \geq \tau$ it is $\Psi_t(S_{\mathscr{B}'}) \subset V$. Let $U = \{x \in R^n : \|x\|_\mathscr{B} < \|\varphi_{-\tau}\|_\mathscr{B}^{-1}\}$. If $x \in U$, then

$$\|\varphi_{-\tau}(x)\|_\mathscr{B} \leq \|\varphi_{-\tau}\|_\mathscr{B} \cdot \|x\|_\mathscr{B} < \|\varphi_{-\tau}\|_\mathscr{B} \cdot \|\varphi_{-\tau}\|_\mathscr{B}^{-1} = 1,$$

i.e. $\|\varphi_{-\tau}(x)\| < 1$, and thus, there exists $s > \tau$, such that $\varphi_{-s}(x) \in S_\mathscr{B}$. From the definition of the mapping h we find that $h(x) = h(\varphi(\varphi(x,-s),s)) = \Psi(h \circ \varphi(x,-s), s)$. Since $\overline{h} \circ \varphi(x,-s) \in S_{\mathscr{B}'}$ and $\Psi_s(S_\mathscr{B}) \subset V$, then $h(x) \in V$, and thus, $h(U) \subset V$. The proof of continuity of the inverse mapping h^{-1} at the point 0 is analogous. Thus, the mapping h is a homeomorphism. Now we shall prove that for arbitrary $(x,t) \in R^n \times R$ the equality

$$h \circ \varphi(x,t) = \Psi(h(x), t) \tag{3.24}$$

is fulfilled. Obviously, from this equality it follows that $h(\gamma_x^\varphi) = \gamma_{h(x)}^\Psi$ for arbitrary $x \in R^n$, where h preserves the orientation of trajectories. Obviously, the equality (3.24) is fulfilled for $x = 0$. If $x \neq 0$ then there exists point $(y,p) \in S_\mathscr{B} \times R$, such that $x = \overline{\varphi}(y,p)$ and for all $t \in R$ it holds that

$$h \circ \varphi(x,t) = h \circ \varphi(y, t+p) = \overline{\Psi} \circ (\overline{h} \times \text{id})(y, t+p) = \overline{\Psi}(\overline{h}(y), t+p)$$
$$= \Psi(\overline{\Psi}(\overline{h}(y), p), t) = \Psi(\overline{\Psi}(\overline{h} \times \text{id}) \circ \overline{\varphi}^{-1}(\overline{\varphi}(y,p)), t)$$
$$= \Psi(h(x), t). \quad \square$$

3.99 Remark. The equivalence of dynamical systems or vector fields defined by the equality (3.24), in the literature referred to as the flow equivalence in general is not suitable for classification of non-linear dynamical systems for several reasons. In fact, if the dynamical system φ has a periodic trajectory γ_x^φ, then $\gamma_{h(x)}^\Psi$ is a periodic trajectory of the dynamical system Ψ with the same period as γ_x^φ. However, the period of the trajectory can be changed by a small change of the vector field. Therefore the above definition of equivalence is not suitable for the study of structural stability of vector fields.

3.100 Theorem. *If the mapping $A \in L(R^n)$ has k eigenvalues with negative real part ($1 \leq k < n$) and $n - k$ eigenvalues with positive real part then the dynamical system $\varphi(x,t) = e^{tA}x$ is globally orbitally topologically equivalent to the dynamical system*

$$\Psi(x,t) = (e^{-t(\text{id}_k)}x_1, e^{t(\text{id}_{n-k})}x_2),$$

$x = (x_1, x_2) \in R^k \times R^{n-k}$ ($\text{id}_k \stackrel{\text{def}}{=} \text{id}_{R^k}$).

Proof

On the basis of Theorem 1.18 we can assume that on R^n there is given such a basis with respect to which the mapping A is of the form $A(x_1, x_2) = (A_1 x_1, A_2 x_2)$, where $(x_1, x_2) \in R^k \times R^{n-k}$, $A_1 \in L(R^k)$ is a mapping all its eigenvalues having a negative real part and $A_2 \in L(R^{n-k})$ is a mapping with all its eigenvalues having a real positive part. Then $\varphi(x,t) = e^{tA}x = ((\exp tA_1)x_1, (\exp tA_2)x_2)$, i.e. $\varphi = \varphi_1 \times \varphi_2$, where $\varphi_1(x_1,t) = (\exp tA_1)x_1$, $\varphi_2(x_2, t) = (\exp tA_2) \cdot x_2$. According to Lemma 3.98 the dynamical system $\varphi_1(\varphi_2)$ is globally orbitally topologically equivalent to the dynamical system $\Psi_1(x_1,t) = (\exp -t(\text{id}_k))x_1$ ($\Psi_2(x_2,t) = (\exp t(\text{id}_{n-k}))x_2$). Let $h_1 : R^k \to R^k$, $h_2 : R^{n-k} \to R^{n-k}$ be the corresponding conjugating homeomorphisms. Since for arbitrary $x = (x_1, x_2) \in R^k \times R^{n-k}$ is $\gamma_x^\varphi = \gamma_{x_1}^{\varphi_1} \times \gamma_{x_2}^{\varphi_2}$, so $h = h_1 \times h_2$ is the conjugating homeomorphism of the dynamical systems φ and $\Psi = \Psi_1 \times \Psi_2$. $\square$

3.101 Remark. The assertion of Theorem 3.100 in fact, means that the differential equation $\dot{x} = Ax$ is globally orbitally topologically equivalent to the differential equation $\dot{x} = (\dot{x}_1, \dot{x}_2) = (-x_1, x_2)$, where $(x_1, x_2) \in R^k \times R^{n-k}$ which, in fact is a system of mutually independent differential equations $\dot{x}_1 = -x_1, \dot{x}_2 = x_2$ that can be written in the form of mutually independent one-dimensional differential equations $\dot{y}_i = -y_i, i = 1, 2, \ldots, k, \dot{y}_j = y_j, j = k+1, k+2, \ldots, n$, where $x_1 = (y_1, y_2, \ldots, y_k), x_2 = (y_{k+1}, y_{k+2}, \ldots, y_n)$.

If A_v is a linear vector field on E_n defined by the mapping $A \in L(E_n)$, then the number of eigenvalues of the mapping A with positive (negative) real parts is denoted by $m^+(A_v)$ $(m^-(A_v))$.

3.102 Lemma. *If the vector field $A_v \in V_L(E_n)$ is hyperbolic then there exists its neighbourhood U in $V_L(E_n)$, such that for arbitrary $B_v \in U$ we have $m^+(B_v) = m^+(A_v)$, and thus, also $m^-(B_v) = m^-(A_v)$.*

Proof

It is sufficient to prove the lemma for $E_n = R^n$. From Theorem 3.81 it follows that there exists an open neighbourhood U of the vector field A_v in $V_L(R^n)$ containing hyperbolic vector fields, only. Since $V_L(R^n)$ is the Banach space, without loss of generality we can assume that U is the sphere with centre at A_v. Let us assume that there exists a vector field $B_v \in U$ such that $m^+(B_v) \neq m^+(A_v)$. Let us define the mapping $D : \langle 0, 1 \rangle \to V_L(R^n)$, $D(s) = sB_v + (1-s)A_v$. Since the mapping D is continuous, eigenvalues of the mapping $D(s)$ depend continuously upon the variable s. Since according to the assumption the mappings $D(0)$ and $D(1)$ have a different number of eigenvalues with positive real parts there must exist $s_0 \in (0, 1)$ for which some of eigenvalues of the mapping $D(s_0)$ have an eigenvalue with zero real part. This, however, is not possible, since $D(\langle 0, 1 \rangle) \subset U$ and all vector fields from U are hyperbolic. □

3.103 Theorem. *Let E_n be an n-dimensional Euclidean space. Then all linear hyperbolic vector fields on E_n are globally structurally stable in $V_L(E_n)$.*

Proof

It is sufficient to prove the theorem for $E_n = R^n$. If $A_v \in HV_L(R^n)$, then according to Lemma 3.102 there exists a neighbourhood U of the vector field A_v in $V_L(R^n)$ such that for arbitrary $B_v \in U$ is $k^+ = m^+(B_v) = m^+(A_v)$ and $k^- = m^-(B_v) = m^-(A_v)$, where $k^+ + k^- = n$. According to Theorem 3.100 both flows φ^{B_v} and φ^{A_v} are globally orbitally topologically equivalent to the same dynamical system $\Psi(x, t) = ((\exp -t(\mathrm{id}_{k^-}))x_1, (\exp t(\mathrm{id}_{k^+}))x_2$, where $x = (x_1, x_2) \in R^{k^-} \times R^{k^+}$, which means that the vector fields B_v and A_v are globally orbitally topologically equivalent. □

Let us now consider the linear parametrized autonomous differential equation in an n-dimensional Euclidean space E_n

$$\dot{x} = A(\mu)x, \tag{3.25}$$

where $A \in C^r(P, L(E_n))$, $1 \leqq r \leqq \infty$ and P is a k-dimensional interval or a k-dimensional C^∞-manifold.

3.104 Definition. *Let $A \in C^r(P, L(E_n))$. The mapping $\varphi : E_n \times P \times R \to E_n$, $\varphi(x, \mu, t) = e^{tA(\mu)}x$ is called a parametrized dynamical system on E_n. If for $\mu \in P$, $A(\mu)_v$ is a linear vector field E_n defined by the mapping $A(\mu)$, then the mapping $A_v : E_n \times P \to T(E_n)$, $A_v(x, \mu) = A(\mu)_v(x)$ is called a parametrized linear vector field on E_n. The set of all parametrized linear vector fields on E_n with the set of parameters P is denoted as $V_L^r(P, E_n)$.*

If $A_v, B_v \in V_L^r(P, E_n)$, $\lambda \in R$, then we define $A_v + B_v = (A+B)_v$ and $\lambda A_v = (\lambda A)_v$. If P is a compact set, then the set $V_L^r(P, E_n)$ with the operations defined like that and with the norm

$$\|A_v\| = \max_{\mu \in P}(\|A(\mu)\|_0 + \sum_{k=1}^{r} \|d_\mu^k A\|_k),$$

where $\|\cdot\|_j$ ($j = 0, 1, \ldots, r$) is a norm on $L^j(E_n)$ is obviously, a Banach space.

3.105 Definition. *We say that the singular point $0 \in E_n$ of the vector field $A \in V_L(E_n)$ is quasi-hyperbolic of degree 1, if the linear mapping A defining this vector field has only one eigenvalue with zero real part of multiplicity 1, excluding its complex conjugate eigenvalue. We say that this singular point is quasi-hyperbolic of degree 2 if it either has two different eigenvalues with zero real part of multiplicity 1 excluding their complex conjugate eigenvalues or only one eigenvalue with zero real part of multiplicity 2, excluding its complex conjugate eigenvalue.*

The following theorem is a direct consequence of Theorem 3.74.

3.106 Theorem. *If P is a k-dimensional interval or a k-dimensional C^∞-manifold and $1 \leqq r \leqq \infty$ then there exists a massive subset $V^0(P, E_n)$ in $V_L^r(P, E_n)$ such that the following assertions hold.*

1. If $k = 1$ and $A \in V^0(P, E_n)$, then either the vector field $A(\mu)_v$ is hyperbolic for all $\mu \in P$ or there exists a set $P_0 \subset P$ consisting of isolated values of the parameter μ for which the point $0 \in E_n$ is the quasi-hyperbolic singular point of the vector field $A(\mu)_v$ of degree 1.

2. If $k = 2$ and $A \in V^0(P, E_n)$, then either the vector field $A(\mu)_v$ is hyperbolic for all $\mu \in P$ or it holds that

a) if P is a C^∞-manifold then there exists a one-dimensional C^r-submanifold P_1 in P and the set $P_2 \subset P_1$ consisting of isolated values of the parameter such that for $\mu \in P_1$ ($\mu \in P_2$) the point $0 \in E_n$ is a quasi-hyperbolic singular point of the vector field $A(\mu)_v$ of degree 1 (degree 2),

b) if P is a two-dimensional interval (it can also be compact), then there exist the sets $Q_1 \subset P$, $Q_2 \subset Q_1$ such that int $(Q_1 \cap P)$ is a one-dimensional C^r-submanifold in R^2, Q_2 consists of isolated values of the parameter, where for $\mu \in Q_1$ ($\mu \in Q_2$) the point $0 \in E_2$ is the quasi-hyperbolic singular point of the vector field $A(\mu)_v$ of degree 1 (degree 2). If the manifold P is compact or P is a compact interval then the set $V^0(P, E_n)$ is also open in $V_L^r(P, E_n)$.

3.107 Example. If $P = \langle -1, 1 \rangle$, $1 \leq r \leq \infty$ and A_v is a parametrized vector field on R^2 defined by the C^r-mapping $A : P \to M(2)$, $\mu \to A(\mu) = (a_{ij}(\mu))$, where $a_{11}(\mu) = a_{22}(\mu) = \mu$ and $a_{12}(\mu) = -a_{21}(\mu) = b \neq 0$, then $A_v \in V^0(P, R^2)$, where the set P_0 from Theorem 3.106 only consists of the one point $\mu = 0$. Let us consider the dynamical system $\varphi(x, t) = e^{tA(\mu)}x$. For $\mu = 0$ the point $0 \in R^2$ is the centre, for $\mu > 0$ it is an unstable focus and for $\mu < 0$ this point is a stable focus of the dynamical system φ. This means that in every one of these cases this dynamical system belongs to another class of equivalence defined by the orbital topological equivalence i.e., by transition of the parameter through the point $\mu = 0$ the topological structure of trajectories is changed. In this case it is obvious, that there is no approximation $\tilde{A}$ of the mapping A in a C^r-topology small enough for the mapping $\tilde{A}$ not to have a pure imaginary eigenvalue for a certain value of the parameter $\mu \in P$. It is even obvious, that there exists a neighbourhood $U(A)$ of the mapping A such that for an arbitrary mapping $\tilde{A} \in U(A)$ the set of such $\mu \in P$ for which the matrix $A(\mu)$ has a pure imaginary eigenvalue, consists of a single point. Obviously, the mapping A belongs to the set $V^0(P, R^2)$, that is an open and dense subset of the set $(V_L^r(P, R^2)$.

Such changes of the topological structure of trajectories of the dynamical system when the parameter is changed, as described in the preceding example for the vector field $A(\mu)_v$ at the point $\mu = 0$, are called bifurcations. Now we shall introduce an exact and much more general definition of the notion of bifurcation.

3.108 Definition. Let M be a set on which a relation of equivalence is defined and let $[x]$ be the class of equivalence defined by the element $x \in M$. Let P be a topological space and $F : P \to M$. The point $\mu_0 \in P$ is called the bifurcation point of the mapping F if for every neighbourhood U of the point μ_0 there exists a $\mu_1 \in U$ such that $[F(\mu_0)] \neq [F(\mu_1)]$.

This change of the class of equivalence $[F(\mu)]$ when the parameter μ is changed is called the bifurcation of the mapping F. The set of bifurcation points of the mapping F is called the bifurcation diagram. Let us denote by $\mathscr{F}$ the set of all mappings from the set P into the set M and let us assume that on $\mathscr{F}$ a topology S is defined such that $(\mathscr{F}, S)$ is the Baire topological space. A bifurcation of some class (type) T of mappings from the space $\mathscr{F}$ is called generic *if the set of mappings from $\mathscr{F}$ for which the bifurcation of class T occurs is massive in $\mathscr{F}$, where the class of bifurcations is defined in the following way. We say that the mappings $F, G \in \mathscr{F}$ have at their bifurcation points $\mu_F, \mu_G \in P$ bifurcations of the same class if there exist open neighbourhoods U, V of the points μ_F and μ_g, respectively, and such a homeomorphism $h : U \to V$ that maps the bifurcation diagram of the mapping F onto the bifurcation diagram of the mapping G, where for arbitrary $\mu \in U$ it is $[F(\mu)] = [G(h(\mu))]$.*

The above definition enables us to speak about bifurcations or generic bifurcations in the spaces $V_L^r(P, E_n), V^r(P, X)$ and $\text{Diff}^r(P, X)$. The notion of the class of bifurcation and the notion of generic bifurcation will be sufficiently explained in Section 5 using an example of 1-parameter systems of vector fields. On the basis of the results presented in Example 3.89 it is not difficult to find generic bifurcations in the space of parametrized linear vector fields $V_L^r(\langle -1, 1 \rangle, R^2)$. Example 3.107, however, shows that though from the mathematical viewpoint these bifurcations are relatively simple, from the practical viewpoint they are not very valuable. In fact, in this example the dynamical system has its centre for $\mu = 0$ which is considerably unstable with respect to small non-linear perturbations of the dynamical system, and thus, it cannot model any real process.

The theory of topological classification of linear dynamical systems and bifurcations of parametrized linear dynamical systems that has been studied in this section is, in fact, an outline of the topic that will be studied throughout the next sections in connection with non-linear dynamical systems. We shall see that compared to linear dynamical systems an analogous theory for dynamical systems is extraordinarily complicated and it is still far from a satisfactory conclusion.

To conclude this section we shall introduce some definition necessary for the classification of non-linear parametrized vector spaces. Let us consider the space $V^r(P, X)$ of parametrized vector fields on the smooth manifold X, where P is a smooth manifold and $1 \leq r \leq \infty$.

3.109 Definition. *We say that the parametrized vector fields $F, G \in V^r(P, X)$ are* germ equivalent *at the point $(x_0, p_0) \in X \times P$ (we write $F \sim_{(x_0, p_0)} G$), if the vector fields $\tilde{F}, \tilde{G} \in V^r(X \times P)$, $\tilde{F}(x, p) = (F(x, p), p)$, $\tilde{G}(x, p) = (G(x, p), p)$, are germ equivalent at the point (x_0, p_0). The class $[F]_{(x_0, p_0)} = \{G \in V^r(P, X) : F \sim_{(x_0, p_0)} G\}$ is called the germ of the* parametrized vector field F at the point (x_0, p_0). The set of all such germs is denoted as $V^r_{(x_0,p_0)}(P, X)$. Two parametrized differential equations defined on an open set $U \times V \subset R^n \times R^k$ (U is the state space and V is the space of parameters) are called germ equivalent at the point $(y_0, \varepsilon_0) \in U \times V$ if the parametrized vector fields on $U \times V$ generated by them are germ equivalent at the point (y_0, ε_0). The class of parametrized vector fields on $U \times V$ germ equivalent at the point (y_0, ε_0) with the parametrized differential equation $\dot{x} = f(x, \varepsilon)$, $f \in C^r(U \times V, R^n)$, is called the germ of this differential equation and we denote it as $[f]_{(y_0,\varepsilon_0)}$. The set of all such germs is denoted as $(PDR)^r_{(y_0,\varepsilon_0)}$.

3.110 Definition. *The germs* $[F]_{(x_0,p_0)} \in V^r_{(x_0,p_0)}(P, X)$, $[G]_{(y_0,q_0)} \in V^r_{(y_0,q_0)}(P, X)$ *are called orbitally C^k-equivalent (orbitally topologically equivalent, if $k = 0$) if there exist $\tilde{F} \in [F]_{(x_0,p_0)}, \tilde{G} \in [G]_{(y_0,q_0)}$, a neighbourhood $U_1 \times V_1$, a neighbourhood $U_2 \times V_2$ of the point (y_0, q_0), a C^k-mapping $h : U_1 \times V_1 \to U_2$ and a C^k-diffeomorphism (homeomorphism) $k : V_1 \to V_2$ such that*

1. $h(x_0, p_0) = y_0, k(p_0) = q_0$;
2. *For every $p \in V$ the mapping $h_p : U_1 \to U_2, h_p(x) = h(x, k(p))$, C^k-diffeomorphism (homeomorphism) mapping the components of connectivity of intersections of trajectories of the vector field F_p with the set U_1 onto components of connectivity of intersections of trajectories of the vector field $G_{k(p)}$ with the set U_2.*

The germ $[H]_{(x_0,p_0)}$ of the mapping $H = (h, k)$ is called the conjugating germ of the C^k-diffeomorphisms (homeomorphisms) of the germs $[F]_{(x_0,p_0)}$ and $[G]_{(y_0,q_0)}$. The germs $[f]_{(x_0,\varepsilon_0)}, [g]_{(y_0,\mu_0)}$ of the parametrized differential equations $\dot{x} = f(x, \varepsilon)$, $\dot{y} = g(y, \mu)$ are called orbitally C^k-equivalent (orbitally topologically equivalent), if the germs $[F]_{(x_0,\varepsilon_0)}$ and $[G]_{(y_0,\mu_0)}$ of the parametrized vector fields F and G created by them are orbitally C^k-equivalent (orbitally topologically equivalent). We shall also say that the above parametrized differential equations are orbitally C^k-equivalent and orbitally topologically equivalent in neighbourhoods of the points $(x_0, \varepsilon_0), (y_0, \mu_0)$, respectively.

3.111 Definition. *We say that the germ $[F]_{(x_0,p_0)} \in V^r_{(x_0,p_0)}(P, X)$ is induced from the germ $[G]_{(x_0,q_0)} \in V^r_{(x_0,q_0)}(Q, X)$, where $F(x, p_0) = G(x, q_0)$, if there exists a neighbourhood U of the point x_0, a neighbourhood V_1 of the point p_0, a neighbourhood V_2 of the point q_0 and a continuous mapping $\Psi : V_1 \to V_2$ such that*

1. $q_0 = \Psi(p_0)$;
2. $G(x, \Psi(p)) = F(x, p)$ *for all* $(x, p) \in U \times V_1$.

3.112 Definition. *The germ $[F]_{(x_0,\varepsilon_0)} \in V^r_{(x_0,\varepsilon_0)}(R^k, R^n)$ is called the k-parametric deformation or k-parametric unfolding of the germ $[v]_{x_0} \in$*

$V_{x_0}^r(R^n)$, if $F(x, \varepsilon_0) = v(x)$ for all x from a certain neighbourhood of the point x_0.

3.113 Definition. *The germ $[G]_{(x_0,\varepsilon_0)} \in V_{(x_0,\varepsilon_0)}^r(R^k, R^n)$ is called the versal deformation or the versal unfolding of the germ $[v]_{x_0} \in V_{x_0}^r(R^n)$, if for an arbitrary natural number p an arbitrary p-parametric deformation of the germ $[v]_{x_0}$ is orbitally topologically equivalent to a germ induced from the germ $[G]_{(x_0,\varepsilon_0)}$. We also say that the germ $[G]_{(x_0,\varepsilon_0)}$ is versal.*

Definitions 3.111–3.113 can be reformulated also for the germs of parametrized differential equations in natural way with the use of parametrized vector fields generated by these differential equations. We shall not introduce individual definitions.

The notions as the germ, orbital equivalence of germs, induced germs and versal deformation of the germ can also be defined in analogous way for diffeomorphisms or parametrized diffeomorphisms, respectively. These definitions will not be introduced separately.

3.114 Example. Let us introduce the germ $[v]_0 \in V_0^r(R)$, represented by the differential equation

$$\dot{y} = y^2 \tag{3.26}$$

and its 1-parametric deformation $[F]_{(0,0)} \in V_{(0,0)}^r(R, R)$ represented by the parametrized differential equation

$$\dot{y} = \mu y + y^2, \tag{3.27}$$

where $\mu \in R$ is a parameter. Let us consider another 1-parametric deformation $[G]_{(0,0)} \in V_{(0,0)}^r(R, R)$ of the germ $[v]_0$ represented by the parametrized differential equation

$$\dot{x} = x^2 - \varepsilon, \quad \varepsilon \in R. \tag{3.28}$$

Let us define the continuous mapping $\varepsilon = \Psi(\mu) = 1/4\mu^2$. This mapping induces from equation (3.28) the germ $[\tilde{G}]_{(0,0)}$ represented by the parametrized differential equation

$$\dot{x} = x^2 - 1/4\mu^2. \tag{3.29}$$

The mapping $x = h(y, \mu) = y + 1/2\mu$ is smooth and for every $\mu \in R$ the mapping $h_\mu : R \to R$, $h_\mu(y) = y + 1/2\mu$ is a C^∞-diffeomorphism transforming the differential equation (3.29) into the form (3.27). This means that the germ $[\tilde{G}]_{(0,0)}$ represented by the parametrized differential equation (3.29) is induced from the germ $[G]_{(0,0)}$ and the germ $[\tilde{G}]_{(0,0)}$ is orbitally equivalent to the germ $[F]_{(0,0)}$. In Chapter 5 we shall prove that the germ $[G]_{(0,0)}$ represented by the differential equation (3.28) is versal (see Theorem 5.18).

3.115 Remark. Let us note that the parametrized differential equations (3.27) and (3.28) are not orbitally topologically equivalent in any neighbourhood of the point $(0,0)$. The differential equation (3.27) has for $\mu < 0$ two singular points, for $\mu = 0$ it has one and for $\mu > 0$ it again has two singular points. The differential equation (3.28) has, however, two singular points for $\varepsilon > 0$. This example shows that the conception of versal deformations has the characteristic of control theory of singularities of differential equations, where the control is governed by the choice of parameters. The analogous theory of versal deformations was at first created for smooth functions of several variables depending on parameters and it is known as the Thom mathematical catastrophe theory. It was formulated by the French mathematician R. Thom and its foundations are explained in [11, 24 and 94] in quite a simple way. In papers [45, 46] with the use of this theory some problems of the theory of bifurcations of parametrized differential equations with certain type of symmetry are solved.

3.6 Grobman–Hartman Theorem

The present section is aimed at the definition of a hyperbolic singular point for nonlinear vector fields on smooth manifolds and description of the topological structure of trajectories in neighbourhoods of such points.

3.116 Lemma. *If φ is a C^r-dynamical system on a smooth manifold X, where $r \geq 2$ and $x \in X$ is its singular point, then for arbitrary $t \in R$ it holds that*

1. $\text{Image}(D_x\varphi_t) \subset T_x(X);$
2. $D_x\varphi_t = e^{tA},$ (3.30)

where
$$A = \frac{d}{ds}(D_x\varphi_s)_{/s=0}.$$

Proof

For arbitrary $t \in R$ we have $\varphi_t(x) = x$, and thus, the mapping $D_x\varphi_t$ maps the space $T_x(X)$ into itself. Now we shall prove assertion 2. The curve $c : R \to L(T_x(X))$, $c(s) = D_x\varphi_s$ is of class C^{r-1} and

$$A = \frac{d}{ds}(D_x\varphi_s)_{/s=0} \in L(T_x(X)).$$

From Theorem 2.31 on the derivation of a composed mapping we obtain that for arbitrary $s,t \in R$ it is $D_x(\varphi_s \circ \varphi_t) = D_x\varphi_s \circ D_x\varphi_t$, and thus,

$$\frac{dc(t)}{dt} = \frac{d}{dt}(D_x\varphi_t) = \frac{d}{ds}(D_x\varphi_{s+t})_{/s=0} = \frac{d}{ds}(D_x(\varphi_s \circ \varphi_t))_{/s=0}$$
$$= \frac{d}{ds}D_x\varphi_s \circ D_x\varphi_t)_{/s=0} = A \circ c(t).$$

This means that c is a solution of the linear differential equation $\dot{u} = A \circ u$ in the Banach space $T_x(X)$ satisfying the initial condition $c(0) = \text{id}$. According to Theorem 3.76 we obtain $c(t) = D_x\varphi_t = e^{tA}$. □

3.117 Lemma. *If $A \in M(n)$, then for arbitrary $\varepsilon > 0$ there exists a matrix $\tilde{A} \in M(n)$ whose eigenvalues are different from each other and $\|A - \tilde{A}\| < \varepsilon$, where $\|\cdot\|$ is a certain norm on $M(n)$.*

Proof

Since the space of matrices $M(n)$ is isomorphic with R^{n^2}, according to Theorem 1.5 all norms on $M(n)$ are equivalent. Therefore it is sufficient to suppose that the norm on $M(n)$ is defined by the relationship

$$\|D\| = \sup_{\|x\| \leq 1} \|D\| \quad \text{for} \quad D \in M(n),$$

where $\|\cdot\|$ is a norm on R^n. Let $C \in M(n)$ be a regular matrix such that $B = CAC^{-1}$ is the Jordan canonical form of the matrix A. Diagonal elements of the Jordan blocks of the matrix B, i.e., eigenvalues of the matrix A can be modified in such a way that we add to them small numbers such that for the matrix $\tilde{B}$ obtained like this is $\|B - \tilde{B}\| < \varepsilon(\|C^{-1}\|\|C\|)^{-1}$. The matrix $\tilde{A} = C^{-1}\tilde{B}C$ has the same eigenvalues as the matrix $\tilde{B}$ and it holds that $\|A - \tilde{A}\| = \|C^{-1}(B - \tilde{B})C\| \leq \|B - \tilde{B}\|\|C^{-1}\|\|C\| < \varepsilon$. □

3.118 Definition. *Let X be a smooth manifold, $r \geq 2$ and φ be a flow of the vector field $F \in V^r(X)$. The mapping $A : T_x(X) \to T_x(X)$, defined by the relationship (3.30) is called the hessian of the vector field F at the singular point $x \in X$ denoted as $\dot{F}(x)$. The eigenvalue μ of the mapping $\dot{F}(x)$ is called the characteristic exponent of the singular point x and the number e^μ, i.e., the eigenvalue of the mapping $D_x\varphi_1$, is called the characteristic multiplier of the singular point x. If $\dim X = n$ and $\mu_1, \mu_2, \ldots, \mu_n$ are characteristic exponents, where $\text{Re}\,\mu_i > 0$, $i = 1, 2, \ldots, k$, $\text{Re}\,\mu_i < 0, i = k+1, k+2, \ldots, n$, then the number k is denoted as $m^+(F, x)$ and the number $n - k$ is denoted as $m^-(F, x)$.*

3.119 Proposition. *Let X be a smooth manifold, $r \geq 2$, (U, α) be an admissible chart on X, $x \in U$ be a singular point of the vector field $F \in V^r(X)$, $y = \alpha(x)$ and f_α be the main part of local representation of the vector field F with respect to the given admissible chart. Then the linear mapping $d_y f_\alpha \in L(R^n)$ is a local representation of the hessian $\dot{F}(x)$.*

Proof

From the properties of the flow φ of the vector field F it follows that there exists an open neighbourhood V of the point x and an open interval $I \subset R$ such that $0 \in I$, $\varphi_t(V) \subset U$ for all $t \in I$. The mapping $\Psi_t = \alpha \circ \varphi_t \circ \alpha^{-1}$ is the local representation of the mapping φ_t, for $t \in I$, the linear mapping $B : T_x(X) \to R^n$, $B([\gamma]_x) = d_0(\alpha \circ \gamma)(1)$ is a coordinate mapping on $T_x(X)$ and it holds that: For arbitrary $v \in R^n$ there exists γ such that $v = d_0(\alpha \circ \gamma)(1)$ and

$$B \circ D_x \varphi_t \circ B^{-1}(v) = B \circ D_x \varphi_t([\gamma]_x) = B([\varphi_t \circ \gamma]_x) = d_0(\alpha \circ \varphi_t \circ \gamma)(1)$$
$$= d_0((\alpha \circ \varphi_t \circ \alpha^{-1}) \circ \alpha \circ \gamma)(1) = d_y \Psi_t \circ d_0(\alpha \circ \gamma)(1) = d_y \Psi_t(v).$$

This means that $d_y \Psi_t$ is a local representation of the mapping $D_x \varphi_t$. From the definition of the mapping $F(x)$ the equality

$$d_y \Psi_t = e^{tB\dot{F}(x)B^{-1}} \tag{3.31}$$

follows. If $z \in \alpha(V)$, then according to Proposition 3.5 $c_z : I \to R^n$, $c_z(t) = \Psi_t(z)$, is a solution of the differential equation $\dot{u} = f_\alpha(u)$ fulfilling the initial condition $c_t(0) = z$. From Theorem 1.48 it follows that the mapping $W : I \to L(R^n)$, $W(t) = d_y \Psi_t$ is a solution of the linear differential equation $\dot{w} = d_y f_\alpha \circ w$ in the Banach space $L(R^n)$, fulfilling the initial condition $W(0) = \text{id}$. From Theorem 3.70 and from the equality (3.31) the equality $d_y f = B \circ \dot{F}(x) \circ B^{-1}$ follows. □

3.120 Definition. *Let X be a smooth manifold and $r \geq 2$. The singular point $x \in X$ of the vector field $F \in V^r(X)$ is called hyperbolic, if the hessian $\dot{F}(x)$ has no eigenvalues with zero real part.*

3.121 Remark. On the basis of Proposition 3.119 we obtain that the definition of the hyperbolicity of a singular point of a linear vector field is in accordance with the definition of hyperbolicity of a singular point of a linear vector field defined on an Euclidean space (see Definition 3.79) and from the definition of the hessian it follows that it is independent of coordinates on X.

3.122 Remark. If $X = R^n$ and the vector field $F \in V^r(R^n)$ is defined by the mapping $F \in C^r(R^n, R^n)$, then in accordance with Proposition 3.119

Vector Fields and Dynamical Systems

instead of the hessian $\dot{F}(x)$ defined only for $r \geq 2$ it is possible to take the mapping $d_x f$, and so it is sufficient in the definition of hyperbolicity of a singular point to assume that $r = 1$.

3.123 Definition. *Let E_n be an n-dimensional Euclidean space. The linear mapping $A \in L(E_n)$ is called hyperbolic if it has no eigenvalue on the unit circle and does not have the zero eigenvalue, i.e., the mapping A has the inverse mapping A^{-1}.*

3.124 Definition. *Let X be a smooth manifold, $r \geq 1$ and $f \in \text{Diff}^r(X)$ or f be a local C^r-diffeomorphism at the point $x \in X$. If $x \in X$ is a fixed point of the diffeomorphism f, then this point is called hyperbolic if the mapping $D_x f : T_x(X) \to T_x(X)$ is hyperbolic. If x is an n-periodic point of the diffeomorphism f, then this point is called hyperbolic if x is a hyperbolic fixed point of the diffeomorphism f^m. If $\dim X = n$ and $\lambda_1, \lambda_2, \ldots, \lambda_n$ are eigenvalues of the mapping $D_x f$, where $|\lambda_i| > 1$, $i = 1, 2, \ldots, k, |\lambda_i| < 1$, $i = k+1, k+2, \ldots, n$, then the number k is denoted as $m^+(f, x)$ and the number $n - k$ is denoted as $m^-(f, x)$.*

3.125 Proposition. *The set $L_H(E_n) \subset L(E_n)$ of all linear hyperbolic mappings is open and dense in $L(E_n)$.*

Proof

Density of the set of all invertible linear mappings from $L(E_n)$ follows directly from Theorem 3.81. Density of the set of all linear mappings having no eigenvalue on the unit circle can be proved with the use of Theorem 3.75 in a way analogous to that presented in the proof of Theorem 3.81. However, a simpler way is to prove this part with the same procedure as introduced in the proof of Lemma 3.117. Openness of the set $L_H(E_n)$ follows from the continuous dependence of eigenvalues of matrices on their elements.

3.126 Lemma. *Let E_n be an n-dimensional Euclidean space over R and for arbitrary eigenvalue λ of the mapping $A \in L(E_n)$ is $|\lambda| < 1$. Then on E_n there exists a norm $\|\cdot\|$, such that*

$$\|A\| = \sup_{\|x\| \leq 1} \|Ax\| < 1.$$

Proof

It is sufficient to prove the lemma for $E_n = R^n$ and $A \in M(n)$. First, let us assume that all eigenvalues $\lambda_1, \lambda_2, \ldots, \lambda_n$ of the matrix A are non-zero different from each other. Let u_j, v_j be the unit eigenvectors of the matrix A and of the matrix A^* transported to A, respectively, that correspond to the eigenvalue λ_j and $\bar{\lambda}_j$, respectively ($j = 1, 2, \ldots, n$). Since $(Ax, y) = (x, A^*y)$ for all $x, y \in C^n$, where $(.,.)$ is the scalar product defined by the basis $e_1, e_2, \ldots, e_n$, where $e_i = (0, 0, \ldots, 0, 1, 0, \ldots, 0)$ (1 being the i-th element), for $i \neq j$ we get: $\lambda_i(u_i, v_j) = (\lambda_i u_i, v_j) = (Au_i, v_j) = (u_i, A^*v_j) = (u_i, \bar{\lambda}_j v_j) = \lambda_j(u_i, v_j)$. This means that $(u_i, v_j) = 0$ for $i \neq j$. Let us define the norm $\|\cdot\|$ on R^n as follows:

$$\|x\| = \left(\sum_{i=1}^{n} (x, v_i)\overline{(x, v_i)}\right)^{1/2}.$$

Since the vectors $u_1, u_2, \ldots, u_n$ form an orthonormal basis in C^n and $(u_i, v_j) = 0$ for $i \neq j$, then $x = \sum_{i=1}^{n}(x, v_i)u_i$. Therefore we get

$$\|Ax\|^2 = \left\|\sum_{i=1}^{n}(x, v_i)\lambda_i u_i\right\|^2 = \sum_{i=1}^{n}(x, v_i)\overline{(x, v_i)}\lambda_i \bar{\lambda}_i \leq \max_{1 \leq i \leq n} |\lambda_i|^2 \|x\|^2.$$

We have proved that

$$\|A\| = \sup_{\|x\| \leq 1} \|Ax\| \leq m(A) < 1,$$

where $m(A) = \max_{1 \leq i \leq n} |\lambda_i|^2$. Now, let us assume that A is an arbitrary matrix, fulfilling the assumptions of the lemma. Let ε be a positive number such that $m(A) + \varepsilon < 1$. From Theorem 3.81 and Lemma 3.117 it follows that for arbitrary positive number δ there exists a matrix $C \in M(n)$ such that $\|A - C\| < \delta$, where the matrix C has all its eigenvalues $v_1, v_2, \ldots, v_n$ non-zero and different from each other. From the continuous dependence of eigenvalues upon coefficients of the matrix it follows that for δ small enough it is $|m(C) - m(A)| < (1/2)\varepsilon$, where $m(C) = \max_{1 \leq i \leq n} |v_i|^2$. From the first part of the proof it follows that on R^n there exists a norm $\|\cdot\|$, such that

$$\|C\| = \sup_{\|x\| \leq 1} \|Cx\| \leq m(C).$$

If $\delta < (1/2)\varepsilon$ then

$$\|A\| \leq \|A - C\| + \|C\| \leq \frac{1}{2}\varepsilon + m(C)$$
$$\leq \frac{1}{2}\varepsilon + |m(C) - m(A)| + m(A) \leq \varepsilon + m(A) < 1. \quad \square$$

3.127 Lemma. *If $0 < s < \infty$, then there exists a number $c > 0$ and a smooth function $\varphi : R \to R$, such that $\varphi(t) = 1$ for all t, for which $|t| \leq 1/4cs^2, \varphi(t) = 0$ for all t, for which $|t| \geq cs^2, 0 < \varphi(t) < 1$ for all t, for which $1/4cs^2 < |t| < cs^2$ and*

$$\left|\frac{d\varphi(t)}{dt}\right| \leq 4s^{-2}$$

for all $t \in R$.

Proof

According to Lemma 2.82 there exists a smooth function $\Psi : R \to R$, such that, $\Psi(t) = 1$ if $|t| \leq 1, \Psi(t) = 0$ if $|t| \geq 4$ and $0 < \Psi(t) < 1$ if $1 < |t| < 4$. Let

$$c = \max_{t \in R} \left|\frac{d\Psi(t)}{dt}\right|.$$

The function $\varphi : R \to R, \varphi(t) = \Psi(4c^{-1}s^{-2}t)$ is obviously smooth and it holds that: $\varphi(t) = 1$ if $|t| \leq 1/4cs^2$, $\varphi(t) = 0$ if $|t| \geq cs^2$ and $0 < \varphi(t) < 1$ if $1/4cs^2 < |t| < cs^2$ and

$$\left|\frac{d\varphi(t)}{dt}\right| = 4c^{-1}s^{-2}\left|\frac{d}{dt}(\Psi(4c^{-1}s^{-2}t))\right| \leq 4s^{-2} \quad \text{for all } t \in R. \quad \square$$

3.128 Lemma. *Let E_n be an n-dimensional Euclidean space, $U \subset E_n$ be an open set containing the point $0 \in E_n, F : U \to E_n$ be a C^r-mapping, $1 \leq r \leq \infty, F(0) = 0, d_0F = 0$ and let Θ be an arbitrary positive number. Then there exist open sets $U_1, U_2 \subset U$ with compact closure containing the point 0 such that $\overline{U}_1 \subset U_2, \overline{U}_2 \subset U$ and a C^r-mapping $G : E_n \to E_n$ such that it holds:*

1. *$F/U_1 = G/U_1$ and $G(x) = 0$ for all $x \in E_n \setminus U_2$;*
2. *$\|d_xG\|_1 \leq K\Theta$ for all $x \in E_n$, where K is a positive number independent of the mapping F and the number Θ.*

Proof

It is sufficient to assume that $E_n = R^n$. Let $\varphi : R \to R$ be a smooth function having the properties introduced in Lemma 3.127. For the number $s > 0$ let us define the sets:

$$U_1 = \left\{x \in R^n : \|x\| < \frac{1}{4}\sqrt{c} \cdot s\right\}, \quad U_2 = \left\{x \in R^n : \|x\| < \sqrt{c} \cdot s\right\}.$$

Let us assume that the number s is so small that $\overline{U}_2 \subset U$. Let us define the mapping $G : R^n \to R^n$, $G(x) = F(x)\varphi(\|x\|^2)$ if $\|x\| \leq \sqrt{c} \cdot s$ and $G(x) = 0$

if $\|x\| \geq \sqrt{c} \cdot s$. Obviously, the mapping G is of class C^r and possesses the property 1. If $\|x\| \leq \sqrt{c} \cdot s$, then

$$d_x G = \varphi(\|x\|^2) d_x F + 2 \frac{d\varphi(t)}{dt}\bigg|_{t=z} \cdot H(x),$$

where $z = \|x\|^2$ and $H : R^n \to L(R^n)$ is a C^{r-1}-mapping such that the matrix representation $[H(x)]$ of the mapping $H(x)$ (with respect to the standard basis of unit vectors, has the i-th column equal to the vector $x_i F(x)^*$, where $x = (x_1, x_2, \ldots, x_n)$. Since $F(0) = 0$ and $d_0 F = 0$, a small number s, can be chosen such that $\|d_x F\|_1 \leq \Theta$ for all x for which $\|x\| \leq \sqrt{c} \cdot s$. Then, from the Mean Value Theorem we get $\|F(x)\| \leq \Theta \|x\|$ if $\|x\| \leq \sqrt{c} \cdot s$ and it holds that:

$$\|d_x G\|_1 \leq \|d_x F\|_1 \varphi(\|x\|^2) + 2 \left|\frac{d\varphi(t)}{dt}\bigg|_{t=z}\right| \cdot \|H(x)\| \leq \Theta + 8s^{-2}\|x\|^2 \Theta \leq K\Theta,$$

where $z = \|x\|^2$, $K = 1 + 8c$. If $\|x\| > \sqrt{c} \cdot s$, then $G(x) = 0$, and thus, $\|d_x G\|_1 \leq K\Theta$ for all $x \in R^n$. □

3.129 Lemma. *If $A \in L(E_n)$ is an invertible mapping and $\Phi : E_n \to E_n$ is a Lipschitz mapping with the Lipschitz constant $\mathrm{Lip}\,(\Phi) < \|A^{-1}\|^{-1}$, then the mapping $A + \Phi$ is invertible, its inverse is a Lipschitz and*

$$\mathrm{Lip}\,((A+\Phi)^{-1}) \leq (\|A^{-1}\|^{-1} - \mathrm{Lip}\,(\Phi))^{-1}. \tag{3.32}$$

Proof

For arbitrary $x, y \in E_n$ it is $\|x - y\| = \|A^{-1}x' - A^{-1}y'\| = \|A^{-1}(x' - y')\| \leq \|A^{-1}\|\|x' - y'\|$, where $x' = Ax$, $y' = Ay$. From this inequality it follows that $\|Ax - Ay\| \geq \|A^{-1}\|^{-1}\|x - y\|$. Therefore $\|(A + \Phi)(x) - (A + \Phi)(y)\| = \|Ax - Ay + \Phi(x) - \Phi(y)\| \geq \|Ax - Ay\| - \|\Phi(x) - \Phi(y)\| \geq \|A^{-1}\|^{-1}\|x - y\| - \mathrm{Lip}\,(\Phi)\|x - y\|$, and so we obtain the inequality

$$\|(A+\Phi)(x) - (A+\Phi)(y)\| \geq (\|A^{-1}\|^{-1} - \mathrm{Lip}\,(\Phi))\|x - y\|. \tag{3.33}$$

From this inequality also injectivity of the mapping $A + \Phi$ follows. If in the inequality (3.33) we replace x, y by the values $(A + \Phi)^{-1}(x)$ and $(A+\Phi)^{-1}(y)$, respectively, we get the inequality equivalent to the inequality (3.32). Now we shall prove surjectivity of the mapping $A+\Phi$. For arbitrary $y \in E_n$ let us solve the equation

$$(A+\Phi)(x) = y. \tag{3.34}$$

If we denote $x_0 = A^{-1}y$ and introduce the new unknown $w = x - x_0$, then from equation (3.34) we get: $(A+\Phi)(x_0+w) = Ax_0+Aw+\Phi(x_0+w) = Ax_0$, i.e., $Aw + \Phi(x_0 + w) = 0$, or also

$$w = -A^{-1} \circ \Phi(x_0 + w). \tag{3.35}$$

Let us define the mapping $F : E_n \to E_n$, $F(w) = -A^{-1} \circ \Phi(x_0 + w)$. Since $\text{Lip}(F) \leq \|A^{-1}\| \cdot \text{Lip}(\Phi) < 1$, the mapping F is contractive and from the Fixed Point Theorem we get that there exists a unique fixed point W of the mapping F. This means that the equation (3.34) has the unique solution $x = x_0 + w$. □

According to Theorem 1.29 the space $C_B^0(E_n, E_m)$ of all continuous and bounded mappings from the n-dimensional Euclidean space E_n into the m-dimensional Euclidean space F_m with the norm $|f| = \max_{x \in E_n} \|f(x)\|$ is a Banach space. The set $C^1 = \{g \in C_B^1(E_n, E_n) : g(0) = 0, d_0 g = 0, g$ as well as dg are bounded mappings$\}$ with the norm $|g|_1 = \sup_{x \in E_n}(\|g(x)\| + \|d_x g\|_1)$ is obviously a closed subspace in $C_B^1(E_n, E_n)$, and thus, it is complete, i.e. a Banach space. If $0 < s < \infty$, then, obviously, the set $C_s^1 = \{g \in C^1 : |g|_1 < s\}$ is open in C^1. The above notations and assertions will be used in the proof of the following lemma.

3.130 Lemma. *Let $A \in L(E_n)$ be an invertible mapping and $0 < a < \|A^{-1}\|^{-1}$. Then it holds that*
  1. for arbitrary $f \in C_a^1$ the mapping $A + f$ is invertible and its inverse mapping is Lipschitz,
  2. the mapping $S : C_a^1 \to C_B^0(E_n, E_n)$, $S(f) = (A + f)^{-1}$ is continuous.

Proof

Proposition 1 is a direct consequence of Lemma 3.129. We shall prove continuity of the mapping S. If $f, g \in C_a^1$ and $x, y \in E_n$ then using the inequality (3.33) for $\Phi = f$ we obtain $\|(A+f)(x) - (A+g)(y)\| \geq \|(A+f)(x) - (A+f)(y)\| - \|(A+f)(y) - (A+g)(y)\| \geq K\|x-y\| - \|f(y) - g(y)\| \geq K\|x-y\| - |f-g|_1$, where $K = (\|A^{-1}\|^{-1} - \text{Lip}(f))^{-1}$. If in this inequality we replace x, y by the values $(A+f)^{-1}(x)$ and $(A+g)^{-1}(x)$, respectively, we obtain the inequality

$$\sup_{x \in E_n} \|(A+f)^{-1}(x) - (A+g)^{-1}(x)\| \leq K^{-1}|f-g|_1,$$

from which continuity of the mapping S follows. □

3.131 Lemma. *If $A \in L_H(E_n) \subset L(E_n)$ ($L_H(E_n)$ is the set of linear hyperbolic mappings), then there exists a number $a > 0$ and a unique continuous mapping $h : C_a^1 \times C_a^1 \to C_B^0(E_n, E_n)$ such that it holds that*

Chapter 3

1. *for arbitrary $f, g \in C_a^1$ is*

$$H \circ (A + f) = (A + g) \circ H, \qquad (3.36)$$

where $H = h(f, g)$,

2. *the mapping $h(f, g)$ is a homeomorphism.*

Proof

Let the mappings $f, g \in C_a^1$ be Lipschitz with $\mathrm{Lip}(f) \leq a$, $\mathrm{Lip}(g) \leq a$ and from Lemma 3.129 it follows that for $a < \|A^{-1}\|^{-1}$ the mappings $A + f$, $A + g$ are invertible. In this case the equation (3.36) is equivalent to the equation

$$H = (A + g) \circ H \circ (A + f)^{-1}, \qquad (3.37)$$

as well as to the equation

$$H = (A + g)^{-1} \circ H \circ (A + f). \qquad (3.38)$$

In the same way as in the proof of Theorem 3.87 with the use of the Jordan form of the matrix A it can be proved that there exist linear subspaces E^s, E^u of the space E_n that are invariant sets of the mapping A, $E_n = E^s \oplus E^u$ and for arbitrary eigenvalue λ of the mapping $A_s = A/E^s$ ($A_u = A/E^u$) is $|\lambda| < 1 (|\lambda| > 1)$. Since for arbitrary eigenvalue μ of the mapping A_u^{-1} is $|\mu| < 1$, according to Lemma 3.126 there exist norms $\|\cdot\|_1$, $\|\cdot\|_2$ on E^s and E_u, respectively, such that

$$\|A_s\|_1 = \sup_{\|x\| \leq 1} \|A_s x\|_1 < 1, \quad \|A_u^{-1}\|_2 = \sup_{\|y\| < 1} \|A_u^{-1} y\|_2 < 1.$$

Further, we shall denote both norms just by $\|\cdot\|$. Let $b = \max(\|A_s\|, \|A_u^{-1}\|)$ and $a < b$. If $g : E_n \to E_n$ and $x \in E_n$, then $g(x) = y_s + y_u$, where $y_s \in E^s$ and $y_u \in E^u$. We shall denote the values y_s and y_u as $[g(x)]_s$ and $[g(x)]_u$, respectively. For every such mapping g let us define two mappings: $g_s : E_n \to E^s$, $g_s(x) = [g(x)]_s$ and $g_u : E_n \to E^u$, $g_u(x) = [g(x)]_u$. Using these notations, let us define the following mappings: $T_s^a : C_B^0(E_n, E_n) \times C_a^1 \times C_a^1 \to C_B^0(E_n, E^s)$, $T_s^a(h, f, g) = h_s - (A + g)_s \circ h_s \circ (A + f)^{-1}$, $T_u^a : C_B^0(E_n, E_n) \times C_a^1 \times C_a^1 \to C_B^0(E_n, E^u)$, $T_u^a(h, f, g) = h_u - (A + g)_u^{-1} \circ h_u \circ (A + f)$, $T^a : C_B^0(E_n, E_n) \times C_a^1 \times C_a^1 \to C_B^0(E_n, E_n)$, $T^a(h, f, g) = (T_s^a(h, f, g), T_u^a(h, f, g))$. From Lemma 3.130 it follows that the mappings T_s^a, T_u^a are continuous, and thus, the mapping T^a is also continuous, where $T^a(\mathrm{id}, 0, 0) = 0$. For arbitrary $(h, f, g) \in C_B^0(E_n, E_n) \times C_a^1 \times C_a^1$ and $k \in C_B^0(E_n, E_n)$ it is $(\partial_1)_{h, f, g} T_s^a(k) = k_s - (A + dg)_s \circ k_s \circ (A + f)^{-1}$, $(\partial_1)_{h, f, g} T_u^a(k) =$

$k_u - (A+dg)_u^{-1} \circ k_u \circ (A+f)$. From Lemma 3.130 it follows that the mappings $(h,f,g) \mapsto (\partial_1)_{(h,f,g)} T_s^a$, $(h,f,g) \mapsto (\partial_1)_{(h,f,g)} T_u^a$ are continuous, and thus, the mapping $(h,f,g) \mapsto (\partial_1)_{(h,f,g)} T^a$ is also continuous. We shall prove that the mapping $(\partial_1)_{(h,0,0)} T^a : k \mapsto (k_s - A_s \circ k_s \circ A^{-1}, k_u - A_u^{-1} \circ k_u \circ A)$ is an isomorphism. If $w \in C_B^0(E_n, E_n)$, then $w = w_s + w_u$ for some $w_s \in C_B^0(E_n, E^s)$, $w_u \in C_B^0(E_n, E^u)$. Since $\|A_s\| < 1$, $\|A_u^{-1}\| < 1$, the mappings $\mathscr{F}_s : C_B^0(E_n, E^s) \to C_B^0(E_n, E^s)$, $\mathscr{F}_s(v) = A_s \circ v \circ A^{-1} + w_s$, $\mathscr{F}_u : C_B^0(E_n, E^u) \to C_B^0(E_n, E^u)$, $\mathscr{F}_u(z) = A_u^{-1} \circ z \circ A + w_u$, are contractive, and thus, from the Fixed Point Theorem we get that each of the mappings $\mathscr{F}_s$, $\mathscr{F}_u$ has a unique fixed point k_s and k_u, respectively, where $\|w_s\| \geq \|k_s\| - \|A_s\|\|k_s\| \geq (1-b)\|k_s\|$, $\|w_u\| \geq \|k_u\| - \|A_u^{-1}\|\|k_u\| \geq (1-b)\|k_u\|$. This means that the equation $(\partial_1)_{(h,0,0)} T^a(k) = w$ has the unique solution $k \in C_B^0(E_n, E_n)$, where $\|k\| \leq (1-b)^{-1}\|w\|$. Thus, the mapping $(\partial_1)_{(h,0,0)} T^a$ is an isomorphism. We have shown that all assumptions of the Implicit Function Theorem are satisfied, which says that there exists the unique continuous mapping $h : C_a^1 \times C_a^1 \to C_B^0(E_n, E_n)$ such that $h(0,0) = \text{id}$ and $T^a(h(f,g), f, g) = 0$ for all $f, g \in C_a^1$, where $a > 0$ is a small enough number. This means that for $H = h(f,g)$, where $f, g \in C_a^1$, the equality (3.36) is satisfied. It is still to be shown that the mapping $H = h(f,g)$, where $f, g \in C_a^1$, is an isomorphism. Let us consider the equation

$$(A+f) \circ q = q \circ (A+g) \tag{3.39}$$

with the unknown $q \in C_B^0(E_n, E_n)$. This equation is of the form (3.36), where instead of g there is the mapping f and f is replaced by g. According to assertion 1 there exists a continuous mapping $q : C_a^1 \times C_a^1 \to C_B^0(E_n, E_n)$, such that for arbitrary $f, g \in C_a^1$ it is

$$(A+f) \circ q(f,g) = q(f,g) \circ (A+g), \tag{3.40}$$

where $a > 0$ is a sufficiently small number. From the definition of the mapping $H = h(f,g)$ we get that

$$H \circ (A+f) = (A+g) H. \tag{3.41}$$

From the equalities (3.40), (3.41) we get the equality $(A+g) \circ H \circ q(f,g) = H \circ (A+f) \circ q(f,g) = H \circ q(f,g) \circ (A+g)$. According to assertion 1 the equation $h \circ (A+g) = (A+g) \circ h$ has the unique solution $h = \text{id}$, and thus, $H \circ q(f,g) = \text{id}$. We shall also prove that $q(f,g) \circ H = \text{id}$. From the equalities (3.40), (3.41) we get that $q(f,g) \circ H \circ (A+f) = q(f,g) \circ (A+g) \circ H = (A+f) \circ q(f,g) \circ H$. From assertion 1 it follows that the equation $h \circ (A+f) = (A+f) \circ h$ has the unique solution $h = \text{id}$, and thus, it is $q(f,g) \circ H = \text{id}$. We have proved that the mapping $H = H(f,g)$ is a homeomorphism and $H^{-1} = q(f,g)$. $\square$

3.132 Hartman Theorem. *Let E_n be an n-dimensional Euclidean space, $V \subset E_n$ be an open neighbourhood of the point $x_0 \in E_n$, $f : V \to E_n$ be a local C^1-diffeomorphism at the point x_0 and the point x_0 be a hyperbolic fixed point of this diffeomorphism. Then there exists an open neighbourhood $U \subset V$ of the point x_0 and a homeomorphism of the set U onto the set $h(U)$ such that*

$$d_{x_0}f \circ h(x) = h \circ f(x) \qquad (3.42)$$

for all $x \in U$.

Proof

Without loss of generality it can be assumed that $x_0 = 0$. Since according to the assumption f is a local C^1-diffeomorphism at the point 0, there exists an open neighbourhood $W \subset V$ of the point 0 such that f/W is a C^1-diffeomorphism. From the Taylor Theorem it follows that if W is an open sphere in E_n having its centre at the point 0 and it is sufficiently small, then $f(x) = d_0f(x) + g(x)$ for all $x \in W$, where $g : W \to E_n$ is a mapping of class C^1, where $g(0) = 0$, $d_0g(0) = 0$. We shall suppose that W is such a neighbourhood of the point 0. According to Lemma 3.128 there exist for an arbitrary small number a open neighbourhoods $U_1, U_2 \subset W$ of the point 0 such that $\overline{U}_1 \subset U_2$, $\overline{U}_2 \subset W$ and a C^1-mapping $G : E_n \to E_n$, such that $g/U_1 = G/U_1$, $G(x) = 0$ for all $x \in E_n \setminus U_2$ and $\|d_xG\|_1 \leq a$ for all $x \in E_n$. This means that $G \in C^1_a$. Let the number a be so small that the assumptions of Lemma 3.131 are fulfilled. Then from this lemma it follows that there exists a unique homeomorphism $H : E_n \to E_n$, such that $d_0f \circ H = H \circ (d_0f + G)$. If $U = U_1$, $h = H/U$, then for arbitrary $x \in U$ we have $d_0f(h(x)) = h(d_0f(x) + g(x)) = h(f(x))$. □

3.133 Lemma. *If $0 \in R^n$ is a hyperbolic singular point of the linear vector field $A_v \in V_L(R^n)$, defined by the mapping $A \in L(R^n)$, then there exists a positive number a and a unique continuous mapping $h : C^1_a \to C^0_B(R^n, R^n)$ such that for arbitrary $R \in C^1_a$ the mapping $H = h(R)$ is a homeomorphism and*

$$H \circ \Psi_t = \varphi_t \circ H \qquad (3.43)$$

for all $t \in R$, where φ is the flow of the vector field $F \in V^1(R^n)$ defined by the differential equation $\dot{x} = Ax + R(x)$ and $\Psi_t : R^n \to R^n$, $\Psi_t(x) = e^{tA}x$.

Proof

Let us consider the differential equations

$$\dot{y} = Ay, \qquad (3.44)$$

$$\dot{y} = Ay + R(y), \qquad (3.45)$$

where $R \in C_a^1$ for some $a > 0$ and let Ψ and φ be dynamical systems generated by solutions of the differential equations (3.44) and (3.45), respectively. Then

$$\Psi_t(x) = e^{tA}x,$$
$$\varphi_t(x) = e^{tA} + g_t^R(x), \qquad (3.46)$$

where

$$g_t^R : R^n \to R^n, g_t^R(x) = \int_0^t e^{(t-s)A} R(\varphi_s(x))\, ds$$

(see the variation of constants formula). Since, $R(0) = 0$, from Theorem 1.41 it follows that $\varphi_t(0) = 0$ for all $t \in R$, and thus, $g_t^R(0) = 0$ for all $t \in R$. The mapping R is, according to the assumption, from the set C_a^1, and thus, it is

$$\|g_t^R(x)\| \leq a \cdot M(t) \qquad (3.47)$$

for all $t \in R, x \in R^n$, where

$$M(t) = \int_0^t e^{(t-s)\|A\|}\, ds.$$

By differentiating the equality (3.46) by x we obtain the equation

$$d_x \varphi_t = e^{tA} + \int_0^t e^{(t-s)A} \circ d_{\varphi_s(x)} R \circ d_x \varphi_s\, ds,$$

yielding the inequality

$$\|d_x \varphi_t\|_1 \leq e^{t\|A\|} + \int_0^t e^{(t-s)\|A\|} \|d_{\varphi_s(x)} R\|_1 \cdot \|d_x \varphi_s\|_1\, ds$$
$$\leq e^{t\|A\|} + a \int_0^t e^{(t-s)\|A\|} \|d_x \varphi_s\|_1\, ds, \text{ i.e. } e^{-t\|A\|}\|d_x \varphi_t\|_1$$
$$\leq 1 + a \int_0^t e^{-s\|A\|} \|d_x \varphi_s\|_1\, ds.$$

Using the Gronwall inequality we obtain the inequality $e^{-t\|A\|}\|d_x \varphi_t\|_1 \leq e^{at}$, i.e.

$$\|d_x \varphi_t\|_1 \leq e^{(a+\|A\|)t} \qquad (3.48)$$

for all $t \in R$. Since

$$d_x g_t^R = \int_0^t e^{(t-s)A} d_{\varphi_s(x)} R \circ d_x \varphi_s \, ds,$$

using the inequality (3.48) we obtain the inequality

$$\|d_x g_t^R\|_1 \leqq a \int_0^t e^{(t-s)\|A\|} \|d_x \varphi_s\|_1 \, ds \leqq a \int_0^t e^{(t-s)\|A\|} e^{(a+\|A\|)s} \, ds$$
$$= e^{t\|A\|}(e^{at} - 1).$$

Thus, we obtain the inequality

$$\|d_x g_t^R\|_1 \leqq \tilde{a}(t) = e^{t\|A\|}(e^{at} - 1) \tag{3.49}$$

for all $t \in R$, $x \in R^n$. Let us now consider the mappings Ψ_1 and φ_1, i.e.

$$\Psi_1(x) = e^A x, \varphi_1(x) = e^A x + g^R(x), \tag{3.50}$$

where $g^R = g_1^R$. If $M = (1/2) \max(M(1), \tilde{a}(1))$, then from the inequalities (3.47), (3.49) we obtain the inequality $|g^R|_1 \leqq a \cdot M$, and this means that $g^R \in C_{aM}^1$. From Lemma 3.131 it follows that for $a > 0$ sufficiently small there exists a unique continuous mapping $h : C_{aM}^1 \to C_B^0(R^n, R^n)$ such that for arbitrary $R \in C_a^1$ the mapping $H = h \circ G(R)$ is a homeomorphism and

$$\Psi_1 \circ H = H \circ \varphi_1. \tag{3.51}$$

For arbitrary $t \in R$ let us define the mapping $H_t : R^n \to R^n$, $H_t = \Psi_t \circ H \circ \varphi_{-t}$, that, obviously, is continuous. Using the equality (3.51) we get $\Psi_1 \circ H_t = \Psi_{t+1} \circ H \circ \varphi_{-t} = \Psi_t \circ \Psi_1 \circ H \circ \varphi_{-t} = \Psi_t \circ H \circ \varphi_{1-t} = \Psi_t \circ H \circ \varphi_{-t} \circ \varphi_1 = H_t \circ \varphi_1$, i.e. $\Psi_1 \circ H_t = H_t \circ \varphi_1$ for all $t \in R$. Since the equation $\Psi_1 \circ h = h \circ \varphi_1$ according to Lemma 3.131 has a unique continuous solution, so $H_t = H$ for all $t \in R$. Therefore from the definition of the mapping H_t we get $H = \Psi_t \circ H \circ \varphi_{-t}$, i.e. $\Psi_t \circ H = H \circ \varphi_t$ for all $t \in R$. □

Let X be a smooth manifold, $r \geq 2, x \in X$ be a singular point of the vector field $F \in V^r(X)$ and $\dot{F}(x)$ be the hessian of this vector field at the point x. Then the dynamical system $\Psi : T_x(X) \times R \to T_x(X)$, $\Psi(w, t) = e^{t\dot{F}(x)} w$ on the tangent space $T_x(X)$ defines the linear vector field $\dot{F}(x)_v \in V_L(T_x(X))$. This vector field is called the linearization of the vector field F at the singular point x. If $X = R^n$ and $F \in V^1(R^n)$ is a vector field defined by the differential equation $\dot{y} = f(y)$, where $f \in C^1(R^n, R^n)$ (in this case it is sufficient to suppose that $r = 1$), then $\dot{F}(x) = d_x f$ and $\Psi(w, t) = e^{t d_x f} w$.

3.134 Grobman–Hartman Theorem. *Let X be a smooth manifold, $r \geq 2$ and $x \in X$ be a hyperbolic singular point of the vector field $F \in V^r(X)$. Then the vector field is orbitally topologically equivalent with its linearization $\dot{F}(x)_v \in V_L(T_x(X))$ at the point $(x, 0_x) \in X \times T_x(X)$, where 0_x is the zero element of the tangent space $T_x(X)$.*

Proof

The assertion of the theorem is of local character, and thus, without loss of generality we can assume that $X = R^n$, $x = 0$, F is defined by the differential equation $\dot{y} = f(y)$, where $f \in C^r(R^n, R^n)$ and $F(x) = d_0 f$. From the Taylor Theorem it follows that there exists an open neighbourhood V of the point 0 and a C^r-mapping $R : V \to R^n$, such that $R(0) = 0$, $d_0 R = 0$ and $f(x) = Ax + R(x)$ for all $x \in V$, where $A = d_0 f$. According to Lemma 3.128 for arbitrary positive number Θ there exist open neighbourhoods U_1, U_2 of the point 0 such that $\tilde{U}_1 \subset U_2$, $\overline{U}_2 \subset V$ and a C^r-mapping $g : R^n \to R^n$, such that $R/U_1 = g/U_1$, $g(x) = 0$ for all $x \in R^n \setminus U_2$ and $\|d_x g\|_1 \leq K\Theta$ for all $x \in R^n$, where K is a positive number independent of R and Θ. Therefore if a is an arbitrary, given number, then the neighbourhood U_1 and the mapping g can be chosen such that $|g|_1 < a$. This means that $g \in C^1_a$. Let us consider the C^r-mapping $\tilde{f} : R^n \to R^n$, $\tilde{f}(x) = Ax + g(x)$ and the differential equations

$$\dot{x} = Ax, \qquad (3.52)$$

$$\dot{x} = f(x) = Ax + R(x), \qquad (3.53)$$

$$\dot{x} = \tilde{f}(x) = Ax + g(x). \qquad (3.54)$$

Let Ψ, φ, $\tilde{\varphi}$ be dynamical systems generated by solutions of the differential equations (3.52), (3.53) and (3.54), respectively. From Lemma 3.133 it follows that there exists a homeomorphism $h : R^n \to R^n$, such that

$$\Psi_t \circ h = h \circ \tilde{\varphi}_t \qquad (3.55)$$

for all $t \in R$. From the continuity of the mapping φ it follows that there exists an open neighbourhood $U_0 \subset U$ of the point $0 \in R^n$ and an interval $I \subset R$ such that for all $x \in U_0, t \in I$ it is $\varphi_t(x) \in U$. Since $f/U_1 = \tilde{f}/U_1$, so $\varphi_t(x) = \tilde{\varphi}_t(x)$ for all $x \in U_0, t \in I$ and from the equality (3.55) we get the equality $\Psi_t \circ h(x) = h \circ \varphi_t(x)$ for all $x \in U_0, t \in I$. □

3.135 Remark. The original Grobman's proof of Theorem 3.134 (see [47, 48]) as well as the proof of Theorem 3.132 by Hartman (see [58, 59]) is technically quite exacting. Proving Theorem 3.132 by solving the functional equation of the type (3.36) makes the proof of Theorem 3.134 much more transparent. The basic idea for this proof was devised by the American mathematician C. C. Pugh (see [128] as well as [74, 76, 119, 147]).

3.136 Theorem on Local Structural Stability. *Let X be a smooth manifold, $r \geq 2$ and x_0 be a hyperbolic singular point of the vector field $F \in V^r(X)$. Then there exists an open neighbourhood $V(F)$ of the vector field F in $V^r(X)$ such that for arbitrary $G \in V(F)$ there exists an open neighbourhood U of the point x_0 such that the vector field G has in U the unique singular point y_0 and the vector fields F and G are orbitally topologically equivalent at the point (x_0, y_0).*

Proof

The assertion of the theorem is of local character, and therefore without loss of generality we can assume that $X = R^n$, $x_0 = 0$ and F is defined by the differential equation $\dot{y} = f(y)$, where $f \in V^r(R^n, R^n)$. From the Taylor Theorem it follows that there exists an open neighbourhood U of the point $0 \in R^n$ and a C^r-mapping $R : R^n \to R^n$, such that $R(0) = 0$, $d_0 R = 0$ and $f(x) = Ax + R(x)$ for all $x \in U$, where $A = d_0 f$. From the Theorem 3.103 it follows that there exists an open neighbourhood $V(F)$ of the vector field F such that for arbitrary vector field $G \in V(F)$ defined by the differential equation $\dot{y} = g(y)$, $g \in C^r(R^n, R^n)$, the mapping $\tilde{A} = d_0 g$ is an isomorphism and from the Inverse Mapping Theorem it follows that if the neighbourhood U is small enough, then the mapping g/U is a C^r-diffeomorphism. Therefore, if $V(F)$ is a sufficiently small neighbourhood of the vector field F, then for arbitrary vector field $G \in V(F)$ and U small enough, the vector field G has the unique singular point y_0 in U. Since $h : R^n \to R^n$, $h(x) = x + y_0$, is a homeomorphism of the set U onto $h(U)$ and obviously, it maps the parts of trajectories of the vector field F situated in U onto parts of trajectories of the vector field G situated in $h(U)$, it is sufficient to assume that $y_0 = 0$. Under this assumption and if U is small enough, then $g(x) = \tilde{A}(x) + \tilde{R}(x)$ for all $x \in U$, where $\tilde{R} : U \to R^n$ is a C^r-mapping such that $\tilde{R}(0) = 0$, $d_0 \tilde{R} = 0$. If the neighbourhood $V(F)$ is sufficiently small, then according to Theorem 3.103 the point 0 is a hyperbolic singular point of the vector field G. From Theorem 3.134 it follows that the vector fields F and G are orbitally topologically equivalent to their linearizations $F(0)_v$ and $G(0)_v$, respectively at the point $(0,0) \in R^n \times R^n$. From Lemma 3.102 it follows that $m^+(F(0)_v) = m^+(G(0)_v)$, and thus, according to Theorem 3.100 there exists a linear vector field $B_v \in V_L(R^n)$ such that both vector fields $F(0)_v$ and $G(0)_v$ are orbitally topologically equivalent to B_v at the point $(0,0) \in R^n \times R^n$. Therefore also the vector fields F and G are orbitally topologically equivalent at the point $(0,0)$. □

3.137 Proposition. *Let X be a smooth manifold, $r \geq 2$, x_0 be a singular point of the vector fields F, $G \in V^r(X)$ that are orbitally C^1-equivalent at the point x_0. Then the vector fields F and G have the same characteristic exponents at the point x_0.*

Proof

Let Ψ and φ be the flows of the vector fields F and G, respectively and h be a C^1-diffeomorphism such that $\Psi_t(x) = h \circ \varphi_t \circ h^{-1}(x)$ for all x from a certain neighbourhood U of the point x_0 and all t from a certain interval containing the point 0. Then with the use of Theorem 2.31 and Lemma 3.116 we get $e^{t\dot{F}(x_0)} = D_{x_0}\Psi_t = A \circ D_{x_0}\varphi_t \circ A^{-1} = A \circ e^{t\dot{G}(x_0)} \circ A^{-1} = e^{tA \circ \dot{G}(x_0) \circ A^{-1}}$, where $A = D_{x_0}h$. This means that $\dot{F}(x_0) = A \circ \dot{G}(x_0) \circ A^{-1}$, and thus, the mappings $\dot{F}(x_0)$ and $\dot{G}(x_0)$ have the same eigenvalues. □

From the preceding proposition it follows that two vector fields with different characteristic exponents at their singular points x_0 and y_0, respectively cannot be orbitally C^1-equivalent at the point (x_0, y_0). A question arises as to whether the equality of characteristic exponents suffices for orbital C^1-equivalence. Let us consider the vector fields F and G defined by the following systems of differential equations

$$\dot{x}_1 = \alpha x_1, \quad \dot{x}_2 = (\alpha - \beta)x_2 + \varepsilon x_1 x_3, \quad \dot{x}_3 = -\beta x_3$$

$$\dot{x}_1 = \alpha x_1, \quad \dot{x}_2 = (\alpha - \beta)x_2, \quad \dot{x}_3 = -\beta x_3,$$

where $\alpha > \beta > 0$ and $\varepsilon \neq 0$. The point $x_0 = 0 \in R^3$ is the hyperbolic singular point of both vector fields. According to Theorem 3.134 these vector fields are orbitally topologically equivalent at the point x_0. Ph. Hartman has proved in his paper [60] (see also [58]) that inspite of the above vector fields having equal linearizations at the point x_0, they are not orbitally C^1-equivalent at the point x_0. This result means also that the vector fields having a certain hyperbolic singular point need not be locally structurally stable at this point with respect to the orbital C^1-equivalence. The following section is devoted to sufficient conditions for orbital C^1-equivalence of vector fields.

3.7 Normal Forms of Differential Equations

In the present section we shall introduce basic methods for finding of normal forms of differential equations, i.e. of suitable representatives of classes of locally orbital C^k-equivalent differential equations at their singular points. With their use the local structure of trajectories of dynamical systems can be studied.

We shall use the following notations: $Z_+^n = Z_0^+ \times Z_0^+ \times \cdots \times Z_0^+$ (n-times), where Z_0^+ is a set of non-negative integers; $|k| = k_1 + k_2 + \cdots + k_n$ for $k = (k_1, k_2, \ldots, k_n) \in Z_+^n$; $Z_+^n(s) = \{k \in Z_+^n : |k| = s\}, s = 0, 1, \ldots$. For $x = (x_1, x_2, \ldots, x_n) \in R^n$ or $x \in C^n$ and $k = (k_1, k_2, \ldots, k_n) \in Z_+^n$ let us define $x^k = x_1^{k_1} x_2^{k_2} \ldots x_n^{k_n}$.

Chapter 3

3.138 Definition. *The mapping*

$$P^s : R^n \to R^m (P^s : C^n \to C^m), \quad P^s(x) = \sum_{|k|=s} a_k x^k,$$

where $a_k \in R^m$ ($a_k \in C^m$), is called a real (holomorphic) vector homogeneous polynomial of degree s, where the sum is taken over all k from the set $Z_+^n(s)$. (This notation for the sum will also be used throughout the book). The set of all real (holomorphic) vector homogeneous polynomials of degree s is denoted by $H^s(n,m)$ and $H^s(n) \stackrel{def}{=} H^s(n,n)(H_c^s(n,m)$ and $H_c^s(n) \stackrel{def}{=} H_c^s(n,m))$.

3.139 Definition. *If $P^s \in H_c^s(2n, n), (u,v) \mapsto P^s(u,v), (u,v) \in C^n \times C^n$, then the expression $P^s(z, \bar{z})$, where $z = (z_1, z_2, \ldots, z_n) \in C^n$, $\bar{z} = (\bar{z}_1, \bar{z}_2, \ldots, \bar{z}_n)$, is called the complex vector homogeneous polynomial of degree s.*

If it will be obvious, which of the above polynomials is concerned, we shall also speak about homogeneous polynomials of degree s.

Let us consider the differential equation

$$\dot{x} = Ax + P(x), \tag{3.56}$$

where $P = (P_1, P_2, \ldots, P_{2n})$ is a C^r-mapping ($r \geq 4$) such that $P(0) = 0$, $d_0 P = 0$ and the matrix $A \in M(2n)$ has all its eigenvalues complex and different from each other. Now we shall present one method for finding the normal forms of the differential equation (3.56) under the above assumptions. Let us note that if the matrix A also has real eigenvalues or eventually multiple then the problem of finding normal forms of the differential equation (3.56) is more difficult. In the conclusion of this section another method will also be presented which is more convenient for solving this more difficult problem.

From Taylor Theorem it follows that $P(x) = \sum_{s=2}^{r} P^s(x) + R(x)$, where $P^s \in H^s(2n)$ ($2 \leq s \leq r$), $R \in C^r$ and $R(x) = o(\|x\|^r)$. Without loss of generality it can be assumed that the matrix A is in Jordan canonical form

$$A = \text{diag}\{A_1, A_2, \ldots, A_{2n}\}, \tag{3.57}$$

where $A_m = (a_{ij}^m) \in M(2)$, $a_{11}^m = a_{22}^m = a_m$, $a_{12}^m = -a_{21}^m = b_m \neq 0$ ($m = 1, 2, \ldots, n$). Obviously, $\lambda_m = a_m + ib_m, \bar{\lambda}_m$, $m = 1, 2, \ldots, n$ are the eigenvalues of the matrix A. If we denote $z_m = x_{2m-1} + ix_{2m}$ ($m = 1, 2, \ldots, n$), where $x = (x_1, x_2, \ldots, x_n)$, then we can define the complex vector homogeneous polynomial $Q^s = (Q_1^s, Q_2^s, \ldots, Q_n^s)$ of degree s, of variables $z, \bar{z}$ with values in C^n, and thus, in the following way. By substituting the expression $x_{2m-1} = (1/2)(z_m + \bar{z}_m)$ and the expression $x_{2m} = (-1/2)i(z_m - \bar{z}_m)$

144

($m = 1, 2, \ldots, n$) in the polynomial P^s, we obtain the complex polynomial $\tilde{P}^s = (\tilde{P}_1^s, \tilde{P}_2^s, \ldots, \tilde{P}_{2n}^s)$ of the variables $z.\bar{z}$ with the values in R^{2n}. Let us define $Q_m^s(z, \bar{z}) = \tilde{P}_{2m-1}^s(z, \bar{z}) + i\tilde{P}_{2m}(z, \bar{z}), m = 1, 2, \ldots, n$. We say that the polynomial Q^s is generated by complexification of the polynomial P^s. Analogously the complexification $Q(z, \bar{z})$ of the mapping P can be generated. The differential equation (3.56) now can be written in the complex form

$$\dot{z} = Bz + \sum_{s=2}^{r} Q^s(z, \bar{z}) + Q(z, \bar{z}), \qquad (3.58)$$

where $B = \text{diag}\{\lambda_1, \lambda_2, \ldots, \lambda_n\}$. The differential equation (3.58) is called the complexification of the differential equation (3.56). The differential equation (3.56) is called the realization of the differential equation (3.58). If the complex conjugate differential equation

$$\dot{\bar{z}} = \overline{Bz} + \sum_{s=2}^{r} \overline{Q^s(z, \bar{z})} + \overline{Q(z, \bar{z})}$$

is added to the differential equation (3.58) and we denote $u_i = z_i, v_i = \bar{z}_i$, $i = 1, 2, \ldots, n$, $u = (u_1, u_2, \ldots, u_n)$, $v = (v_1, v_2, \ldots, v_n)$, we get a complex system of differential equations whose right sides depend only upon u and v, and not upon $\bar{u}$ and $\bar{v}$. This means that the problem of the normal forms of differential equations of the form (3.46) can be solved by solving this problem for the complex system of differential equations of the form (3.58), where the mappings Q^s and Q depend only on z, being independent of $\bar{z}$. Then, let us consider the differential equation of the form

$$\dot{z} = Bz + \sum_{s=2}^{r} H^s(z) + H(z), \qquad (3.59)$$

where $B = \text{diag}\{\lambda_1, \lambda_2, \ldots, \lambda_n\}$,

$$H^s(z) = \sum_{|k|=s} h_k^s z^k, \qquad (3.60)$$

$z = (z_1, z_2, \ldots, z_n) \in C^n$, $h_k^s \in C^n$, $\lambda_1, \lambda_2, \ldots, \lambda_n \in C$.

3.140 Definition. *The eigenvalues $\lambda_1, \lambda_2, \ldots, \lambda_n$ of the matrix $A \in M(n)$ or $A \in M_c(n)$ are called resonant if there exists an n-tuple of the numbers $k = (k_1, k_2, \ldots, k_n) \in Z_+^n$ such that for some $r \in \{1, 2, \ldots, n\}$ the equality*

$$\lambda_r = (k, \lambda), \qquad (3.61)$$

holds, where (k, λ) is the scalar product of the vectors k and $\lambda = (\lambda_1, \lambda_2, \ldots, \lambda_n)$, i.e., $(k, \lambda) = \sum_{j=1}^{n} k_j \lambda_j$. The equality (3.61) is called the resonance. The number s is called the order of resonance. If for no eigenvalue of the matrix A and no $k \in Z^n$ the equality (3.61) is fulfilled, then we say that eigenvalues of the matrix A are non-resonant.

3.141 Example. If $\lambda_1 = 5$, $\lambda_2 = 1$ and $\lambda_3 = 2$ are eigenvalues of the matrix A, then $\lambda_1 = \lambda_2 + 2\lambda_3$ is the resonance of order 3.

3.142 Example. Let the matrix A have two eigenvalues λ_1, λ_2 such that $\lambda_1 + \lambda_2 = 0$. Then $\lambda_1 = (m+1)\lambda_1 + m\lambda_2, \lambda_2 = m\lambda_1 + (m+1)\lambda_2$ for an arbitrary natural number n. This case, obviously occurs when the matrix A has a pair of complex conjugated pure imaginary eigenvalues. If $A \in M(2)$ then the resonances introduced here are all the possible resonances.

3.143 Definition. Let $A \in M_c(n)$ and for $r \in \{1, 2, \ldots, n\}$ the resonance condition (3.61) be fulfilled. Let v_r be an eigenvector of the matrix A corresponding to the eigenvalue λ_r. Then the vector homogeneous polynomial of the form $\alpha v_r z^k$, where $z = (z_1, z_2, \ldots, z_n), \alpha \in C$, is called the resonant term of degree $s = |k|$. The sum of finite number of resonant terms of degree s is called the resonant polynomial of degree s.

3.144 Example. Let the matrix $A \in M_c(2)$ have the eigenvalues $\lambda_1 = ib$, $\lambda_2 = -ib$, $b \neq 0$. Then from the resonances introduced in Example 3.142 we get that $z_j(z_1 z_2)^m v_j$, $j = 1, 2,$, where v_1, v_2 are eigenvectors of the matrix A corresponding to the eigenvalues λ_1 and λ_2, respectively, and are the resonant terms of degree $2m+1$.

3.145 Theorem. Let a differential equation of the form (3.59) be given. Then for arbitrary natural number K, where $2 \leq K \leq r$, there exists a mapping $\Phi : C^n \to C^n$, $\Phi = \mathrm{id} + \sum_{s=2}^{K} R^s$, where $R^s \in H_c^s(n)$, transforming the differential equation (3.59) into the form

$$\dot{u} = Bu + \sum_{s=2}^{K} W^s(u) + \tilde{H}(u), \qquad (3.62)$$

where $\tilde{H}(u) = o(\|u\|^K)$ and $W^s \in H_c^s(n)$, $s = 2, 3, \ldots, K$, are the resonant polynomials of degree s.

Proof

The vectors $e_1 = (1, 0, \ldots, 0)^*$, $e_2 = (0, 1, 0, \ldots, 0)^*$, $\ldots$, $e_n = (0, 0, \ldots, 0, 1)^*$ are the eigenvectors of the matrix B corresponding to the eigenvalues $\lambda_1, \lambda_2, \ldots, \lambda_n$. The mapping H^s ($2 \leq s \leq r$) can be written in the form

$$H^s(z) = \sum_{|k|=s} \sum_{i=1}^{n} h_{ki}^s e_i z^k, \qquad (3.63)$$

where $h_{ki}^s \in C$. Let us introduce the new variable u through the mapping

$$z = \Phi_2(u) = u + R^2(u), \qquad (3.64)$$

$$R^2(u) = \sum_{|k|=2} \sum_{i=1}^{n} \alpha_{ki}^2 e_i u^k, \qquad (3.65)$$

where $\alpha_{ki}^2 \in C$. From the Inverse Mapping Theorem it follows that there exists an inverse mapping to Φ_2 having the form

$$u = \Phi_2^{-1}(z) = z - R^2(z) + o(\|z\|^2). \qquad (3.66)$$

If using the form of the differential equation (3.59), we get $\dot{u} = \dot{z} - d_z R^2(\dot{z}) + \cdots = Bu + BR^2(u) - d_u R^2 \circ Bu + H^2(u) + \ldots$, where the three dots stand for the terms of the order higher than 2. Thus, we obtain at the differential equation

$$\dot{u} = Bu + \sum_{s=2}^{r} \tilde{H}^s(u) + \tilde{H}(u), \qquad (3.67)$$

where $\tilde{H}(u) = o(\|u\|^r)$, $\tilde{H}^s \in H_c^s(n)$ ($s = 2, 3, \ldots, r$), where

$$\tilde{H}^2(u) = H^2(u) + BR^2(u) - d_u R^2 \circ Bu. \qquad (3.68)$$

Now let us look for numbers $\alpha_{ki}^2 \in C$ such that the vector polynomial $\tilde{H}^2$ contains resonant terms only. Obviously it holds that

$$Be_i u^k - d_u(e_i u^k) \circ Bu = [\lambda_1 - (k, \lambda)] e_i u^k \qquad (3.69)$$

for $i = 1, 2, \ldots, n$. If $\lambda_i - (k, \lambda) = 0$, then let us choose $\alpha_{ki}^2 = 0$ and if this resonance condition is not fulfilled, then let us put $\alpha_{ki}^2 = (\lambda_i - (k, \lambda))^{-1} h_{ki}^2$ ($i = 1, 2, \ldots, n$). Then the polynomial $\tilde{H}^2$ will contain only the resonant terms, and thus, the theorem is proved for $K = 2$. Further, we can proceed using the method of induction. Let us assume, that the differential equation (3.59) is yet in such a form that all polynomials H^s, $s = 2, 3, \ldots, K - 1$ are resonant. Let us introduce the new variable u through the mapping

$z = \Phi_K(u) = u + R^K(u)$, where $R^K \in H_c^K(n)$. When eliminating the non-resonant terms of degree $K > 2$ we can proceed in the same way as in the case of $K = 2$. In fact, the mapping Φ_K does not change the terms of order lower than K and at analogous choice of coefficients of the polynomials R^K as in case of $K = 2$, this mapping only keeps the resonant terms of degree K. Thus for every $s \in \{2, 3, \ldots, K\}$, the transformation $\Phi_s = \text{id} + R^s$, is obtained, where $R^s \in H_c^s(n)$ eliminating the non-resonant terms of degree s, where the terms of order lower than s are not changed, at all. The mapping Φ to be found is of the form $\Phi = \Phi_K \circ \Phi_{K-1} \circ \cdots \circ \Phi_2$. □

3.146 Corollary. *If the matrix B has all its eigenvalues non-resonant, then for an arbitrary natural number K, $2 \leq K \leq r$ there exists a mapping*

$$\Phi = \text{id} + \sum_{s=2}^{K} R^s,$$

where $R^s \in H_c^s(n)$, $2 \leq s \leq K$, transforming the differential equation (3.59) into the form $\dot{u} = BU + o(\|u\|^K)$.

3.147 Remark. From Corollary 3.146 it follows that if the matrix B has all its eigenvalues non-resonant and the mapping H from the right side of the differential equation (3.59) is analytical, then the differential equation (3.59) can be linearized through the mapping Φ that is expressed in the form of a formal infinite power series. An analogous situation occurs in the case when there exist the resonant eigenvalues of the matrix B. Then through the mapping Φ which is of the form of a formal infinite power series, all the non-resonant terms of any high order can be eliminated. However, very often, power series like this, are divergent. Inspite of this, the present method is very efficient. The differential equation, obtained by this formal procedure, can in many cases be at least orbitally C^0-equivalent to the original differential equation.

The fundamental idea of the above methods of finding the normal forms comes from H. Poincaré. Poincaré's method was further developed by A. M. Ljapunov [91], H. Dulac [34], G. D. Birkhoff [14], C. L. Siegel, J. M. Moser [137], A. D. Brjuno [22] and J. N. Bibikov [13]. There many references to the literature dealing with these problems can be found.

3.148 Definition. *The differential equation of the form (3.62) is called the normal form (of order K) of the differential equation (3.59). Let the differential equation*

$$\begin{bmatrix} \dot{u} \\ \dot{v} \end{bmatrix} = \begin{bmatrix} B & 0 \\ 0 & \bar{B} \end{bmatrix} \begin{bmatrix} u \\ v \end{bmatrix} + \sum_{s=2}^{r} \begin{bmatrix} W_1^s(u,v) \\ W_2^s(u,v) \end{bmatrix} + \begin{bmatrix} H_1(u,v) \\ H_2(u,v) \end{bmatrix},$$

where $W^s = (W_1^s, W_2^s)$ $(s = 2, 3, \ldots, r)$ is a resonance polynomial of degree s and $H_i(u,v) = o\left(\|u,v\|^r\right)$, $i = 1, 2$, is a normal form of the system of differential equations consisting of the differential equation (3.58) and of its complex conjugate differential equation. Then the differential equation

$$\dot z = Bz + \sum_{s=2}^r W_i^s(z,\overline{z}) + H_1(z,\overline{z})$$

is called the normal form of the differential equations (3.58).

Let us note that every normal form in the sense of the preceding definition need not always be the simplest representation of the given differential equation, that would enable better determination of the topological structure of trajectories of the given differential equation than other representations. The notion of the best representation of a differential equation, in general, is very difficult to define. Moreover, normal forms in the sense of this definition are not defined uniquely. In definite cases it is always to be judged what form of normal form it is advantageous to choose.

The derivation of the normal form introduced above, is similar to the way introduced in [10].

From the construction of complexification of the differential equation (3.56) and from the construction of the normal form of the differential equation presented in the proof of Theorem 3.145 the following theorem can be obtained.

3.149 Theorem. *Let the matrix $A \in M(2n)$ have all eigenvalues complex, different from each other and $P : R^{2n} \to R^{2n}$ is a C^r-mapping, where $2 \leq r \leq \infty$. Then the realization of the normal form of order r of the complexification of the differential equation (3.56) is the normal form of order r of the differential equation (3.56).*

3.150 Theorem. *Let $U_1 \times U_2$ be an open neighbourhood of the point $(0,0) \in R^2 \times R$, $P = (P_1, P_2) \in C^r(U_1 \times U_2, R^2)$ and $A \in C^r(U_2, M(2))$, $A(\varepsilon) = (a_{ij}(\varepsilon)), \varepsilon \in U_2, 2 \leq r < \infty$ (or P, A be analytical mappings), where $P(0, \varepsilon) = 0$ for all $\varepsilon \in U_2, d_{(0,0)}^s P = 0, s = 1, 2, \ldots, r$ and the matrix $A(0)$ has the eigenvalues $\lambda_1^0 = ib, \lambda_2^0 = -ib, b \neq 0$. Then the following assertions hold.*

1. There exists an open neighbourhood $V_1 \times V_2 \subset U_1 \times U_2$ of the point $(0,0)$ a C^r-mapping (an analytical mapping) $\Phi : V_1 \times V_2 \to R^2$ and C^r-mappings (analytical mappings) $\lambda_i : V_2 \to R$, $i = 1, 2$, such that $\lambda_1(0) = 0, \lambda_2(0) = b$, where for arbitrary $\varepsilon \in V_2$ $\lambda(\varepsilon) = \lambda_1(\varepsilon) + i\lambda_2(\varepsilon), \overline{\lambda(\varepsilon)}$ are eigenvalues of the matrix $A(\varepsilon)$ and the system of differential equations

$$\begin{aligned}\dot x_1 &= a_{11}(\varepsilon)x_1 + a_{12}(\varepsilon)x_2 + P_1(x,\varepsilon), \\ \dot x_2 &= a_{21}(\varepsilon)x_1 + a_{22}(\varepsilon)x_2 + P_2(x,\varepsilon),\end{aligned} \qquad (3.70)$$

is orbitally C^∞-equivalent at the point $0 \in R^2$ with the system of differential equations

$$\dot{y}_1 = \varepsilon\dot{\lambda}_1(0)y_1 - by_2 + \sum_{2m+n=2}^{r-1}(b_{1mn}y_1 - b_{2mn}y_2)\varepsilon^n(y_1^2+y_2^2)^m + \tilde{P}_1(y,\varepsilon),$$

(3.71)

$$\dot{y}_2 = by_1 + \dot{\lambda}_1(0)y_2 + \sum_{2m+n=2}^{r-1}(b_{2mn}y_1 + b_{2mn}y_2)\varepsilon^n(y_1^2+y_2^2)^m + \tilde{P}_2(y,\varepsilon),$$

where the numbers $b_{1mn}, b_{2mn} \in R$ are independent of ε, $\tilde{P}(0,0) = 0$, $d^s_{(0,0)}\tilde{P} = 0$, $s = 1, 2, \ldots, r$, $\tilde{P} = (\tilde{P}_1, \tilde{P}_2)$,

$$\dot{\lambda}_1(0) = \frac{d\lambda_1(\varepsilon)}{d\varepsilon}\bigg|_{\varepsilon=0},$$

where $\Phi_\varepsilon : V_2 \to R^2$, $\Phi_\varepsilon(x) = \Phi(x,\varepsilon)$ is the conjugating diffeomorphism.
2. The mapping $\Psi : R^2 \to R^2$, $\Psi(\varrho, \Theta) = (\varrho\cos\Theta, \varrho\sin\Theta)$, transforms the system (3.71) into the form

$$\dot{\varrho} = \varrho(\dot{\lambda}_1(0)\varepsilon + \sum_{2m+n=2}^{r-1} c_{mn}\varepsilon^n\varrho^{2m} + Q_1(\varrho,\Theta,\varepsilon)),$$

$$\dot{\Theta} = -b + Q_2(\varrho,\Theta,\varepsilon),$$

(3.72)

where $\Theta_i(\varrho,\Theta,\varepsilon) = \varrho^{-1}\tilde{Q}_i(\varrho,\Theta,\varepsilon)$, $i = 1, 2$,

$$\tilde{Q}_1(\varrho,\Theta,\varepsilon) = \cos\Theta F_1(\varrho,\Theta,\varepsilon) - \sin\Theta F_2(\varrho,\Theta,\varepsilon),$$
$$\tilde{Q}_2(\varrho,\Theta,\varepsilon) = \cos\Theta F_2(\varrho,\Theta,\varepsilon) + \sin\Theta F_1(\varrho,\Theta,\varepsilon),$$
$$F_j(\varrho,\Theta,\varepsilon) = P_j(\varrho\cos\Theta, \varrho\sin\Theta, \varepsilon),$$

$j = 1, 2$, if $\varrho \neq 0$ and $Q_i(0,\Theta,\varepsilon) = 0$ ($i = 1, 2$) for arbitrary Θ, ε and $c_{01} \neq 0$.
3. The mappings Q_1, Q_2 are of the class C^{r-1} (analytical).

Proof

From Lemma 3.65 it follows that there exists an open set $V_2 \subset U_2, 0 \in V_2$ and C^r-mappings (analytical mappings) $\lambda_i : V_2 \to R$, $i = 1, 2$, $C : V_2 \to M(2)$ such that $\lambda_1(0) = 0$, $\lambda_2(0) = b$, $\lambda(\varepsilon) = \lambda_1(\varepsilon) + i\lambda_2(\varepsilon)$, $\overline{\lambda(\varepsilon)}$ are eigenvalues of the matrix $A(\varepsilon)$, the matrix $C(\varepsilon)$ is regular and $\tilde{A}(\varepsilon) = C(\varepsilon)A(\varepsilon)(C(\varepsilon))^{-1}$ is the Jordan canonical form of the matrix $A(\varepsilon)$ for all $\varepsilon \in V_2$. The linear transformation $y = C(\varepsilon)x$ transforms the differential equation defined by the system (3.70) into the same form, where in place of the matrix $A(\varepsilon)$ the matrix $\tilde{A}(\varepsilon)$ is standing. Thus, it is sufficient to assume that $a_{11}(\varepsilon) = a_{22}(\varepsilon) = \lambda_1(\varepsilon), a_{21}(\varepsilon) = -a_{12}(\varepsilon) = \lambda_2(\varepsilon)$ for all $\varepsilon \in V_2$. The complexification of the differential equation corresponding to the system (3.70) is of the form

$$\dot{z} = (ib)z + \sum_{s=2}^{r} \sum_{|(k,p,q)|=s} a_{kpq}^{s} \varepsilon^k z^p \bar{z}^q + H(z,\bar{z},\varepsilon), \qquad (3.73)$$

where $a_{kpq}^s \in C$ and they are independent of ε, $H(0,0,0) = 0$, $d^s_{(0,0,0)} H = 0$, $s = 1,2,\ldots,r$, from which we obtain also the differential equation

$$\dot{\bar{z}} = -(ib)\bar{z} + \sum_{s=2}^{r} \sum_{|(k,p,q)|=s} \overline{a_{kpq}^{s}} \varepsilon^k z^p \bar{z}^q + \overline{H(z,\bar{z},\varepsilon)}. \qquad (3.74)$$

The system of equations (3.73), (3.74) can be written in the form

$$\dot{Z} = BZ + \sum_{s=2}^{r} \sum_{|(k,p)|=s} (h_{1kp}^s (w,0)^* + h_{2kp}^s (0,\bar{w})^*) \varepsilon^k Z^p + \tilde{H}(Z,\varepsilon), \quad (3.75)$$

where $B = \mathrm{diag}\{ib, -ib\}, (w,0)^*, (0,\bar{w})^*$ are eigenvectors of the matrix B corresponding to the eigenvalues $\lambda_1^0 = ib$, and $\lambda_2^0 = -ib$, respectively; the numbers $h_{1kp}^s, h_{2kp}^s \in C$ do not depend on ε, $h_{2kp}^s = \overline{h_{1kp}^s}$ and $\tilde{H}(0,0) = 0, d^s_{(0,0)} \tilde{H} = 0, s = 1,2,\ldots,r$. All resonances are of the form $\lambda_1^0 = (m+1)\lambda_1^0 + m\lambda_2^0, \lambda_2^0 = m\lambda_1^0 + (m+1)\lambda_2^0, m \in Z_0^+$. According to Theorem 3.145 there exists a C^∞-diffeomorphism $u = \Phi(Z)$ that transforms the differential equation (3.75) with $\varepsilon = 0$ into the form

$$\dot{u} = Bu + \sum_{m=1}^{r-1} (h_{1m} u_1 (u_1 u_2)^m (w,0)^* + h_{2m} u_2 (u_1 u_2)^m (0,\bar{w})^*) + F(u),$$

where $u = (u_1, u_2) \in C^2, u_2 = \bar{u}_1, h_{2m} = \bar{h}_{1m}, F(0) = 0, d_0^s F = 0, s = 1,2\ldots,r$. Therefore the diffeomorphism Φ transforms the parametrized differential equation (3.75) into the form

$$\dot{u} = Bu + \sum_{2m+n=2}^{r-1} (h_{1mn} \varepsilon^n u_1 (u_1 u_2)^m (w,0)^* + h_{2mn} \varepsilon^n u_2 (u_1 u_2)^m (0,\bar{w})^*)$$
$$+ \hat{H}(u,\varepsilon),$$

where $u = (u_1, u_2) \in C^2, u_2 = \bar{u}_1, h_{2mn} = \bar{h}_{1mn}, \hat{H}(0,0) = 0, d^s_{(0,0)} \hat{H} = 0, s = 1,2,\ldots,r$. For $v = u_1$ from this equation the complex differential equation of the form

151

Chapter 3

$$\dot{v} = (ib)v + \sum_{2m+n=2}^{r-1}(h_{1mn}w\varepsilon^n v|v|^{2m}) + G(v,\bar{v},\varepsilon), \qquad (3.76)$$

is obtained, where $G(0,0,0) = 0$, $d^s_{(0,0,0)}G = 0$, $s = 1, 2, \ldots, r$. If $v = y_1 + iy_2$ and $y = (y_1, y_2)$, then the realization of the differential equation (3.76) can be written in the form (3.71). Since Lemma 3.65 and Theorem 3.145 are still valid for analytical mappings, all the above considerations are also valid for A, P analytical. Thus assertion 1 is proved. If $y_1 = \varrho \cos\Theta$, $y_2 = \varrho \sin\Theta$, then

$$\dot{y}_1 = \dot{\varrho}\cos\Theta - \varrho\dot{\Theta}\sin\Theta = \varepsilon\dot\lambda_1(0)\varrho\cos\Theta - b\varrho\sin\Theta$$
$$+ \varrho\cos\Theta\left(\sum_{2m+n=2}^{r-1}b_{1mn}\varepsilon^n\varrho^{2m}\right) + \tilde{P}_1(\varrho\cos\Theta, \varrho\sin\Theta, \varepsilon),$$

$$\dot{y}_2 = \dot{\varrho}\sin\Theta + \varrho\dot{\Theta}\cos\Theta = b\varrho\cos\Theta + \varepsilon\dot\lambda_1(0)\varrho\sin\Theta$$
$$+ \varrho\sin\Theta\left(\sum_{2m+n=2}^{r-1}b_{2mn}\varepsilon^n\varrho^{2m}\right) + \tilde{P}_2(\varrho\cos\Theta, \varrho\sin\Theta, \varepsilon).$$

By multiplying the first equation by $\cos\Theta$, the second one by $\sin\Theta$ and adding them together, we obtain the differential equation for ϱ. By multiplying the first equation by $\sin\Theta$, the second one by $\cos\Theta$ subtracting one from the other, we get the differential equation for Θ. These equations are of the form

$$\dot\varrho = \varrho\left(\dot\lambda_1(0)\varepsilon + \sum_{2m+n=2}^{r-1}c_{mn}\varepsilon^n\varrho^{2m}\right) + \tilde{Q}_1(\varrho,\Theta,\varepsilon),$$
$$\varrho\dot\Theta = -b\varrho + \tilde{Q}_2(\varrho,\Theta,\varepsilon), \qquad (3.77)$$

where $c_{mn} \in R$, $\tilde{Q}_1, \tilde{Q}_2$ are C^r-functions (or analytical functions), defined in assertion 2. Since $\tilde{P}_j(y,\varepsilon) = a_j(y,\varepsilon)y_1 + b_j(y,\varepsilon)y_2$, $j = 1, 2$, where

$$a_j(y,\varepsilon) = \int_0^1 \frac{\partial\tilde{P}_j(ty,\varepsilon)}{\partial y_1}\,dt, \quad b_j(y,\varepsilon) = \int_0^1 \frac{\partial\tilde{P}_j(ty,\varepsilon)}{\partial y_2}\,dt,$$

the functions Q_1, Q_2 defined in assertion 2 are of class C^{r-1} and $\tilde{Q}_j(\varrho,\Theta,\varepsilon) = \varrho Q_j(\varrho,\Theta,\varepsilon)$, $j = 1, 2$. After dividing the second equation of the system (3.77) by ρ we obtain the system in the form (3.72). □

3.151 Remark. From the proof of Theorem 3.150 it can be seen that if the right sides of system (3.70) are analytical, then this system can be transformed with the use of a formal power series (that, in general, need not be convergent) into the form (3.71), where on the right sides of this system the sums range from $2m + 1 = 2$ to ∞. In polar coordinates this system is of the form (3.72), where the sums on the right sides of this system are taken from $2m + n = 2$ to ∞, and thus, the first equation for ϱ does not depend upon Θ, at all.

The method of calculation of normal forms of differential equations that will be introduced now, is by the Dutch mathematician F. Takens (see [151]). The present method is also very efficient in the case when the matrix of linearization of the differential equation has multiple eigenvalues and the calculations of the coefficients of normal forms are, in fact, simple. In fact, the Takens method is bound to the methods of Lie algebras, and thus, we shall start our explanation with the definition of the Lie algebra.

3.152 Definition. *Let V be a vector space over R, or over C and $f : V \times V \to V$ be a bilinear mapping (its values $f(x,y)$ will be denoted as $[x, y]$) possessing the following properties:*

1. $[x, y] + [y, x] = 0$, for all $x, y \in V$;
2. $[[x, y], z] + [[y, z], x] + [[z, x], y] = 0$ for all $x, y, z \in V$.

The vector space V with the operation $[x, y]$ possessing the properties 1 and 2 is called the Lie algebra. The element $[x, y]$ is called the commutator or the Lie bracket of the elements x, y.

3.153 Example. If $f, g \in C^\infty(R^n, R^n)$ then we define the mapping $[f, g] \in C^\infty(R^n, R^n)$ so that $[f, g](x) = (d_x f) \circ g(x) - (d_x g) \circ f(x)$. The vector space $C^\infty(R^n, R^n)$ with the operation $[f, g]$ is the Lie algebra.

3.154 Example. Let F, G be C^∞-vector fields on R^n,

$$F(x) = \sum_{i=1}^{n} f_i(x) \partial/\partial x_i, G(x) = \sum_{i=1}^{n} g_i(x) \partial/\partial x_i, x \in R^n.$$

Let us define the vector field $[F, G] \in V^\infty(R^n)$,

$$[F, g](x) = \sum_{i=1}^{n} h_i(x) \partial/\partial x_i,$$

where

$$h_i(x) = \sum_{j=1}^{n} \left(\frac{\partial g_i(x)}{\partial x_j} f_j(x) - \frac{\partial f_i(x)}{\partial x_j} g_j(x) \right),$$

$i = 1, 2, \ldots, n$. The vector field $V^\infty(R^n)$ with the operation $[F, G]$ is the Lie algebra.

Chapter 3

Let us consider the differential equation

$$\dot{x} = f(x), \qquad (3.78)$$

where $f \in C^r(R^n, R^n)$, $2 \leq r \leq \infty$. Let φ be the flow of another differential equation

$$\dot{y} = g(y), \qquad (3.79)$$

where $g \in C^r(R^n, R^n)$. For arbitrary $t \in R$ the mapping $\varphi_t : R^n \to R^n$, $\varphi_t(y) = \varphi(y, t)$, is the C^r-difeomorphism transforming the differential equation (3.78) into the form

$$\dot{y} = f_t(y) = (\varphi_t)_* f(y), \qquad (3.80)$$

where $(\varphi_t)_* f(y) = (d_z \varphi_t) f(z)$, $z = (\varphi_{-t})(y)$.

3.155 Lemma. *Let φ be the flow of the differential equation (3.79), $f, g \in C^r(R^n, R^n)$, $2 \leq r \leq \infty$ and $f_t : R^n \to R^n$, $f_t(y) = (\varphi_t)_* f(y)$, where $t \in R$. Then the mapping f_t is of the class C^{r-1} and*

$$\frac{df_t}{dt} = [f_t, g], \; f_0 = f, \qquad (3.81)$$

where $[f_t, g]$ is the Lie bracket of the mappings f_t and g.

Proof

For arbitrary $z \in R^n$ there is

$$\frac{d}{dt} f_t(\varphi_t(z)) = \frac{\partial f_t(\varphi_t(z))}{\partial t} + (d_{\varphi_t(z)} f_t) \frac{d\varphi_t(z)}{dt}. \qquad (3.82)$$

It follows from the definition of the mapping f_t that $f_t(\varphi_t(z)) = (d_z \varphi_t) f(z)$ and by derivation of this equality we obtain

$$\frac{d}{dt} f_t(\varphi_t(z)) = d_z \left(\frac{d\varphi_t}{dt}\right) \circ f(z). \qquad (3.83)$$

From the equalities (3.82), (3.83) we obtain

$$\frac{\partial f_t(\varphi_t(z))}{\partial t} = d_z \left(\frac{d\varphi_t}{dt}\right) \circ f(z) - (d_{\varphi_t(z)} f_t) g(\varphi_t(z)). \qquad (3.84)$$

As

$$\frac{d\varphi_t}{dt} = g \circ \varphi_t,$$

thus
$$d_z\left(\frac{d\varphi_t}{dt}\right) = d_z(g \circ \varphi_t) = (d_{\varphi_t(z)}g) \circ (d_z\varphi_t).$$

Therefore from equality (3.84) we obtain the equality

$$\frac{\partial f_t(\varphi_t(z))}{\partial t} = (d_{\varphi_t(z)}g) \circ (d_z\varphi_t)f(z) - (d_{\varphi_t(z)}f_t) \circ g(\varphi_t(z))$$
$$= (d_{\varphi_t(z)}g) \circ f_t(\varphi_t(z)) - (d_{\varphi_t(z)}f_t) \circ g(\varphi_t(z)).$$

As $z \in R^n$ is an arbitrary point and φ_t is a diffeomorphism, it follows from the last equality that for arbitrary $y \in R^n$ it holds that

$$\frac{df_t}{dt} = (d_y g) \circ f_t(y) - (d_y f_t) \circ g(y) = [f_t, g](y)$$

which means that equality (3.81) holds, while from definition of the mapping f_t it is obvious that $f_0 = f$. □

We can look at equality (3.81) as a linear differential equation whose solution u fulfilling the initial condition $u(0) = f$ is such that its value $u(t)$ is the right side of the differential equation (3.80) equal to f_t. Of course, this solution is dependent on the choice of the mapping g. By proper choice of the mapping which defines the diffeomorphism transforming the differential equation (3.78) to the form (3.80) a relatively simple normal form of the differential equation (3.78) can often be found. This Takens idea is realized similarly in the case of the above-mentioned method of successive elimination of non-resonant terms from the right side of the differential equation. The elimination of non-resonant terms of degree s from the right side of the differential equation (3.78) can be done by proper choice of the mapping $g \in H^3(n)$, i.e. the right side of the differential equation (3.79).

If the mappings g and f are such as in Lemma 3.155 then from the equality (3.81) we obtain the equality

$$\frac{d(T^m(f_t))}{dt} = T^m([f_t, g]), \qquad (3.85)$$

for an arbitrary natural number $m \leq r-1$ ($T^m(h)(y)$ is the Taylor polynomial of degree m of the mapping h at the point 0). If $g = g^m \in H^m(n)$, i.e. g^m is a real vector homogeneous polynomial of degree m, then $[T^m(f_t), g^m]$ is a vector homogeneous polynomial of degree greater than m and therefore $T^m([f_t, g^m]) = T^m([T^m(f_t), g^m])$. The equality (3.85) for $g = g^n$ we can write in the form

$$\frac{d(T^m(f_t))}{dt} = T^m([T^m(f_t), g^m]). \qquad (3.86)$$

Chapter 3

Let us consider the differential equation

$$\frac{dU}{dt} = T^m([U, g^m]) \qquad (3.87)$$

with the initial condition

$$U(0) = T^m(f). \qquad (3.88)$$

3.156 Lemma. *Let $m \in N$, $m \geq 2$. Then the initial value problem (3.87), (3.88) has a unique solution $U : R \to P^m(n)$, $U(t) = T^m(f) + t[T^1(f), g^m]$ ($P^m(n)$ is the space of vector polynomials of degree m with coefficients from R^n).*

Proof

The state space of the differential equation (3.87) is a finite dimensional vector space $P^m(n)$. From Theorem 1.41 it follows that the initial value problem (3.87), (3.88) has a unique solution U. As the differential equation (3.87) is linear, according to the Theorem 3.76 this solution is defined on R, i.e. $U: R \to P^m(n)$. Let $V: R \to P^m(n)$, $V(t) = T^m(f) + t[T^1(f), g^m]$. Then from the properties of Lie brackets we obtain $[V(t), g^m](x) = [T^m(f) + t[T^1(f), g^m], g^m](x) = [T^m(f), g^m](x) + t[[T^1(f), g^m], g^m](x)$. The vector polynomial $[[T^1(f), g^m], g^m](x)$ is a homogeneous polynomial of degree $2m - 1$ and $T^m(f)(x) = T^1(f)(x) + R_2(x)$, where $R_2(0) = 0$, $d_0 R_2 = 0$. Therefore there is $[V(t), g^m](x) = [T^1(f), g^m](x) + R_m(x)$, where $R_m(0) = 0$, $d_0 R_m = 0$, $s = 1, 2, \ldots, m$. Thus, we obtain that

$$T^m([V(t), g^m]) = [T^1(f), g^m] = \frac{dV(t)}{dt},$$

where $V(0) = T^m(f)$. It means that $V = U$. □

For each natural number $m \geq 2$ let us define the mapping $L^m : H^m(n) \to H^m(n)$, and denote $B^m(n) = \text{Image } L^m$.

3.157 Lemma. *If the assumptions of Lemma 3.155 are fulfilled where, $g = g^n \in H^m(n)$, then*

$$T^m(f_1) - T^m(f) = [T^1(f), g^m] \in B^m(n). \qquad (3.89)$$

Proof

From Lemma 3.156 and from the inequality (3.86) the equality $T^m(f_t) = T^m(f) + t[T^1(f), g^m]$ follows and therefore for $t = 1$ we obtain the equality (3.86).

3.158 Takens Theorem. *Let $f \in C^r(R^n, R^n)$, $2 \leq r \leq \infty$, $L^m : H^m(n) \to H^m(n)$, $L^m(h) = [T^1(f), h]$, where $m \geq 2$ is a natural number. Let $B^m(n) = \text{Image } L^m$ and $W^m(n)$ be a space complementary to $B^m(n)$, i.e. $H^m(n) = B^m(n) + W^m(n)$ (not necessarily the direct sum). Then for arbitrary q, $1 \leq q \leq r$, the differential equation (3.78) is orbitally C^∞-equivalent at the point $0 \in R^n$ to the differential equation*

$$\dot{y} = T^1(f)(y) + w_2(y) + w_3(y) + \cdots + w_q(y) + R_q(y), \quad (3.90)$$

where $w_m \in W^m(n)$, $m = 2, 3, \ldots, q$ and $T^q(R_q)(y) \equiv 0$.

Proof

Proof of Theorem 3.158 for $q < \infty$. We shall proceed by induction with respect to q. For $q = 1$ the proof is trivial. Let us assume that the differential equation (3.78) is in a certain neighbourhood of the point 0 in the form

$$\dot{x} = T^1(f)(x) + w_2(x) + w_3(x) + \cdots + w_{q-1}(x) + R_{q-1}(x), \quad (3.91)$$

where $w_m \in W^m(n), m = 2, 3, \ldots, q-1, T^{q-1}(R_{q-1}(x) = 0$. If follows from both Lemmas 3.155 and 3.157 that if $g^q \in H^q(n)$ and φ is the flow of the differential equation $\dot{y} = g^q(y)$, then C^∞-diffeomorphism $\Phi = \varphi_1$ transforms the differential equation (3.91) into the form

$$\dot{y} = f_1(y) = T^1(f)(y) + w_2(y) + \cdots + w_{q-1}(y) + [T^1(f), g^q](y) + R_q(y),$$

where $T^{q-1}(R_q)(y) \equiv 0$. The mapping R_q can be written in the form $R_q = w_q + b_q + R_{q+1}$, where $w_q \in W^q(n)$, $b_q \in B^q(n)$. $T^q(R_{q+1})(y) \equiv 0$. It is sufficient to choose $g^q \in H^q(n)$ so that $b_q = -[T^1(f), g^q]$ and we obtain the differential equation (3.90).

Proof of Theorem 3.158 for $q = \infty$. In the preceding proof we proved that for an arbitrary natural number j there exists a C^∞-diffeomorphism Φ_j transforming the differential equation (3.78) in a sufficiently small neighbourhood of the point 0 into the form (3.90) where $q = j$. Let us denote Ψ_j a C^∞-diffeomorphism transforming the differential equation (3.90) for $q = j - 1$ into the form (3.90) for $q = j$, i.e. Ψ_j eliminates all the terms of degree j which do not belong to the subspace $W^j(n)$ where all the terms of degree smaller than j remain unchanged. For this mapping there is $T^{j-1}(\Psi) = \text{id}$. As $\Phi_j = \Psi_j \circ \Phi_{j-1}$, so $T^{j-1}(\Phi_j) = T^{j-1}(\Phi_{j-1})$. The sequence of mappings $\{\Phi_j\}_{j=1}^\infty$ defines a formal infinite power series $\lim_{j \to \infty} T^j(\Phi_j)(x)$. It follows from the Borel Theorem that there exists a C^∞-diffeomorphism Φ such that

$$T(\Phi)(x) = \lim_{j \to \infty} T^j(\Phi_j)(x)(T(\Phi)(x)$$

is the formal Taylor series of the mapping Φ at the point 0). This diffeomorphism transforms the differential equation (3.78) into the form

$$\dot{y} = T^1(f)(y) + \sum_{s=2}^{\infty} w_s(y) + R_\infty(y), \tag{3.92}$$

where $w_s \in W^s(n)$ for all $s \geqq 2$ and $T(R_\infty)(x) \equiv 0$. □

Let us consider the parametrized differential equation

$$\dot{x} = f(x, \varepsilon), \tag{3.93}$$

where $f \in C^\infty(R^n \times R^k, R^n)$, $\varepsilon \in R^k$ is a parameter. This differential equation can be written in the form

$$\begin{bmatrix} \dot{\varepsilon} \\ \dot{x} \end{bmatrix} = \begin{bmatrix} 0 \\ f(x, \varepsilon) \end{bmatrix} = F(x, \varepsilon), \tag{3.94}$$

which is a differential equation with the state variable (x, ε). For an arbitrary natural number m let us define the set

$$H_0^m(n+k) = \Big\{ P \in H^m(n+k) : P(x) = \sum_{|\alpha|=m} a_\alpha x^\alpha,$$

$$a_\alpha = (0, \tilde{a}_\alpha) \in R^k \times R^n \Big\}.$$

The set $H^m(n+k)$ is obviously a subspace of the vector space $H^m(n+k)$. Let us denote

$$\pi_m : H_0^m(n+k) \to H^m(n+k, n), \quad \pi_m \left(\sum_{|\alpha|=m} (0, \tilde{a}_\alpha) x^\alpha \right) = \sum_{|\alpha|=m} \tilde{a}_\alpha x^\alpha.$$

From the form of the mapping F on the right side of the differential equation (3.94) it follows that the mapping $L^m : H_0^m(n+k) \to H^m(n+k)$, $L_0^m(h) = [T^1(F), h]$ maps $H_0^m(n+k)$ to $H_0^m(n+k)$. Let $B^m(n+k) = \text{Image } L_m^0$ and $W_0^m(n+k)$ be a complementary space to $B_0^m(n+k)$ in the space $H_0^m(n+k)$.

3.159 Theorem. *Let $f \in C^r(R^n \times R^k, R^n)$, $2 \leqq r \leqq \infty$. Then for an arbitrary q, $1 \leqq q \leqq r$ there exists a neighbourhood $U \times V$ of the point $0 \in R^n \times R^k$ and a C^∞-mapping $\Phi : U \times V \to R^n$ such that for arbitrary $\varepsilon \in V$ the differential equation (3.93) is orbitally C^∞-equivalent at the point $0 \in R^n$ to the differential equation*

$$\dot{y} = T^1(f)(y, \varepsilon) + w_2(y, \varepsilon) + \cdots + w_q(y, \varepsilon) + R_q(y, \varepsilon), \tag{3.95}$$

where $w_m \in \pi_m(W_o^m(n+k))$, $m = 2, 3, \ldots, q$, $T^q(R_q)(y, \varepsilon) \equiv 0$, where $\Phi_\varepsilon : U \to R^n$, $\Phi_\varepsilon(x) = \Phi(x, \varepsilon)$, is the conjugating diffeomorphism.

Proof

We can follow the same procedure as for the proof of the Theorem 3.158 where the role of the mapping L^m, spaces $H^m(n)$, $B^m(n)$ and $W^m(n)$ is replaced by the mappings L^m and spaces $H_0^m(n+k)$, $B_0^m(n+k)$ respectively $W_0^m(n+k)$. As for $g \in H_0^m(n+k)$ there is $L_0^m(g) \in H_0^m(n+k)$. The sequence of C^∞-diffeomorphism $\{\Phi_j\}_{j=1}^q$ constructed analogously to that in the proof of Theorem 3.158, does not change the differential equation $\dot{\varepsilon} = 0$ and eliminates all the polynomials of degree smaller than $q+1$ which do not belong to $W_0^m(n+k)$ ($m = 2, 3, \ldots, q$). When omitting the differential equation $\dot{\varepsilon} = 0$ we obtain the differential equation (3.95). □

3.160 Remark. If the assumptions of Theorem 3.150 are fulfilled, where the mappings P and A are of the class C^∞, then all the assertions of this theorem are also valid for $r = \infty$. The proof of this assertion can be done by the method of formal power series when using the Borel Theorem.

3.8 Poincaré Mapping

In the preceding sections we dealt with the questions regarding the local classification of the vector fields in neighbourhoods of their singular points. However, the methods we have used here, have in fact broader possibilities of application. At the first view it seems that the topological structure of trajectories in neighbourhoods of periodic trajectories is of global character. The aim of this section is to show that it is not so and that the problem of the local topological classification of vector fields in neighbourhoods of their periodic trajectories can be reduced to the problem of the local topological classification of diffeomorphism in neighbourhoods of their fixed and periodic points.

The following theorem gives an idea about the local structure of trajectories of vector fields in neighbourhoods of their regular points and thereby also information about the local structure of trajectories in a neighbourhood of a sufficiently small piece of the given periodic trajectory. To get information on the local structure of the trajectories in a neighbourhood of the whole periodic trajectory we shall prove later with the help of this theorem its global version.

3.161 Local Flow Box Theorem. *Let X be a smooth manifold of dimension n and $x_0 \in X$ be a regular point of the vector field $F \in V^r(X)$, where $1 \leq r \leq \infty$. Then there exists an admissible chart (U, ∞) on X such that it holds that*

1. $x_0 \in U$, $\alpha(x_0) = 0 \in R^n$, $\alpha(U) = \{(t, z_1, z_2, \ldots, z_{n-1}) \in R \times R^{n-1} : |t| < 1, |z_i| < 1, i = 1, 2, \ldots, n-1\}$,

2. *if f_α is the main part of the local representation of the vector field F with respect to the chart (U, ∞), then $f_\alpha(t, z) = (1, 0, \ldots, 0) \in R^n$ for all $(t, z) \in \alpha(U)$.*

Proof

Let (V, β') be an admissible chart on X such that $x_0 \in V$, $\beta'(x_0) = 0 \in R^n$ and $F_{\beta'}(y) = (y, f_{\beta'}(y))$ is the local representation of the vector field F with respect to this chart. As x_0 is a regular point of the vector field F, thus $f_{\beta'}(0) = q \neq 0$. Let $A : R^n \to R^n$ be a regular linear mapping such that $Aq = e_1 = (1, 0, \ldots, 0)$ and $\beta = A \circ \beta'$. Then (V, β) is an admissible chart on X, where $\beta(x_0) = 0$. If $F_\beta(y) = (y, f_\beta(y))$ is the local representation of the vector field F with respect to this chart, then $f_\beta(y) = A f_{\beta'}(A^{-1}y)$ and obviously $f_\beta(0) = Aq = e_1$. Let $\Psi_\beta : W_0 \times \langle -\varepsilon_0, \varepsilon_0 \rangle \to W = \beta(V)$ be a local C^r-flow of the vector field F_β, where $W_0 = \{y = (y_1, y_2, \ldots, y_n) : |y_i| < \delta, i = 1, 2, \ldots, n\} \subset W$, $\delta > 0$, $\varepsilon_0 > 0$. As $f_\beta(0) = e_1$ thus $\Sigma = \{y = (y_1, y_2, \ldots, y_n) \in R^n : y_1 = 0\}$ is obviously a local transversal at the point $0 \in R^n$ to the trajectory of the vector field F passing through this point. If $U_\delta = \{z = (z_1, z_2, \ldots, z_{n-1}) \in R^{n-1} : |z_i| < \delta, i = 1, 2, \ldots, n-1\}$, then obviously $W_0 \cap \Sigma = \{0\} \times U_\delta$. Let us define the mapping $g : \langle -\varepsilon_0, \varepsilon_0 \rangle \times U_\delta \to W$, $g(t, z) = \Psi_\beta(0, z, t)$. The mapping g is of the class C^r and from the first property of the vector field (see Definition 3.13) we obtain that $g(0, z) = z$ for all $z \in U_\delta$. Therefore

$$d_{(0,0)} g(s, v) = \frac{\partial \Psi_\beta(0, 0)}{\partial t} s + \frac{\partial g(0, 0)}{\partial z} = e_1 s + (0, v) = (s, v)$$

for all $(s, v) \in R \times R^{n-1}$, i.e. $d_{(0,0)} g = \mathrm{id}$. From the Inverse Mapping Theorem it follows that for $\varepsilon > 0$ small enough the mapping g is a C^r-diffeomorphism of the set $\tilde{U}_\varepsilon = (-\varepsilon, \varepsilon) \times U_\varepsilon$ onto the set $V_\varepsilon = g(\tilde{U}_\varepsilon)$, where $U_\varepsilon = \{z = (z_1, z_2, \ldots, z_{n-1}) : |z_i| < \varepsilon, i = 1, 2, \ldots, n-1\}$. If $w \in V_\varepsilon$ and $g^{-1}(w) = (s, u) \in \tilde{U}_\varepsilon$, then $g^{-1}(\Psi_\beta(w, t)) = g^{-1}(\Psi_\beta(g(s, u), t)) = g^{-1}(\Psi_\beta(\Psi_\beta(0, u, s), t))) = g^{-1}(\Psi_\beta(0, u, s+t)) = g^{-1} \circ g(s+t, u) = (s+t, u)$ for $|s|, |t|$ small enough. It means that the mapping $g^{-1} : V_\varepsilon \to \tilde{U}_\varepsilon$ maps the trajectories of the vector field F_β to the trajectories of the vector field $\tilde{F}_\varepsilon(y) = (y, \tilde{f}_\varepsilon(y))$, where $\tilde{f}_\varepsilon(y) = (1, 0, \ldots, 0)$ for all $y \in \tilde{U}_\varepsilon$. Let $\tilde{U}_1 = \{(t, z_1, z_2, \ldots, z_{n-1}) \in R \times R^{n-1} : |t| < 1, |z_i| < 1, i = 1, 2, \ldots, n-1\}$. The mapping $h : \tilde{U}_1 \to U_\varepsilon$, $h(y) = \varepsilon y$ is obviously a C^r-diffeomorphism and the mapping $h^{-1} \circ g^{-1} : V_\varepsilon \to \tilde{U}_1$ is a C^r-diffeomorphism, mapping the trajectories of the vector fields F_β onto the trajectories of the vector field $\tilde{F}_1(y) = (y, \tilde{f}_1(y))$, where $\tilde{f}_1(y) = (1, 0, \ldots, 0)$ for all $y \in \tilde{U}_1$. If we choose $U = \beta^{-1}(g(\tilde{U}_\varepsilon))$ and $\alpha = h^{-1} \circ g^{-1} \circ \beta$, then for the chart (U, α) the assertions of the theorem are valid, where $f_\alpha = \tilde{f}_1$. □

3.162 Definition. *Let X, P be smooth manifolds. We say that the vector field $F \in V^r(X \times P)$ is generated by the parametrized vector field $G \in V^r(P, X)$, if $F(x, p) = (G(x, p), 0_p)$ for all $(x, p) \in X \times P$, where 0_p is the zero element of the tangent space $T_p(P)$.*

3.163 Lemma. *Let X, P be smooth manifolds, (U, α), (V, β) be admissible charts on X, and P, respectively and let $F \in V^r(X \times P)$ be a vector field generated by the parametrized vector field $G \in V^r(P, X)$. Then the local representation $F_{\alpha \times \beta}$ of the vector field F with respect to the chart $(U \times V, \alpha \times \beta)$ is of the form $F_{\alpha \times \beta}(y, \mu) = (G_{\alpha \times \beta}(y, \mu), 0, 0, \ldots, 0)$, where $G_{\alpha \times \beta}$ is the local representation of the parametrized vector field G with respect to the given charts and $(0, 0, \ldots, 0) \in R^k$.*

The proof of the above lemma is trivial. Let us only note that if $g_{\alpha \times \beta}$ is the main part of the local representation of the parametrized vector field G, then $(y, \mu) \to (g_{\alpha \times \beta}(y, \mu), 0)$, where $0 \in R^k$, is the main part of the local representation of the vector field F. Therefore according to Proposition 3.5 c is an integral curve of the vector field F iff $(\alpha \times \beta) \circ c$ is a solution of the system of differential equation

$$\dot{\mu} = 0 \in R^k, \dot{y} = g_{\alpha \times \beta}(y, \mu). \tag{3.96}$$

3.164 Theorem. *Let P, X be smooth manifolds, $\dim X = n$, $\dim P = k$, $G \in V^r(P, X)$, $1 \leq r \leq \infty$ and x_0 be a regular point of the vector field G_{p_0}, where $p_0 \in P$. Then there exists an admissible chart (W, h) on $X \times P$ such that the following assertions hold.*

1. $(x_0, p_0) \in W$, $h(x_0, p_0) = (0, 0) \in R^n \times R^k$, $W = W_1 \times W_2$, where $W_1 \subset X$, $W_2 \subset P$ are open sets, $h = h_1 \times h_2$, $h_1 : W_1 \times W_2 \to R^n$, $h_2 : W_2 \to R^k$.

2. For each $p \in W_2$ the mapping $h_{1p} : W_1 \to R^n, h_{1p}(x) = h_1(x, p)$, C^r-diffeomorphism of the set W_1 onto the set $I^n = I \times I \times \cdots \times I$ (n-times), where $I = (-1, 1)$.

3. The mapping h_2 is the C^r-diffeomorphism of the set W_2 onto the set I^k.

4. The main part of the local representation of the vector field G with respect to the chart (W, h) is of the form $g_h(t, z, \mu) = (1, 0, \ldots, 0) \in R^n$ for all $(t, z, \mu) \in I \times I^{n-1} \times I^k$.

Chapter 3

Sketch of the proof.

Let F be a vector field on $X \times P$ generated by the parametrized vector field G. Then the point (x_0, p_0) is obviously a regular point of the vector field F, and therefore in its sufficiently small neighbourhood the Local Flow Box Theorem can be used. By repeating individual steps of the proof of this theorem for the vector field F, which has a specific form, we can be persuaded with the help of Theorem 3.165 that the chart (W, h) can be chosen in such a way that all the assertions of the theorem may be valid. $\square$

3.165 Definition. *Let γ be a trajectory of the vector field $F \in V^r(X)$. The set $\Gamma \subset \gamma$ is called the arc of the trajectory γ with the end points $p, q \in \gamma$ if the following conditions are fulfilled.*

1. Γ is not a periodic trajectory, i.e. $\Gamma \neq \gamma$ if γ is a periodic trajectory.

2. There exists a number a, $0 < a < \infty$, such that $\Gamma = \varphi_p(\langle 0, a \rangle)$, where φ is the flow of the vector field F and $q = \varphi_p(a)$.

3.166 Definition. *Let Γ be the arc of a trajectory γ of the vector field $F \in V^r(X)$. The pair (U, h) is called a tubular neighbourhood of the arc Γ if*

1. U is an open neighbourhood of the arc;

2. h is a C^r-diffeomorphism of the set U onto the set $I^n = I \times I^{n-1}$, where $I = (-1, 1)$, $I^{n-1} = I \times I \times \cdots \times I$ $((n-1)$-times$)$;

3. h maps connected parts of intersections of trajectories of the vector field F with the set U onto connected parts of intersections of trajectories of the vector fields G, where $G(x, y) = (x, y, g(x, y)) = (x, y, 1, 0, \ldots, 0) \in R \times R^{n-1} \times R^n$ for all $(x, y) \in I \times I^{n-1}$;

4. $h(U \cap \gamma) = I \times \{0\} \subset I \times I^{n-1}$.

3.167 Global Flow Box Theorem. *Let X be a smooth manifold of dimension n and Γ be an arc of the trajectory γ of the vector field $F \in V^r(X)$, where $1 \leq r < \infty$. Then there exists a tubular neighbourhood (U, h) of the arc Γ.*

Proof

Let $\Gamma = \varphi_q(\langle 0, a \rangle)$, where φ is the flow of the vector field F, $q \in \Gamma$, $a > 0$. Then for $\varepsilon > 0$ sufficiently small the mapping $\varphi_q / \langle -\varepsilon, a + \varepsilon \rangle$ is injective. As the set $\tilde{\Gamma} = \varphi_q(\langle -\varepsilon, a + \varepsilon \rangle)$ is compact, it follows from Theorem 3.161 that there exist numbers $-\varepsilon = a_1 < a_2 < \cdots < a_{k+1} = a + \varepsilon$ and admissible charts (U_i, h_i), $i = 1, 2, \ldots, k$ on X such that for each $i \in \{1, 2, \ldots, k\}$ is $\Gamma = \varphi_q(\langle a_i, a_{i+1} \rangle) \subset U_i$ and (U_i, h_i) is the tubular neighbourhood of the arc Γ_i where $U_i \cap U_{i+1} \neq \emptyset$, $U_j \cap U_i = \emptyset$ for $j \neq i-1, i+1$. Let $p_1 = \varphi_q(-\varepsilon)$

162

and $I_\delta^{n-1} = \{(0, y_1, y_2, \ldots, y_{n-1}) \in I \times I^{n-1} : |y_i| < \delta, i = 1, 2, \ldots, n-1\}$.
The tubular neighbourhood (U_1, h_1) can be chosen in such a way that $h_1(p_1) = (0, 0)$. Then for $\delta > 0$ sufficiently small $\Sigma_{p_1}^\delta = h_1^{-1}(I_\delta^{n-1}) \subset U_1$ is a local transversal to $\tilde{\Gamma}$ at the point p_1. For arbitrary $p \in \tilde{\Gamma}$ there obviously exists $t \in \langle 0, a + 2\varepsilon \rangle$, such that $\varphi_t(p_1) = p$. The mapping $\varphi_t : X \to X$ is a C^r-diffeomorphism, $D_p \varphi_t : T_{p_1}(X) \to T_p(X)$ is an isomorphism and $D_p \varphi_t(T_{p_1}(\tilde{\Gamma})) \subset T_p(\tilde{\Gamma})$. Therefore if $\delta > 0$ is small enough then for arbitrary $p \in \tilde{\Gamma}$ is $\Sigma_p^\delta = \varphi_t(\Sigma_{p_1}^\delta) \subset \cup_{i=1}^k U_i$ and Σ_p^δ is a local transversal to $\tilde{\Gamma}$ at the point $p = \varphi_t(p_1)$. Therefore from the compactness of the set $\tilde{\Gamma}$ and from the properties of the flow φ it follows that if $\delta > 0$ is small enough then for arbitrary $p \in \tilde{\Gamma}, \Sigma_p^\delta$ is a transversal to the vector field F i.e. for each $x \in \Sigma_p^\delta$ the trajectory of the vector field F passing through the point x crosses Σ_p^δ transversally at the point x. Let $\tilde{U} = \cup_{p \in \tilde{\Gamma}} \Sigma_p^\delta$. It follows from the properties of the tubular neighbourhoods $(U_i, h_i), i = 1, 2, \ldots, k$, that if $\delta > 0$ is small enough then $\Sigma_p^\delta \cap \Sigma_{\tilde{p}}^\delta = \emptyset$, if $p \neq \tilde{p}$ and $\tilde{U}$ is the neighbourhood of the arc $\tilde{\Gamma}$. Let us define the mapping $\pi_1 : \tilde{U} \to \tilde{\Gamma}$, $\pi_1(z) = p$, where $z \in \Sigma_p^\delta$ and $\pi_2 : \tilde{U} \to \Sigma_{p_1}^\delta$, $\pi_2(z) = \varphi_{-t}(z)$, where $z \in \Sigma_p^\delta, p = \varphi_t(p_1)$. As Σ_p^δ is a C^r-submanifold at X for each $p \in \tilde{\Gamma}$, the mapping π_1 is of the class C^r. The mapping π_2 is of the class C^r because φ is of the class C^r. Let $g_1 : \tilde{\Gamma} \to \langle -1, 1 \rangle$ and $g_2 : \Sigma_{p_1}^\delta \to I^{n-1}$ be some C^r-diffeomorphisms. Let us define the mapping $\tilde{h} : \tilde{U} \to \langle -1, 1 \rangle \times I^{n-1}, h(z) = (g_1 \circ \pi_1(z), g_2 \circ \pi_2(z))$, which is obviously of the class C^r. The inverse mapping to $\tilde{h}$ is given by the relation $\tilde{h}^{-1}(s, u) = \varphi_{g(s)}(g_2^{-1}(u))$, where $g(s) = g_1^{-1}(s)$. As the mappings $\varphi, g_1^{-1}, g_2^{-1}$ are of the class C^r, $\tilde{h}^{-1}$ is also of the class C^r and thus $\tilde{h}$ is a C^r-diffeomorphism. From the construction of the mapping $\tilde{h}$ it follows that if γ_x is a trajectory of the vector field F passing through the point $x \in \Sigma_{p_1}^\delta$, then $\tilde{h}(\gamma_x \cap \tilde{U}) = \langle -1, 1 \rangle \times \{g_2(x)\}$. It means that (U, h), where $U = \text{int } \tilde{U}, h = \tilde{h}/U$ is a tubular neighbourhood of the arc Γ. $\square$

Let us consider a C^r-vector field F on a smooth manifold X of dimension n where $1 \leq r < \infty$. Let us assume that this vector field has a periodic trajectory γ passing through the point $x_0 \in X$. As the point x_0 is a regular point of the vector field F, $F(x_0)$ is a non-zero element of the tangent space $T_{x_0}(X)$ generating the one-dimensional subspace $\langle F(x_0) \rangle = \{kF(x_0) : k \in R\}$ in $T_{x_0}(X)$ which, in fact, is the tangent space $T_{x_0}(\gamma)$ to γ at the point x_0. Let $G(x_0)$ be an algebraic complement to $T_{x_0}(\gamma)$, i.e. $T_{x_0}(\gamma) \oplus G(x_0) = T_{x_0}(X)$ where $T_{x_0}(\gamma) \cap G(x_0) = \{0\}$. There obviously exists a submanifold Σ at X of dimension $n-1$ such that $T_{x_0}(\Sigma) = G(x_0)$. It means that $T_{x_0}(\gamma) \oplus T_{x_0}(\Sigma) = T_{x_0}(X)$ and thus Σ is a local transversal to γ at the point x_0. The existence of the local transversal to γ at the point x_0 also follows from Theorem 3.161.

3.168 Theorem. *Let X be a smooth compact n-dimensional manifold, P be a smooth compact k-dimensional manifold and let a parametrized vector* field $F \in V^r(P, X)$ be given, where $1 \leq r < \infty$. Let us assume that the vector field $F_{p_0} \in V^r(X)$, where $p_0 \in P$, has τ_0-periodic trajectory γ passing through the point x_0. Then there exists a local transversal Σ to γ at the point x_0, an open neighbourhood $B_0 \times U_0 \times V_0 \times W_0$ of the point (F, x_0, p_0, τ_0) in $V^r(P, X) \times \Sigma \times P \times R$ and just one C^r-function $\tau : B_0 \times U_0 \times V_0 \to W_0$ such that $\tau(F, x_0, p_0) = \tau_0$ and $\varphi^G(x, p, t) \in \Sigma$ for $(G, x, p) \in B_0 \times U_0 \times V_0$ iff $t = \tau(G, x, p)$, φ^G is a parametrized flow for G.

Proof

According to Theorem 3.29 $V^r(P, X)$ is a Banach space. Let us define the mapping $\Phi : X \times V^r(X, P) \times P \times R \to X, \Phi(x, G, p, t) = \varphi^G(x, p, t)$. This mapping is a parametrized dynamical system on X with the set of parameters $\tilde{P} = V^r(X, P) \times P$ (dim $\tilde{P} = \infty$). From Propositions 3.5, 3.18 and from Theorem 1.48 it follows that the mapping Φ is of the class C^r. From Theorem 3.161 (Local Flow Box Theorem) it follows that there exists an admissible chart (U, h) on X such that $x_0 \in U, h(x_0) = 0 \in R^n, h(U) = I^n = I \times I \times \cdots \times I$ (n-times) and the main part f_h of the local representation of the vector field F_{p_0} with respect to this chart is of the form $f_h(t, z) = (1, 0, \ldots, 0) \in R^n$ for all $(t, z) \in I \times I^{n-1}$. The set $\tilde{\Sigma} = \{(t, z) \in I \times I^{n-1} : t = 0\}$ is obviously a transversal to the vector field f_h and $\Sigma = h^{-1}(\tilde{\Sigma})$ is a local transversal to γ at the point x_0. Let $h = (h_1, h_2)$, where $h_1 : I^n \to R, h_2 : I^n \to R^{n-1}$. From the continuity of the mapping Φ it follows that there exists an open neighbourhood $B_0 \times U_0 \times V_0 \times W_0$ of the point (F, x_0, p_0, τ_0) in $V^r(P, X) \times \Sigma \times P \times R$ such that the mapping $\Psi : B_0 \times U_0 \times V_0 \times W_0 \to R, \Psi(G, x, p, t) = h_1 \circ \varphi^G(x, p, t)$ is defined. As $x_0 \in \Sigma$, γ is a periodic trajectory of the vector field F_{p_0} with period τ_0 thus $\Psi(F, x_0, p_0, \tau_0) = 0$ and from the definition of the mapping Ψ it follows that

$$\frac{\partial}{\partial t}[\Psi(F, x_0, p_0, \tau_0)] = h_1 \circ F(x_0, p_0) = 1.$$

From the Implicit Function Theorem it follows that for the neighbourhoods B_0, U_0, V_0 sufficiently small there exists a unique C^r-function $\tau : B_0 \times U_0 \times V_0 \to R$ such that $\tau(F, x_0, p_0) = \tau_0$ and $h_1 \circ \varphi^G(x, p, t) = 0$ for $(G, x, p) \in B_0 \times U_0 \times V_0$ iff $t = \tau(G, x, p)$. It means that $\varphi^G(x, p, t) \in \Sigma$ iff $t = \tau(G, x, p)$.

3.169 Corollary. *Let X be a smooth n-dimensional compact manifold and let the vector field $F \in V^r(X)$ be given, where $1 \leq r < \infty$. Let us assume that F has a τ_0-periodic trajectory γ passing through the point x_0. Then there exists a local transversal Σ to γ at the point x_0, an open neighbourhood $B_0 \times U_0 \times W_0$ of the point (F, x_0, τ_0) in $V^r(X) \times \Sigma \times R$ and a unique C^r-function $\tau : B_0 \times U_0 \to W_0$ such that $\tau(F, x_0) = \tau_0$ and $\varphi^G(x, t) \in \Sigma$ for $(G, x) \in B_0 \times U_0$ iff $t = \tau(G, x)$, where φ^G is the flow of the vector field G.*

3.170 Definition. *Let the assumptions of Corollary 3.169 be fulfilled. Then for each $G \in B_0 \subset V^r(X)$ the mapping $\pi_G : U_0 \to \Sigma, \pi_G(x) = \varphi^G(x, \tau(G, x))$ is defined. This mapping is called the Poincaré mapping attached to F and its periodic trajectory γ (in short the Poincaré mapping). We shall also denote it as $\pi_G[F, \gamma, \Sigma, x_0, U_0]$ while, instead of $\pi_F[F, \gamma, \Sigma, x_0, U_0]$ we shall write either π or $\pi[F, \gamma, \Sigma, x_0, U_0]$.*

3.171 Definition. *Let the assumptions of Theorem 3.168 be fulfilled. Then for each $G \in B_0 \subset V^r(P, X)$ the mapping $H_G : U_0 \times V_0 \to \Sigma, H_G(x, p) = \varphi^G(x, p, \tau(G, x, p))$ is defined. This mapping is called the parametrized Poincaré mapping attached to F and the periodic trajectory γ of the vector field F_{p_0} (in short the parametrized Poincaré mapping). We shall also denote it as $H_G[F, \gamma, \Sigma, x_0, p_0, U_0, V_0]$ while, instead of $H_F[F, \gamma, \Sigma, x_0, p_0, U_0, V_0]$ we shall write either H or $H[F, \gamma, \Sigma, x_0, p_0, U_0, V_0]$.*

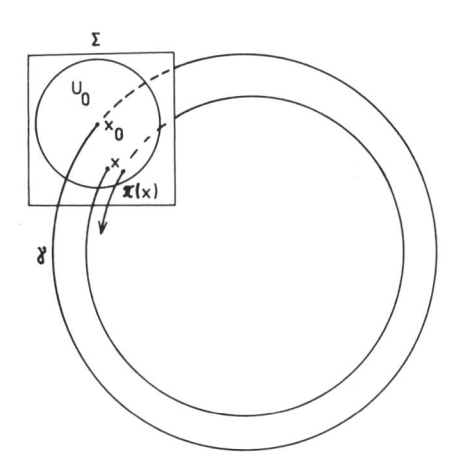

Fig. 28. Poincaré mapping.

3.172 Remark. Let the assumptions of Corollary 3.169 be fulfilled. Inspite of the fact that the vector field F has a periodic trajectory, the vector field $G \in B_0 \subset V^r(X)$ need not have a periodic trajectory. However, the Poincaré mapping π_G can be defined in the sense of the Definition 3.171. In the literature this Poincaré mapping is often defined only under the condition that G has a periodic trajectory.

3.173 Remark. Let $\pi_G = \pi_G[F, \gamma, \Sigma, x_0, U_0]$ be a Poincaré mapping, i.e. $\pi_G(x) = \varphi^G(x, \tau(G, x))$ for $x \in U_0$. Let us assume that the k-th iteration of the mapping π_G is defined. From the third property of the flow we obtain that $\pi_G^2(x) = \varphi^G(\pi_G(x), \tau(G, \pi_G(x))) = \varphi^G(\varphi^G(x, \tau(G, x)), \tau(G, \pi_G(x))) = \varphi^G(x, \tau(G, x) + \tau(G, \pi_G(x)))$. It can be proved by induction that the k-th iteration of the mapping π_G is of the form

$$\pi_G^k(x) = \varphi^G(x, \tau(G, x) + \sum_{i=1}^{k-1} \tau(G, \pi_G^i(x))).$$

As $\pi_F(x_0) = x_0, \tau(F, x_0) = \tau_0$ from the preceding equality and from the continuity of the mappings $(G, x, t,) \mapsto \varphi^G(x, t), (G, x) \mapsto \pi_G(x), (G, x) \mapsto \tau(G, x)$ it follows that if the mapping π_G has a k-periodic point x sufficiently close to x_0 then the vector field G, sufficiently close to F, has a T-periodic trajectory γ while its period T is approximately equal to τ_0.

3.174 Theorem. *Let X be a smooth manifold, γ be a periodic trajectory of the vector field $F \in V^r(X)$, where $1 \leq r < \infty$, passing through the point x_0. Then there exists a local transversal Σ to γ at the point x_0, an open neighbourhood $B_0 \times U$ of the point (F, x_0) in $V^r(X) \times \Sigma$ such that it holds that*

1. for each $G \in B_0$ the Poincaré mapping $\pi_G = \pi_G[F, \gamma, \Sigma, x_0, U]$ is defined;

2. the mapping π_G is of the class C^r;

3. π_G is the C^r-diffeomorphism of the set U onto the set $\pi_G(U)$.

Proof

Assertions 1 and 2 result from Corollary 3.169. The inverse mapping to π_G is $\pi_G^{-1} = \pi_{-G}[-F, \gamma, \Sigma, x_0, V]$, where $V = \pi_G(U)$. If $B_0 \times U$ is a sufficiently small neighbourhood of the point (F, x_0), then according to assertion 2 the mapping $\pi_{-G}[-F, \gamma, \Sigma, x_0, V]$ is of the class C^r. It means that π_G^{-1} is of the class C^r and thus π_G is the C^r-diffeomorphism of the set U onto the set $\pi_G(U)$. □

3.175 Theorem. *Let X be a smooth manifold, γ be a periodic trajectory of the vector field $F \in V^r(X)$, where $1 \leq r < \infty$. Then there exist local transversals Σ_1, Σ_2 to γ at the point $x_1 \in \gamma$, $x_2 \in \gamma$, an open neighbourhood B_0 of the vector field F in $V^r(X)$, an open neighbourhood U_1 of the point x_1 in Σ and an open neighbourhood U_2 of the point x_2 in Σ_2 such that for each $G \in B_0$ the Poincaré mappings $\pi_G^1 = \pi_G[F, \gamma, \Sigma_1, x_1, U_1]$, $\pi_G^2 = \pi_G[F, \gamma, \Sigma_2, x_2, U_2]$ are defined and these mappings are C^r-conjugated diffeomorphisms.*

Proof

From Theorem 3.174 it follows that there exist open neighbourhoods B_0, U_1, U_2 of the mapping F, of the point x_1 and of the point x_2, respectively, such that the mappings π_G^1, π_G^2 are defined and both of them are

C^r-diffeomorphisms. We must prove their C^r-conjugacy. First of all, let us assume that $x_1 \neq x_2$. Let Γ be the arc of the trajectory γ with the end points x_1, x_2. As the mapping $(G, x, t) \mapsto \varphi^G(x, t)$ is of the class C^r, for an arbitrary C^r-neighbourhood W of the arc Γ there exists a neighbourhood B_0 of the vector field F in $V^r(X)$ such that an arbitrary vector field $G \in B_0$ has a trajectory γ^G crossing transversally both Σ_1 and Σ_2 at the points $x_1^G \in \Sigma_1$, respectively $x_2^G \in \Sigma_2$ where the arc Γ^G of the trajectory γ^G with the end points x_1^G and x_2^G lies completely in the neighbourhood W. According to Theorem 3.167 there exists a tubular neighbourhood (U, h) of the arc Γ^G. Let $V_1 = \Sigma_1 \cap U$, $V_2 = \Sigma_2 \cap U$, $g : V_1 \to V_2$ be the projection along the trajectories of the vector field F and $\tilde{g} : h(V_1) \to h(V_2)$ be the projection along the trajectories of the local representation of the vector field G with respect to the chart (U, h). From the form of this local representation we obtain that $\tilde{g}$ is a C^∞-diffeomorphism. From the definition of the mappings g and $\tilde{g}$ it follows that $g = h^{-1} \circ \tilde{g} \circ h$, which means that g is the C^r-diffeomorphism. For arbitrary $x \in V_1$ there is obviously $\pi_G^2 \circ g(x) = g \circ \pi_G^1(x)$ and thus the diffeomorphisms π_g^1, π_G^2 are C^r-conjugated where g is their conjugating diffeomorphism. If $x_1 = x_2$ then instead of Theorem 3.167 it is sufficient to use Theorem 3.161 in the same way as we have used Theorem 3.167 in the case $x_1 \neq x_2$. $\square$

The Poincaré mapping $\pi_G = \pi_G[F, \gamma, \Sigma, x_0, U]$ is in fact, a local recurrent mapping derived from the flow φ^G of the vector field $G \in V^r(X)$ (see the Definition 3.23). In Section 3.1 we dealt with the question of the construction of vector field $G \in V^r(X)$ to a given C^r-diffeomorphism g so that this diffeomorphism may represent a recurrent mapping derived from the flow of the vector field G. In fact, we managed to construct such a vector field (see Proposition 3.24) but only in this sense: while the diffeomorphism g is defined on a smooth manifold X of dimension n, the vector field G is defined on a C^r-manifold $\tilde{X}$ of dimension $n + 1$ where the recurrent mapping g derived from the flow of the vector field G is defined on a C^r-submanifold Σ in $\tilde{X}$ which is a global transversal to G and $\tilde{g}$ is a C^r-diffeomorphism conjugated with g. Further we shall formulate a theorem on the existence of such vector field G of the class C^r to an arbitrarily small C^r-approximation of the given Poincaré mapping π but defined on the same manifold X on which the flow defining the Poincaré mapping is defined. We say that the vector field G realizes the C^r-approximation of the Poincaré mapping. Proposition 3.24 is obviously of different character and for the study of the so-called generic bifurcations which we shall deal with in the Chapter 5, it is not applicable.

3.176 Definition. *Let P, X be smooth manifolds, $p_0 \in P$, $G \in V^r(P, X)$, where $1 \leq r < \infty$, γ is a periodic trajectory of the vector field G passing through the point $x_0 \in X$, $M_{p_0} \subset P$ is an open neighbourhood of the point p_0 and $N_\gamma \subset X$ is an open neighbourhood of the trajectory γ. Then we* define a set: $V_G^r(M_{p_0}, N_\gamma) = \{\tilde{G} \in V^r(P,X) : \tilde{G}(x,p) = G(x,p)$ for all $(x,p) \in X \times P \setminus (N_\gamma \times M_{p_0})\}$. In addition let Σ be a local transversal to γ at the point x_0 and $U \subset \Sigma$, $V \subset P$ be open sets such that $x_0 \in U$, $P_0 \in V$ and the Poincaré mapping $H = H[G,\gamma,\Sigma,x_0,p_0,U,V]$ is defined. Let $K \subset U$, $L \subset V$ be open sets such that $\overline{K} \subset U$, $\overline{L} \subset V$, $x_0 \in K$, $p_0 \in L$. Then we define the set: $Z_H^r(U[K], V[L]) = \{\tilde{H} \in C^r(U \times V, \Sigma) : \tilde{H}(x,p) = H(x,p)$ for all $(x,p) \in U \times V \setminus (K \times L)\}$.

3.177 Theorem. *Let X, P be smooth compact manifolds, $G \in V^r(P,X)$, where $1 \leq r < \infty$, $p_0 \in P$ and γ is a periodic trajectory of the vector field G_{p_0} passing through the point $x_0 \in X$. Then there exists a local transversal Σ to γ at the point x_0 such that the parametrized Poincaré mapping $H = H[G,\gamma,\Sigma,x_0,p_0,U,V]$ is defined, where $U \times V$ is an open neighbourhood of the point (x_0,p_0) in $\Sigma \times P$ and it holds that: there exists an open neighbourhood K of the point x_0 in Σ, $\tilde{K} \subset U$, an open neighbourhood $M_{p_0}, L \subset P$ of the point p_0, $\tilde{L} \subset M_{p_0}$, an open neighbourhood $N_\gamma \subset X$ of the trajectory γ, an open neighbourhood $W^r(H)$ of the mapping H in $Z_H^r(U[K], V[L])$ and a continuous mapping $\varkappa : W^r(H) \to V_G^r(M_{p_0}, N_\gamma)$ such that for each $\tilde{H} \in W^r(H)$, the parametrized vector field $\tilde{G} = \varkappa(\tilde{H})$ is such that the parametrized Poincaré mapping $H_{\tilde{G}} = H_{\tilde{G}}[G,\gamma,\Sigma,x_0,U,V]$ is defined and $H_{\tilde{G}} = \tilde{H}$.*

3.178 Definition. *Let X be a smooth manifold, γ be a periodic trajectory of the vector field $F \in V^r(X)$, where $1 \leq r < \infty$, passing through the point $x_0 \in X$ and let N be an open neighbourhood of this trajectory. Let us define the set: $V^r(X[N_\gamma]) = \{\tilde{F} \in V^r(X) : \tilde{F}(x) = F(x)$ for all $x \in X \setminus N_\gamma\}$. In addition let Σ be a local transversal of the vector field F at the point x_0 and $U \subset \Sigma$ be an open neighbourhood of the point x_0 in Σ such that the Poincaré mapping $\pi = \pi[F,\gamma,\Sigma,x_0,U]$ is defined. Let $\hat{K}K \subset U$ be an open neighbourhood of the point x_0 in Σ such that $K \subset U$. Let us define the set $Z^r(U[K]) = \{\tilde{\pi} \in C^r(U,\Sigma) : \tilde{\pi}(x) = \pi(x)$ for all $x \in U \setminus K\}$.*

The consequence of Theorem 3.177 is the following theorem.

3.179 Theorem on Realization of C^r-Approximation of Poincaré Mapping. *Let X be a smooth manifold, γ is a periodic trajectory of the vector field $F \in V^r(X)$, where $1 \leq r < \infty$, passing through the point $x_0 \in X$. Then there exists a local transversal Σ to γ at the point x_0 such that the Poincaré mapping $\pi = \pi[F,\gamma,\Sigma,x_0,U]$ is defined and it holds that: there exists an open neighbourhood $K \subset U$ of the point x_0 in Σ, $\hat{K} \subset U$, an open neighbourhood $N_\gamma \subset X$ of the trajectory γ, an open neighbourhood $W^r(\pi)$ of the mapping π in $Z^r(X[N_\gamma])$ and a continuous mapping $\Psi : W^r(\pi) \to V_F^r(X[N_\gamma])$ such that for each $\tilde{\pi} \in W^r(\pi)$ the vector field $\tilde{F} = \Psi(\tilde{\pi})$ is such that the Poincaré mapping $\pi_{\tilde{F}}[F,\gamma,\Sigma,x_0,U]$ is defined and $\pi_{\tilde{F}} = \tilde{\pi}$.*

The detailed proof of Theorem 3.177 can be found in [106]. This proof is large, technically quite difficult and therefore we do not present it. In [122] there is a construction of the vector field which realizes only a special linear approximation of the given Poincaré mapping. This result is sufficient for the study of generic properties of the vector fields but not for the study of the generic bifurcation which will be dealt with in Chapter 5.

3.180 Remark. Theorem 3.179 guarantees the existence of only a local transversal Σ to the periodic trajectory γ for which the assertion on the realization of approximation of the Poincaré mapping $\pi = \pi[F, \gamma, \Sigma, x_0, U]$ holds. If Σ_1 is another local transversal to γ at the point $x_1 \in \gamma$ then according to Theorem 3.179 there exists an open neighbourhood U_1 of the point x_1 in Σ such that the Poincaré mapping $\pi_1 = \pi[F, \gamma, \Sigma_1, x_1, U_1]$ is C^r-conjugated with π. It means that Theorem 3.179 is sufficiently general in view of its application to the study of the topological structure of trajectories of vector fields in neighbourhoods of their periodic trajectories.

The above properties of the Poincaré mapping enable us to define well the hyperbolicity of periodic trajectories of vector fields.

3.181 Definition. *Let γ be a periodic trajectory of the vector field $F \in V^r(X)$, Σ be a local transversal to γ at the point $x \in \gamma$ and $\pi = \pi[F, \gamma, \Sigma, x, U]$ be a Poincaré mapping. The periodic trajectory γ is called hyberbolic, if $|\lambda| \neq 1$ for all eigenvalues λ of the mapping $D_x\pi : T_x(\Sigma) \to T_x(\Sigma)$. The eigenvalues of the mapping $D_x\pi$ are called characteristic multipliers of the periodic trajectory γ. If $\dim X = n$ and $\lambda_1, \lambda_2, \ldots, \lambda_{n-1}$ are characteristic multipliers, where $|\lambda_i| > 1$, $i = 1, 2, \ldots, k$, $|\lambda_i| < 1$, $i = k+1, k+2, \ldots, n$, then we denote the number k as $m^+(F, \gamma)$ and the number $n - k - 1$ as $m^-(F, \gamma)$.*

From Theorem 3.175 it follows that if $\pi_1 = \pi[F, \gamma, \Sigma, x, U], \pi_2 = \pi[F, \gamma, \Sigma_1, y, V]$ are two Poincaré mappings then the mappings $D_x\pi_1$ and $D_x\pi_2$ have the same eigenvalues and thus the definition of the hyperbolicity of a periodic trajectory does not depend on the choice of the Poincaré mapping.

3.182 Theorem. *Let X be a smooth n-dimensional manifold, γ be a τ-periodical trajectory of the vector field $F \in V^r(X)$, where $1 \leq r < \infty$, φ be the flow of the vector field F, Σ be a local transversal to γ at the point $x \in \gamma$ and $\pi = \pi[F, \gamma, \Sigma, x, V]$ be a Poincaré mapping. Then the mapping $D_x\varphi_\tau : T_x(X) \to T_x(X)$ has eigenvalues $\lambda_1 = 1, \lambda_2, \ldots, \lambda_n$, where $\lambda_2, \lambda_3, \ldots, \lambda_n$ are eigenvalues of the mapping $D_x\pi : T_x(\Sigma) \to T_x(\Sigma)$ and the eigenvector of the mapping $D_x\varphi_\tau$ corresponding to the eigenvalue $\lambda_1 = 1$ equals $F(x)$.*

Proof

If $t \in R$ then $\varphi(x,t) \in \gamma$ and therefore $\varphi(\varphi(x,t),\tau) = \varphi(x, t + \tau) = \varphi(x,t)$. Thus we obtain that $\varphi_\tau \circ \varphi_x = \varphi_x$. When using a Theorem 2.31 on the derivation of the composite mapping we obtain $F(x) = F(\varphi_x(\tau)) = D_\tau \varphi_x(e_\tau) = D_\tau(\varphi_\tau \circ \varphi_x)(e_\tau) = D_x\varphi_\tau \circ D_\tau\varphi_x(e_\tau) = D_x\varphi_\tau(F(x))$, where e_τ is the unit vector in $T_\tau(R)$, i.e.

$$D_x\varphi_\tau(F(x)) = F(x). \tag{3.97}$$

Thus we obtain that $F(x)$ is an eigenvector of the mapping $D_x\varphi_\tau$ corresponding to the eigenvalue $\lambda_1 = 1$. Let (U, α) be an admissible chart on X such that $x \in U$ and $\pi(V) \subset U$. According to Proposition 2.30 the mapping $D\varphi_\tau : T(X) \to T(X)$ has a local representation $(D\varphi_\tau)_\alpha = T_\alpha \circ D\varphi_\tau \circ T_\alpha^{-1} : \alpha(U) \times R^n \to \alpha(U) \times R^n$, $(D\varphi_\tau)_\alpha(y, u) = (\Psi_\alpha(y), d_y\Psi_\alpha(u))$, where $(\pi_X^{-1}(U), T_\alpha)$ is the chart on $T(X)$ derived from the chart (U, α) (see the proof of Proposition 2.26) and $\Psi_\alpha = \alpha \circ \varphi_\tau \circ \alpha^{-1}$. It means that the mapping $D_x\varphi_\tau$ has the local representation $(D_x\varphi_\tau)_\alpha = d_z\Psi_\alpha$, where $z = \alpha(x)$. From Proposition 2.24 it follows that λ is an eigenvalue of the mapping $D_x\varphi_\tau$ iff it is an eigenvalue of the Jacobi matrix $[d_z\Psi_\alpha]$. From equality (3.97) we obtain $F_\alpha(z) = T_\alpha \circ F \circ \alpha^{-1}(z) = T_\alpha \circ D_x\varphi_\tau \circ F(\alpha^{-1}(z)) = T_\alpha \circ D_x\varphi_\tau T_\alpha^{-1}(F_\alpha(z)) = (D_x\varphi_\tau)_\alpha(F_\alpha(z))$. From this equality it follows that $f_\alpha(z) = d_z\Psi_\alpha(f_\alpha(z))$, where f_α is the main part of the local representation F_α of the vector field F, which means that $(f_\alpha(z))^*$ is an eigenvector of the matrix $[d_z\Psi_\alpha]$ corresponding to the eigenvalue $\lambda = 1$. As the eigenvalues of the matrix $[d_z\Psi_\alpha]$ do not depend on the choice of the chart (U, α), without loss of generality, we can assume on the basis of Theorem 3.161 that the chart (U, α) is such that $f_\alpha(y) = (1, 0, \ldots, 0) \in R^n$ for all $y \in \alpha(U)$. Then

$$[d_z\Psi_\alpha] = \begin{bmatrix} 1 & A \\ 0 & B \end{bmatrix},$$

where $A \in M(1, n-1)$, $B \in M(n-1, n-1)$, $0 = 0_{n-1,1} \in M(n-1, 1)$. From Theorem 3.175 it follows that eigenvalues of the mapping $D_x\pi$ do not depend on the choice of the local transversal to γ at the point x, and thus without loss of generality, we can also assume that $\Sigma = \alpha^{-1}(\{y = (y_1, y_2, \ldots, y_n) : y_1 = 0\} \cap \alpha(U))$. Let us define the mapping $\pi_\alpha : \alpha(\Sigma \cap V) \to \alpha(U)$, $\pi_\alpha = \alpha \circ \pi \circ \alpha^{-1}$. Then the mapping $d_z\pi_\alpha$, where $z = \alpha(x)$, is a local representation of the mapping $D_x\pi$ and $[d_z\pi_\alpha] = B$. It means that the mapping $D_x\pi$ has the same eigenvalues as the matrix B. $\square$

3.183 Remark. Let the assumptions of Theorem 3.182 be fulfilled, where $X = R^n$. Let $F(x) = (x, f,(x))$, i.e. the flow φ of the vector field F is defined by the differential equation $\dot{x} = f(x)$. Then $\Phi(t) = [d_x\varphi_t]$ is the solution of the differential equation $\dot{Y} = A(t)Y$, where $A(t) = [d_{\varphi_t(x)}f]$

where $\Phi(0) = I_n$. As γ is a τ-periodic trajectory and $x \in \gamma$, the matrix $A(t)$ is periodic with the period τ. From the Floquet Theorem it follows that there exists a constant matrix $D \in M(n)$ such that $\Phi(\tau) = e^{\tau D}$. The eigenvalues of the matrix D are called characteristic exponents of the periodic trajectory γ (compare with the definition of the characteristic exponents of the singular points — Definition 3.118).

4 INVARIANT MANIFOLDS

In this chapter we shall present theorems on so-called invariant manifolds playing the same role in the local topological classification of vector fields as the invariant subspaces of the linear dynamical systems for their global topological classification which we studied in Section 3.5 of the preceding chapter.

4.1 Stable and Unstable Manifolds

In the preceding chapter we have analysed the topological structure of the trajectories of vector fields in neighbourhoods of their hyperbolic singular points which as it has turned out depends to a substantial extent upon characteristic exponents of these singular points. Introduction of this section will be devoted to a similar analysis of the diffeomorphisms in neighbourhoods of their fixed points.

4.1 Definition. *Let X be a smooth manifold. The diffeomorphism $f \in \text{Diff}^r(X)$, where $r \geq 1$, is called contractive (expansive) in its fixed point $x \in X$ if $|\lambda| < 1$ ($|\lambda| > 1$) for an arbitrary eigenvalue λ of the mapping $D_x f$. If $X = E_n$ is an n-dimensional Euclidean space then the linear diffeomorphism $A \in L(E_n)$ is called contractive (expansive) if it is contractive (expansive) at the point $0 \in E_n$.*

4.2 Lemma. *If E_n is an n-dimensional Euclidean space, then for a linear mapping $A \in L(E_n)$ the following assertions hold.*

1. If A is a contractive diffeomorphism, then $\lim_{n \to \infty} A^n x = 0$ for arbitrary $x \in E_n$.

2. If A is an expansive diffeomorphism, then $\lim_{n \to \infty} A^{-n} x = 0$ for arbitrary $x \in E_n$.

Proof

Let A be a contractive diffeomorphism. According to Lemma 3.126 on E_n there exists a norm $\|\cdot\|$ such that $\|A\| = \sup_{\|x\|\leq 1}\|Ax\| < 1$. Since $\|A^n x\| \leq \|A\|^n \|x\|$, obviously, $\lim_{n\to\infty} A^n x = 0$ for arbitrary $x \in E_n$, and thus, assertion 1 holds. The number λ is an eigenvalue of the mapping A iff λ^{-1} is an eigenvalue of the mapping A^{-1}, and thus, assertion 2 follows from assertion 1. □

4.3 Theorem. *Let X be a smooth manifold of dimension n and $x \in X$ a fixed point of the diffeomorphism $f \in \mathrm{Diff}^r(X)$, where $r \geq 1$, then the following assertions hold.*

1. If the diffeomorphism f is contractive at the point x then there exists such a neighbourhood U of the point x, that $\lim_{n\to\infty} f^n(y) = x$ for arbitrary $y \in U$.

2. If the diffeomorphism f is expansive at the point x, then there exists a neighbourhood U of the point x such that $\lim_{n\to\infty} f^{-n}(y) = x$ for arbitrary $y \in U$.

Proof

The theorem is of local character and thus, it will suffice to assume that $X = R^n$. Therefore both assertions follow from the Hartman Theorem and Lemma 4.2.

The following theorem follows directly from Theorem 1.18.

4.4 Theorem. *If E_n is an n-dimensional Euclidean space and $A \in L(E_n)$ then there exist linear subspaces E^s, E^c, E^u of the space E_n that are invariant sets of the mapping A, i.e. $A(E^s) \subset E^s$, $A(E^c) \subset E^c$, $A(E^u) \subset E^u$ and it holds that*

1. the space E_n is the direct sum of the spaces E^s, E^c and E^u, i.e. $E_n = E^s \oplus E^c \oplus E^u$;

2. A/E^s is contractive and A/E^u is an expansive diffeomorphism. If the mapping A is hyperbolic then $E^c = \{0\}$.

4.5 Definition. *Let X be a smooth manifold. If x is a fixed point of the diffeomorphism $f \in \mathrm{Diff}^r(X)$ then the set $W^s(x) = W^s(f,x) = \{y \in X : \lim_{n\to\infty} f^n(y) = x\}$ ($W^u(x) = W^u(f,x) = \{y \in X : \lim_{n\to\infty} f^{-n}(y) = x\}$) is called the stable (unstable) set of the diffeomorphism f at the point. If U is an open neighbourhood of the point then the set $W^s_{\mathrm{loc}}(x) = W^s_{\mathrm{loc}}(f,x) = \{y \in U : \lim_{n\to\infty} f^n(y) = x\}$ ($W^u_{\mathrm{loc}}(x) = W^u_{\mathrm{loc}}(f,x) = \{y \in U : \lim_{n\to\infty} f^{-n}(y) = x\}$) is called the local stable (local unstable) set of the diffeomorphism f at the point x. If x is a periodic point of the diffeomorphism f with the minimal period p, then the set $W^s(x) = W^s(f^p,x)$ ($W^u(x) = W^u(f^p,x)$) is called the* stable (unstable) set of the diffeomorphism f at the point x and the set $W^s_{\text{loc}}(x) = W^s_{\text{loc}}(f^p, x)(W^u_{\text{loc}}(x) = W^u_{\text{loc}}(f^p, x))$ is called a local stable (local unstable) set of the diffeomorphism f at the point x.

4.6 Definition. *Let X be a smooth manifold and $x \in X$ be a singular point of the vector field $F \in V^r(X)$. The set $W^s(x) = W^s(F, x) = \{y \in X : \lim_{t \to \infty} \varphi^F_t(y) = x\}$ ($W^u(x) = W^u(F, x) = \{y \in X : \lim_{t \to -\infty} \varphi^F_t(y) = x\}$) is called the stable (unstable) set of the vector field F at the point x, where φ^F is a flow of the vector field F. If a neighbourhood U of the point x is given then the set $W^s_{\text{loc}}(x) = W^s_{\text{loc}}(F, x) = W^s(F/U, x)$ ($W^u_{\text{loc}}(x) = W^u_{\text{loc}}(F, x) = W^u(F/U, x)$) is called the local stable (local unstable) set of the vector field F at the point x. If γ is the periodic trajectory of the field F then the set $W^s(\gamma) = W^s(F, \gamma) = \{x \in X : \omega(x) = \gamma\}$ ($W^u(\gamma) = W^u(F, \gamma) = \{x \in X : \alpha(x) = \gamma\}$) is called the stable (unstable) set of the periodic trajectory γ, where $\omega(\gamma)(\alpha(\gamma))$ is the ω-limit (α-limit) set of the periodic trajectory γ. If an open neighbourhood V of the trajectory γ is given then the set $W^s_{\text{loc}}(\gamma) = W^s_{\text{loc}}(F, \gamma) = W^s(F/V, \gamma)$ ($W^u_{\text{loc}}(\gamma) = W^u_{\text{loc}}(F, \gamma) = W^u(F/V, \gamma)$) is called the local stable (local unstable) set of the periodic trajectory γ.*

4.7 Theorem on Invariant Manifolds of Periodic Points. *If X is a smooth n-dimensional manifold and $x \in X$ is a hyperbolic periodic or fixed point of the diffeomorphism $f \in \text{Diff}^r(X)$, where $r \geq 1$ then the following assertions hold.*

1. *The sets $W^s(f, x)$, $W^u(f, x)$ are immersive C^r-submanifolds in X, where $\dim W^s(f, x) = m^-(f, x)$ and $\dim W^u(f, x) = m^+(f, x)$ (see Definition 3.124).*

2. *There exists a neighbourhood U of the point x such that the following assertions hold*

   a) *the sets $W^s_{\text{loc}}(f, x) = \{y \in U : \lim_{n \to \infty} f^n(y) = x\}$, $W^u_{\text{loc}}(f, x) = \{y \in U : \lim_{n \to \infty} f^{-n}(y) = x\}$ are C^r-submanifolds in X,*

   b) *$T_x(W^s_{\text{loc}}(f, x)) = E^s_x$, $T_x(W^u_{\text{loc}}(f, x)) = E^u_x$, where $E^s_x(E^u_x)$ is a stable (unstable) subspace of the linear diffeomorphism $D_x f$ (and thus, $W^s_{\text{loc}}(f, x) \overline{\pitchfork}_x W^u_{\text{loc}}(f, x))$.*

4.8 Theorem on Invariant Manifolds of Singular Points. *If X is a smooth n-dimensional manifold and $x \in X$ is a hyperbolic singular point of the vector field $F \in V^r(X)$, where $r \geq 1$ then the following assertions hold.*

1. *The sets $W^s(F, x)$, $W^u(F, x)$ are immersive C^r-submanifolds in X, where $\dim W^s(F, x) = m^-(f, x)$ and $\dim W^u(F, x) = m^+(f, x)$ (see Definition 3.118).*

2. *There exists an open neighbourhood U of the point $x \in X$ such that the following assertions hold*

a) the sets $W^s_{loc}(F,x) = W^s(F/U,x)$, $W^u_{loc}(F,x) = W^u(F/U,x)$ are C^r-submanifolds in X,

b) $T_x(W^s_{loc}(F,x)) = E^s_x$, $T_x(W^u_{loc}(F,x)) = E^u_x$, where $E^s_x(E^u_x)$ is a stable (unstable) subspace of the hessian $F(x)$ (and thus,

$$W^s_{loc}(F,x) \pitchfork_x W^u_{loc}(F,x)).$$

4.9 Example. Let us consider the system $\dot{x} = x$, $\dot{y} = -y + x^2$. Let us denote by W^s, W^u the stable and unstable set of this system, respectively, (of the vector field defined by it) at the point $0 \in R^2$ and E^s, E^u the stable and unstable set, respectively, of its linearization at the point 0. Obviously, $W^s = \{(x,y) : x = 0\} = E^s$ and $E^u = \{(x,y) : y = 0\}$. The trajectories of the above system are the graphs of solutions of the differential equation

$$\frac{dy}{dy} = -\frac{y}{x} + x$$

and out of them only those of which the invariant manifold W^u is composed are to be selected. The general solution of this equation is of the form

$$y_c(x) = \frac{x^2}{3} + \frac{c}{x},$$

where c is a constant. Let $W^u = \{(x,y) : y = h(x)\}$. Since according to Theorem 4.8 it is $T_0(W^u) = E^u$, then

$$h(0) = \frac{dh(0)}{dx} = 0.$$

This means that $h(x) = y_0(x) = x^2/3$, i.e., $W^u = \{(x,y) : y = x^2/3\}$ (Fig. 29).

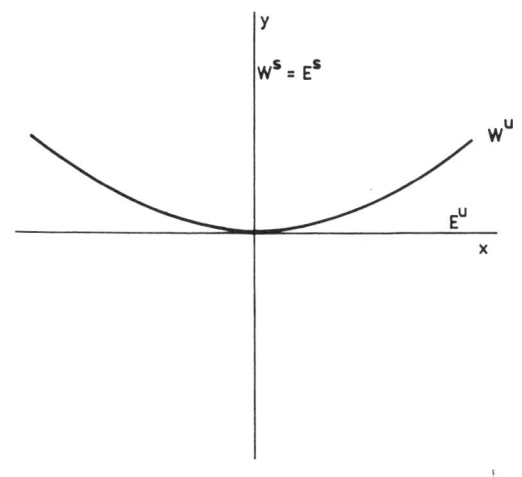

Fig. 29.

4.10 Theorem on the Invariant Manifolds of Periodic Trajectories. *If X is a smooth n-dimensional manifold and γ is a hyperbolic periodic trajectory of the vector field $F \in V^r(X)$, where $r \geq 1$, then the following assertions hold.*

1. *The sets $W^s(F,\gamma)$, $W^u(F,\gamma)$ are immersive C^r-submanifolds in X, where $W^s(F,\gamma) = m^-(F,\gamma)$ and $\dim W^u(F,\gamma) = m^+(F,\gamma)$ (see Definition 3.181).*

2. *There exists an open neighbourhood V of the trajectory γ such that the following assertions hold*

   a) *the sets $W^s_{loc}(F,\gamma) = W^s_{loc}(F/V,\gamma)$, $W^u_{loc}(F,\gamma) = W^u(F/V,\gamma)$ are C^r-submanifolds in X,*

   b) $W^s_{loc}(F,\gamma) \overline{\pitchfork}_\gamma W^u_{loc}(F,\gamma)$ *(see Fig. 30).*

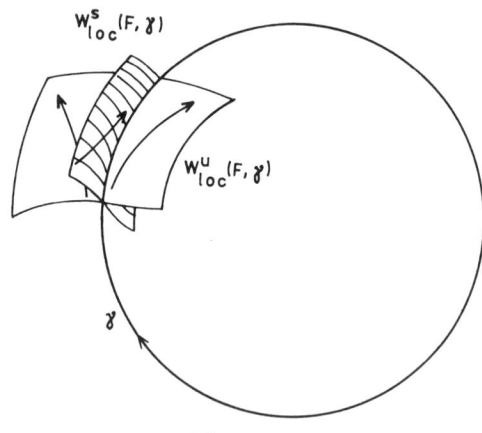

Fig. 30.

On the basis of the above theorems, the stable (unstable) invariant sets of periodic and singular points or of periodic trajectories will be called the stable (unstable) manifolds.

The first proof of existence of the local stable manifold at least of the class C^0 for diffeomorphisms was by J. Hadamard [52] whose ideas were later further developed in the papers of O. Perron [125, 126]. In the work [143] by S. Sternberg is presented the proof of differentiability of a local invariant manifold for diffeomorphisms. In the work by S. Smale [140] the first proof of the existence of global invariant manifolds is presented (see also [2]). The proof by M. Irvin [76, 77] of Theorem 4.7 is the first of the proofs of this theorem that is based on the use of the Implicit Function Theorem in the Banach space. A more geometrical proof of this theorem can be found in [66]. The problem of existence of invariant manifolds for vector fields has been treated by a series of mathematicians. First results in this respect have been achieved by N. N. Bogoljubov and V. Mitropolskii in [20] and further developed in [33, 35, 54, 79, 87, 88, 126].

From the above theorems on invariant manifolds we shall prove only assertion 2 of Theorems 4.7 and 4.8 on local existence of invariant manifolds. The proof of the global existence of invariant manifolds from diffeomorphism can be found in the book [67]. Complete proofs of Theorems 4.8 and 4.10 can be found, e.g., in [2].

Since assertion 2 of Theorem 4.7 is of local character, in its proof it is sufficient to assume that $X = R^n$ and $x = 0$. Let us denote $B_\varepsilon = B_\varepsilon(0) = \{y \in R^n : \|y\| \leq \varepsilon\}$, $W_\varepsilon^s(f, 0) = \{y \in B_\varepsilon : f^n(y) \in B_\varepsilon \text{ for all } n \geq 0\}$, $W_\varepsilon^u(f, 0) = \{y \in B_\varepsilon : f^{-n}(y) \in B_\varepsilon \text{ for all } n \geq 0\}$.

4.11 Proposition. *If $0 \in R^n$ is a hyperbolic fixed point of the diffeomorphism $f \in \text{Diff}^r(R^n)$, where $r \geq 1$, then there exists $\varepsilon > 0$ such that $W_\varepsilon^s(f, 0) = W^s(f/B_\varepsilon, 0)$ and $W_\varepsilon^u(f, 0) = W^u(f/B_\varepsilon, 0)$.*

Proof

The mapping f can be written in the form

$$f(x) = Ax + \Phi(x), \tag{4.1}$$

where $A = d_0 f$, $\Phi \in C^r$, $\Phi(0) = 0$, $d_0 \Phi = 0$. Since 0 is a hyperbolic fixed point of the diffeomorphism, from the Hartman Theorem it follows that there exists a neighbourhood U of the point 0, number $\varepsilon > 0$ and homeomorphism $h : B_\varepsilon \to U$, such that $h(0) = 0$ and

$$h \circ f(x) = A \circ h(x) \tag{4.2}$$

for all $x \in U$. From Theorem 4.4 it follows that if $x \in U$ and $A^n x \in U$ for all $n \geq 0$, then $x \in E^s$, where E^s is a stable subspace of the mapping A. According to Lemma 4.2 for $x \in U$ $\lim_{n \to \infty} A^n x = 0$. If $y \in W_\varepsilon^s(f, 0)$, then $f^n(y) \in B_\varepsilon$ for all $n \geq 0$ and from the equality (4.2) we get that $h \circ f^n(y) = A^n \circ h(y)$, and thus $A^n \circ h(y) \in U$ for all $n \geq 0$. Therefore $\lim_{n \to \infty} A^n \circ h(y) = 0$. From the equality (4.2) it follows that $f^n(y) = h^{-1} \circ A^n \circ h(y)$, and thus, $\lim_{n \to \infty} f^n(y) = h^{-1}(0) = 0$, i.e., $y \in W^s(f/B_\varepsilon, 0)$. Thus, we obtain that $W_\varepsilon^s(f, 0) \subset W^s(f/B_\varepsilon, 0)$. The inverse inclusion is obvious and thus, $W_\varepsilon^s(f, 0) = W^s(f/B_\varepsilon, 0)$. The equality $W_\varepsilon^u(f, 0) = W^u(f/B_\varepsilon, 0)$ can by proved analogously. □

4.12 Proposition. *Let $0 \in R^n$ be a hyperbolic fixed point of the diffeomorphism $(f \in \text{Diff}^r(R^n))$, where $r > 1$, and E^s (E^u) be a stable (unstable) invariant subspace of the mapping $A = d_0 f$. Then there exists a number*

$\varepsilon > 0$ and a C^r-diffeomorphism $h : B^s_\varepsilon \to E^u$, where $B^s_\varepsilon = E^s \cap B_\varepsilon$, such that $W^s_\varepsilon(f,0) = \text{graph}(h)$ and $T_0(W^s_\varepsilon(f,0)) = E^s$.

This proposition will be proved, in the same way as used in [122]. In another paper [66] a different proof of this proposition is presented based on a suitable use of the Implicit Function Theorem for Lipschitz mappings and with the use of this method a more general theorem on the existence of invariant manifolds of so called hyperbolic sets of diffeomorphisms will be proved (a hyperbolic periodic point is a special type of hyperbolic set of diffeomorphism).

If the diffeomorphism $f \in \text{Diff}^r(R^n)$ fulfils the assumptions of Proposition 4.12, then $R^n = E^s \oplus E^u$. We shall use the notations $x_s = \pi_s(x), x_u = \pi_u(x)$ for $x \in R^n$, $g^s = \pi_s \circ g$ and $g^u = \pi_u \circ g$ for $g : R^n \to R^n$, where $\pi_s : R^n \mapsto E^s$, $\pi_u : R^n \to E^u$ are projections. Let us suppose that $\varepsilon > 0$ is a small number such that on B_ε $f \in \text{Diff}^r(R^n)$ has the form (4.1) on B_ε, where

$$\sup_{x \in B_\varepsilon} \|d_x \Phi\| < \delta \tag{4.3}$$

for some $\delta > 0$. The mappings $A^s = A/E^s$, $(A^u)^{-1}/E^u$, where $A^u = A/E^u$ are contractive, and thus, according to Theorem 3.124 the norms $\|\cdot\|_1, \|\cdot\|_2$ on E^s and E^u, respectively, can be chosen such that

$$\|A^s\|_1 < a < 1, \qquad \|(A^u)^{-1}\|_2 < a < 1. \tag{4.4}$$

Let us define on R^n the norm $\|x\| = \max(\|x_x\|_1, \|x_u\|_2)$. If $\varepsilon > 0$ is a small enough number, then the number δ in the inequality (4.3) can be chosen so that

$$0 < \delta < \frac{1}{2}(a^{-1} - 1). \tag{4.5}$$

To be able to prove Proposition 4.12 the following lemma will be also needed.

4.13 Lemma. *Let ε, δ, a be positive numbers for which the inequalities (4.3), (4.4), (4.5) are fulfilled and $f^m(x) \in B_\varepsilon$, $f^m(x') \in B_\varepsilon$ for all $m \geq 0$, where $x, x' \in R^n$ are such that $x_s = x'_s$. Then $x_u = x'_u$.*

Proof

Let $x = (x_s, x_u) \in B_\varepsilon$ and $x' = (x'_s, x'_u) \in B_\varepsilon$ are the points such that $\|x_s - x'_s\| \leq \|x_u - x'_u\|$. First, let us prove that $\|f^s(x) - f^s(x')\|_1 \leq \|f^u(x) - f^u(x')\|_2$. From the definition of the norm $\|\cdot\|$ and from (4.4) we obtain that

$$\|x - x'\| = \|x_u - x'_u\|_2 = \|(A^u)^{-1}[A^u x_u - A^u x'_u]\|_2$$
$$\leq a\|A^u x_u - A^u x'_u\|_2 = a\|\pi_u Ax - \pi_u Ax'\|_2, \quad \text{i.e.}$$

Invariant Manifolds

$$\|\pi_u Ax - \pi_u Ax'\|_2 \geq a^{-1}\|x - x'\|. \tag{4.6}$$

With the use of (4.3), (4.6) and from the Mean Value Theorem we obtain that

$$\|f^u(x) - f^u(x')\|_2 = \|\pi_u Ax - \pi_u Ax' + \Phi^u(x) - \Phi^u(x')\|_2$$
$$\geq \|\pi_u Ax - \pi_u Ax'\|_2 - \|\Phi^u(x) - \Phi^u(x')\|_2$$
$$\geq (a^{-1} - \delta)\|x - x'\|, \quad \text{i.e.}$$

$$\|f^u(x) - f^u(x')\|_2 \geq (a^{-1} - \delta)\|x - x'\|. \tag{4.7}$$

From the inequalities (4.6) and (4.7) the inequality

$$\|f^s(x) - f^s(x')\|_1 \leq (a + \delta) \cdot (a^{-1} - \delta)^{-1}\|f^u(x) - f^u(x')\|_2$$

follows. As $a < 1$, from inequality (4.5) it follows that the inequality $(a + \delta)(a^{-1}-\delta) < 1$ and therefore we have $\|f^s(x)-f^s(x')\|_1 \leq \|f^u(x)-f^u(x')\|_2$. From this inequality, from both the definition of norm $\|\cdot\|$ and (4.7) follows the inequality $\|f(x) - f(x')\| \geq (a^{-1} - \delta)\|x - x'\|$. Obviously, it can be proved by induction that there is $\|f^m(x) - f^m(x')\| \geq (a^{-1}-\delta)^m\|x-x'\|$ for arbitrary $m \in N$. As $a^{-1} - \delta > 1$, thus it follows from the last inequality that $\lim_{m\to\infty} \|f^m(x) - f^m(x')\| = \infty$ for $x_u \neq x'_u$ which is contrary to the assumption of the lemma. Therefore the equality $x_u = x'_u$ has to be valid. □

According to Lemma 4.13 for arbitrary $x_s \in B^s_\varepsilon$ the set $(\pi_s)^{-1}(x_s) \cap W^s_\varepsilon(f, 0)$ consists of a single point. Thus, we obtain the mapping $h : B^s_\varepsilon \to E^u$ for which there is $W^s(f, 0) = \text{graph}(h)$. However, the information on the order of differentiability of the mapping h does not follow from this lemma. In the proof of Proposition 4.12 we show that such a mapping can be defined as a solution of some implicit equation of the class C^r which can be solved with the help of Implicit Function Theorem which means that this mapping is also of the class C^r.

Proof of Proposition 4.12

We can assume that f is of the form (4.1). Let us assume that ε, δ and a be such positive numbers that the inequalities (4.3), (4.4) and (4.5) are fulfilled. If $z \in W^s_\varepsilon(f, 0)$, then $f^m(z) \in B_\varepsilon$ for all $m \geq 0$ and therefore we have $f^m(z) = ((f^m(z))_s,$

$$(f^m(z))_u) = \left((f^m(z))_s, (A^u)^m \left[z_u + \sum_{i=0}^{m-1}(A^u)^{-1-i}\Phi^u(f^i(z))\right]\right).$$

179

As $f^m(z) \in B_\varepsilon$ for all $m \geq 0$ and the mapping A^u is expansive, we obtain that

$$\lim_{m \to \infty} \sum_{i=0}^{m-1} (A^u)^{-1-i} \Phi^u(f^i(z)) = -z_u, \qquad (4.8)$$

and therefore we can write z in the form

$$z = \left(z_s, -\sum_{i=0}^{\infty} ((A^u)^{-1-i} \Phi^u(f^i(z))) \right).$$

If we put $f^m(z)$ instead of z in the last equality, we obtain

$$f^m(z) = \left((A^s)^m z + \sum_{i=0}^{m-1} (A^s)^{m-1-i} \Phi^s(f^i(z)), -\sum_{i=m}^{\infty} (A^u)^{m-1-i} \Phi^u(f^i(z)) \right). \qquad (4.9)$$

We prove that for each $x \in B_\varepsilon^s$ there exists a unique sequence $\{\alpha(x)(m)\}_{m=0}^{\infty}$ of elements from R^n such that

$$\lim_{m \to \infty} \alpha(x)(m) = 0 \qquad (4.10)$$

and for all $m \geq 0$ the equality

$$\alpha(x)(m) = \left((A^s)^m x + \sum_{i=0}^{m-1} (A^s)^{m-1-i} \Phi^s(\alpha(x)(i)), \right.$$
$$\left. -\sum_{i=m}^{\infty} (A^u)^{m-1-i} \Phi^u(\alpha(x)(i)) \right) \qquad (4.11)$$

holds (compare to (4.9)). For $m = 0$ we obtain that

$$\alpha(x)(0) = \left(x, -\sum_{i=0}^{\infty} (A^u)^{-1-i} \Phi^u(\alpha(x)(i)) \right)$$

and let us define the mapping

$$h : B_\varepsilon^s \to E^u, \qquad h(x) = -\sum_{i=0}^{\infty} (A^u)^{-1-i} \Phi^u(\alpha(x)(i)). \qquad (4.12)$$

From (4.10) it follows that $\alpha(x)(m+1) = (A+\Phi)(a(x)(m)) = f(a(x)(m))$. We obtain for $m = 0$ that $\alpha(x)(1) = f(\alpha(x)(0)) = f(x, h(x))$ and by induction we obtain that $\alpha(x)(m) = f^m(x, h(x))$ for all $m > 0$. From (4.10) it follows that $\lim_{m \to \infty} f^m(x, h(x)) = 0$. It means that if $z \in \text{graph}(h)$, then $\lim_{m\to\infty} f^m(z) = 0$. If $z \in B_\varepsilon^s$, then it follows from Lemma 4.13 that $z \in \text{graph}(h)$. As $f(\text{graph}(h)) \subset \text{graph}(h)$, we obtain that $W_\varepsilon^s(f,0) = \text{graph}(h)$. It remains to be proved that there exists just one sequence $\{\alpha(x)(m)\}_{m=0}^\infty$ which has the properties of (4.10) (4.11) for each $x \in B_\varepsilon^s$ and that the mapping h, defined by relation (4.12), is of the class C^r. Let us denote

$$G = \left\{\alpha = \{\alpha(m)\}_{m=0}^\infty : \alpha(m) \in R^n, m \geq 0, \lim_{m\to\infty} \alpha(m) = 0\right\} \text{ and}$$

$$G_\varepsilon = \{\alpha = \{\alpha(m)\}_{m=0}^\infty : \alpha(m) \in B_\varepsilon \text{ for all } m \geq 0\}.$$

The space G with the norm $\|\alpha\| = \sup_{m\geq 0} \|\alpha(m)\|$ is a Banach space. The set G_ε is an open subset of the Banach space

$$H = \left\{\alpha = \{\alpha(m)\}_{m=0}^\infty : \alpha(m) \in R^n, m \geq 0, \sup_{m\geq 0} \|\alpha(m)\| < \infty\right\}.$$

Let us define the mapping $F : B_\varepsilon^s \times G_\varepsilon \to G$, $F(x,\alpha) = \beta$, $x \in B_\varepsilon^s$, $\alpha = \{\alpha(m)\}_{m=0}^\infty \in G_\varepsilon$, $\beta = \{\beta(x)(m)\}_{m=0}^\infty$, where $\beta(x)(m) = \alpha(m) - \gamma(x)(m)$, $m \geq 0$, $\gamma(x)(m) = (\gamma_1(x)(m), \gamma_2(m))$,

$$\gamma_1(x)(m) = (A^s)^m x + \sum_{i=0}^{m-1}(A^s)^{m-1-i}\Phi^s(\alpha(i)),$$

$$\gamma_2(m) = \sum_{i=m}^{\infty}(A^u)^{m-1-i}\Phi^u(\alpha(i))$$

(compare $\beta(x)(m)$ to (4.11)). For arbitrary $x \in B_\varepsilon^s$ and arbitrary $j \in N$ such that $0 \leq j \leq m$ there is

$$\|\gamma_1(x)(m)\| \leq a^m\varepsilon + a^{m-1}\|\Phi^s(\alpha(0))\| + \cdots + a^{m-j}\|\Phi^s(\alpha(m-j-1))\|$$
$$+ \cdots + \|\Phi^s(\alpha(m-1))\|$$
$$\leq a^m\varepsilon + a^{m-j}(1-a)^{-1}b + (1-a)^{-1}\sup_{i\geq j}\|\Phi^s(\alpha(i))\|,$$

where $b = \sup_{z\in B_\varepsilon^s}\|\Phi^s(z)\|$. Let $\varkappa$ be an arbitrary positive number. As $\lim_{i\to\infty} \alpha(i) = 0$ there exists $j \in N$ such that $\sup_{i\geq j}\|\Phi^s(\alpha(i))\| < \varkappa/2$. As $a < 1$ there is $\tilde{m} \in N$ such that $a^m\varepsilon + a^{m-j}(1-a)^{-1}b < \varkappa/2$ for all $\tilde{m} \geq \tilde{m}$. Thus, we obtain that for arbitrary $\varkappa > 0$ there exists $\tilde{m} \in N$ such that for all $m \geq \tilde{m}$ there is $\|\gamma_1(x)(m)\| < \varkappa$ which means that $\lim_{m\to\infty}\gamma_1(x_m)(m) = 0$ for all $x \in B_\varepsilon^s$. As

$$\lim_{m\to\infty}\|\gamma_2(m)\| \leq \lim_{m\to\infty}(1-a)^{-1}\sup_{i\geq m}\|\Phi^u(\alpha(i))\| = 0,$$

thus $\lim_{m\to\infty} \beta(x)(m) = 0$ for all $x \in B_\varepsilon^s$ and so $\beta \in G$. From C^r-differentiability of the mapping Φ the C^r-differentiability of the mapping F follows (we leave the proof to the reader, see [122]) and it holds $(\partial_2)_{(x,\alpha)}F(\alpha') = \alpha' - \beta'(x)$, where

$$\beta'(x)(m) = (\beta_1(x)(m), \beta_2(x)(m)),$$

$$\beta_1(x)(m) = \sum_{i=0}^{m-1}(A^s)^{m-1-i}d_{\alpha(i)}\Phi^s(\alpha'(i)),$$

$$\beta_2(x)(m) = -\sum_{i=m}^{\infty}(A^u)^{m-1-i}d_{\alpha(i)}\Phi^u(\alpha'(i)).$$

Thus, we obtain that $F \in C^r$, $F(0,0) = 0$ and $(\partial_2)_0 F = \text{id}_G$. According to the Implicit Function Theorem for $\varepsilon > 0$ small enough there exists a C^r-mapping $\alpha : B_\varepsilon^s \to G$ such that $\alpha(0) = 0$, $F(x, \alpha(x)) = 0$ for all $x \in B_\varepsilon^s$. As $\alpha(x)(0) = (x, h(x))$ for all $x \in B_\varepsilon^s$ and α is of the class C^r, so $h \in C^r$ and thereby the first part of Proposition 4.12 is proved. As $W_\varepsilon^s(f, 0) = \text{graph}(h)$ it is sufficient to prove that $d_0 h = 0$. We define the mapping $\gamma : C^r(B_\varepsilon^s, G) \to C^r(B_\varepsilon^s, R^n)$, $\gamma(\Psi)(x) = \Psi(x)(0)$. Then $h = \pi_2 \circ \gamma \circ \alpha$, where $\pi_2 : R^n \to E^u$ is the projection. From the implicit equation $F(x, \alpha(x)) = 0$ we obtain that $d_0\alpha(v) = -(\partial_1)_{(0,0)}F(v) = -u$, where $u \in G$, $u(m) = (-(A^s)^m v, 0)$ for $m \geq 0$ and $v \in E^s$. As the mappings π_2 and γ are linear, for arbitrary $v \in E^s$ there is $d_0 h(v) = \pi_2 \circ \gamma \circ d_0 \alpha(v) = \pi_2((A^s)^0 v, 0)$ i.e. $d_0 h = 0$. $\square$

4.14 Proposition. *Let the asumptions of Proposition 4.12 be fulfilled. Then there exists a number $\varepsilon > 0$ and a C^r-diffeomorphism $k : B_\varepsilon^u \to E^s$ ($B_\varepsilon^u = E^u \cap B_\varepsilon$) such that $W_\varepsilon^u(f, 0) = \text{graph}(k)$ and $T_0(W_\varepsilon^u(f, 0)) = E^u$.*

Proof

As $W^u(f, 0) = W_\varepsilon^s(f^{-1}, 0)$ and the unstable invariable subspace of the mapping $d_0 f$ is equal to the stable invariable subspace of the mapping $d_0(f^{-1})$, Proposition 4.15 is a direct consequence of Proposition 4.12. $\square$

Assertion 2 of Theorem 4.7 follows directly from Proposition 4.12 and from Proposition 4.14.

Proof of Assertion 2 of Theorem 4.8

The assertion is of local character and therefore it is sufficient to assume that $X = R^n$. From the definition of hyperbolicity of a singular point and from the equality (3.30) it follows that x is a fixed point of the diffeomorphism φ_1, where φ_1 is the flow of the vector field F. We prove that $W^s(\varphi_1/U, x) = W^s(F/U, x)$ for a sufficiently small neighbourhood U

of the point x. The inclusion $W^s(F/U, x) \subset W^s(\varphi_1/U, x)$ is obvious. Let $y \in W^s(\varphi_1/U, x)$. When we use the Mean Value Theorem we obtain

$$\|\varphi_t(y) - x\| = \|\varphi_t(y) - \varphi_t(x)\| = \|\varphi_{t-[t]}(\varphi_{[t]}(y)) - \varphi_{t-[t]}(\varphi_{[t]}(x))\|$$
$$\leq \sup_{z \in U} \|d_z\varphi_{t-[t]}\|\|\varphi_{[t]}(y) - \varphi_{[t]}(x)\| \leq K\|\varphi_{[t]}(y) - x\|$$

for arbitrary $t \geq 0$, where U is a neighbourhood of the point x, small enough, $[t]$ is the integral part of the number t and $K > 0$. As $y \in W^s(\varphi_1/U, x)$, thus $\lim_{[t] \to \infty} \|\varphi_{[t]}(y) - x\| = 0$, thus $\lim_{t \to \infty} \|\varphi_t(y) - x\| = 0$. It means that $y \in W^s(F/U, x)$. The equality $W^u(F/U, x) = W^u(\varphi_1/U, x)$ can be proved analogously. Thus, assertion 2 of Theorem 4.8 is a consequence of assertion 2 of Theorem 4.7. □

4.15 Remark. If x is a hyperbolic periodic, or fixed point of the diffeomorphism $f \in \text{Diff}^r(X)$, then the invariant sets $W^s(f, x)$, $W^u(f, x)$ can be expressed by means of local invariant manifolds such as: $W^s(f, x) = \cup_{n \geq 0} f^{-n}(W^s_{\text{loc}}(f, x))$, $W^u(f, x) = \cup_{n \geq 0} f^n(W^u(f, x))$. Analogously, for the invariant sets of vector fields at the singular points $W^s(F, x) = \cup_{t \geq 0} \varphi_{-t}(W^s_{\text{loc}}(F, x))$, $W^u(F, x) = \cup_{t \geq 0} \varphi_t(W^u(F, x))$.

4.2 Centre Manifolds

4.16 Example. Let us consider a plane dynamical system defined by the system of differential equations

$$\dot{x} = -x, \quad \dot{z} = z^2 \qquad (4.13)$$

the singular point $(0,0)$ of which obviously, is not hyperbolic. The flow defined by this system is of the form

$$\varphi(x_0, z_0, t) = (e^{-t}x_0, -z_0(tz_0 - 1)^{-1}). \qquad (4.14)$$

Trajectories of the system (4.13) are solutions of the differential equation

$$\frac{dx}{dz} = -\frac{x}{z^2},$$

the general solution of which is of the form $x = ke^{z^{-1}}$, where k is a constant. Let us define the function

$$u_k(z) = ke^{z^{-1}} \quad \text{for } z < 0, \quad u_k(z) = 0 \quad \text{for } z \geq 0 \qquad (4.15)$$

for $k \in R$. This function is of the class C^r and therefore $M(k) = $ graph$(u_k) = \{(x,z) \in R^2 : x = u_k(z)\}$ is a C^∞-submanifold in R^2. Let $(x_0, z_0) \in M(k)$, i.e. $x_0 = u_k(z_0)$. If $z_0 \geq 0$, then $x_0 = u_k(z_0) = 0$ and therefore $\varphi_t(x_0, z_0) = (0, -z_0(tz_0 - 1)^{-1}) \in M(k)$ for all t such that $|t| < z_0^{-1}$. If $z_0 < 0$, then $x_0 = u_k(z_0) = ke^{z_0^{-1}}$, and thus, $e^{-t}x_0 = ke^{-t}e^{z_0^{-1}} = ke^{(-z_0(tz_0-1)^{-1})^{-1}}$, i.e. $\varphi_t(x_0, z_0) \in M(k)$ for all t such that $|t| < -z_0^{-1}$. We have proved that for every real number k, $M(k)$ is a local invariant manifold of the dynamical system φ in a neighbourhood of the point $(0,0)$ (Fig. 31), where the manifold $M(0)$ is obviously a global invariant manifold.

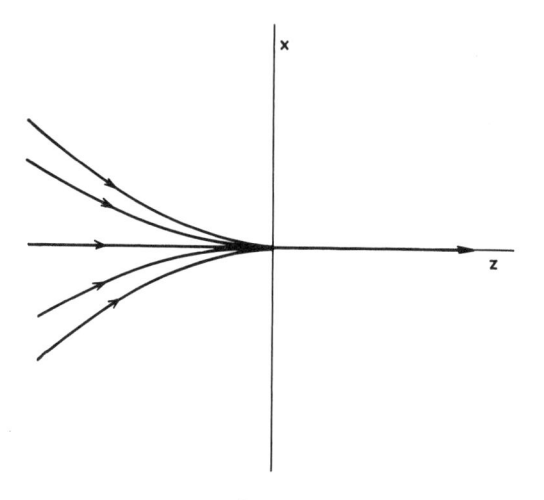

Fig. 31.

Let us consider the system of differential equations

$$\begin{aligned} \dot{x} &= A^- x + P(x,y,z), \\ \dot{y} &= A^+ y + Q(x,y,z), \\ \dot{z} &= A^0 z + R(x,y,z), \end{aligned} \quad (4.16)$$

where $x \in R^m$, $y \in R^n$, $z \in R^p$, $A^- \in M(m)$, $A^+ \in M(n)$, $A^0 \in M(p)$, the matrix $A^-(A^+, A^0)$ has all its eigenvalues with the negative (positive, zero) real part, P, Q, R are C^r-mappings, where $r \geq 1$ on an open set $W \subset R^m \times R^n \times R^p$, $P(0) = 0$, $Q(0) = 0$, $R(0) = 0$, $d_0 P = 0$, $d_0 Q = 0$, $d_0 R = 0$.

4.17 Definition. Let $k \in N$, $0 \leq k \leq r$ and there exist C^k-mappings $y = v^-(x)$, $z = w^-(x)$; $x = v^+(y)$, $z = w^+(y)$; $x = v^0(z)$, $y = w^0(z)$; $x = h^{0+}(y,z)$, $y = h^{0-}(x,z)$, having at the origin both zero values and zero derivatives (if $k \geq 1$) such that $W^s = \{(x, y, z) : y = v^-(x), z = w^-(x)\}$, $W^u = \{(x, y, z) : x = v^+(y), z = w^+(y)\}$, $W^c = \{(x, y, z) : x = v^0(z), y = w^0(z)\}$, $W^{cu} = \{(x, y, z) : x = h^{0+}(y, z)\}$, $W^{cs} = \{(x, y, z) : y : h^{0-}(x, z)\}$ are invariant sets of the system (4.16). Then W^s, W^u, W^c, W^{cu}, W^{cs} is called the stable, unstable, centre, centre unstable and centre stable invariant C^k-manifold (global), respectively, of the system (4.16). If these sets are defined in a certain small neighbourhood of the origin only, and they are just local invariant sets of the system (4.16) in this neighbourhood then they are called the stable, local, unstable etc. invariant C^k-manifold of the system (4.16) in the neighbourhood of the point 0 and we denote them in the same way as their corresponding global invariant manifold but with the index loc.

4.18 Remark. In Example 4.16 we have shown that the system (4.13) has an infinite number of centre C^∞-manifold, and thus, the manifolds $M(k)$, where k is an arbitrary real number. Let us note that out of all these manifolds it is just one, namely $M(0)$, where if $(x(t), z(t))$ is the solution of the system (4.13) and $(x(t), z(t)) \in M(k)$ for all $t \in R$, then $\sup_{t \in R} |x(t)| < \infty$. Another important property of these centre manifolds is that for arbitrary $k \in R$ it is

$$u_k(0) = \frac{d^j u_k(0)}{dz^j} = 0$$

for arbitrary natural number j, i.e., all centre manifolds $M(k)$, $k \in R$ have at the point $(0,0)$ a tangent of arbitrary high order. We shall see that both the above properties of these centre manifolds are of general character.

4.19 Remark. By analogy as we have defined stable, unstable and centre manifolds of singular points of differential equations the stable, unstable and centre manifolds of non-hyperbolic periodic or fixed points of diffeomorphisms can be defined. These definitions will not be defined separately.

In general, the non-linear system (4.16) need not have a global flow which reduces the chances of finding invariant manifolds. To prove the existence of local invariant manifolds of system (4.16) in a neighbourhood of the point 0 it is advantageous instead of this system to consider another system identical to system (4.16) in a certain neighbourhood of the point 0 and outside this neighbourhood it is defined in such a way that it has a global flow. This new system will obviously have in a small enough neighbourhood of the origin the same local invariant manifolds as the original one. Naturally, these systems can have different global invariant manifolds (if they exist for both the systems). Now we shall introduce one of the possible ways to define a modified system like this.

Let P, Q, R be a C^r-mapping from the right side of the system (4.16) defined on an open set $D \subset R^m \times R^n \times R^p$ containing the origin and let

Θ be an arbitrary positive number. Then according to Lemma 3.128 there exist the C^r-mappings $\tilde{P}$, $\tilde{Q}$, $\tilde{R}$ defined on $R^m \times R^n \times R^p$ and open sets D_1, D_2 containing the origin such that

1. $\overline{D}_1 \subset \overline{D}_2 \subset D$;
2. $\tilde{P}/D_1 = P/D_1$, $\tilde{Q}/D_1 = Q/D_1$, $\tilde{R}/D_1 = R/D_1$ and all the mappings $\tilde{P}$, $\tilde{Q}$, $\tilde{R}$ having zero values on the set $R^m \times R^n \times R^p \setminus D_2$;
3. $\|d_z \tilde{P}\|_1 \leq q\Theta$, $\|d_z \tilde{Q}\|_1 \leq q\Theta$, $\|d_z \tilde{R}\|_1 \leq q\Theta$ for all $z \in R^m \times R^n \times R^p$, where q is a positive constant independent both of the mappings P, Q, R and the number Θ. Thus, we arrive at the system

$$\begin{aligned} \dot{x} &= A^- x + \tilde{P}(x, y, z), \\ \dot{y} &= A^+ y + \tilde{Q}(x, y, z), \\ \dot{z} &= A^0 z + \tilde{R}(x, y, z), \end{aligned} \qquad (4.17)$$

called the modification of the system (4.16). From the property 3 of the mappings $\tilde{P}, \tilde{Q}, \tilde{R}$ we get that $\tilde{P}, \tilde{Q}, \tilde{R} \in \mathrm{Lip}(q\Theta)$. This means that the mappings $\tilde{P}, \tilde{Q}, \tilde{R}$ can be chosen as globally Lipschitz with an arbitrary small Lipschitz constant. From Theorem 1.42 it follows that there exists a global C^r-flow of the system (4.17).

Now we shall find an implicit relationship that must be fulfilled by the C^k-mappings $x = h_1(z)$, $y = h_2(z)$, defining the C^k-centre manifold W^c of the system (4.17) if such a manifold exists (in Definition 4.17 it is $h_1 = v^0$, $h_2 = w^0$). Let

$$\varphi(x, y, z, t) = (X(x, y, z, t), Y(x, y, z, t), Z(x, y, z, t))$$

be a C^r-flow of the system (4.17). From the invariance of the set W^c and from the variation of constants formula we get that

$$\frac{dz(t)}{dt} = A^0 z(t) + \tilde{R}(h(z(t)), z(t)), \qquad (4.18)$$

$$h_1(z(t)) = x(t) = e^{A^-(t-t_0)} h_1(z) + \int_{t_0}^{t} e^{-A^-(t-s)} \tilde{P}(h(z(s)), z(s))\, ds, \qquad (4.19)$$

$$h_2(z(t)) = y(t) = e^{A^+(t-t_0)} h_2(z) + \int_{t_0}^{t} e^{A^+(t-s)} \tilde{Q}(h(z(s)), z(s))\, ds \qquad (4.20)$$

for all $t, t_0 \in R$, where $h = (h_1, h_2)$, $x(t) = X(h(z), z, t,)$, $y(t) = Y(h(z), z, t)$, $z(t) = Z(h(z), z, t)$, $z \in R^p$, $t \in R$. From the definition of the mapping $z = z(t)$ and from the equality (4.18) we get that the mapping $\Psi_h : R^p \times R \to R^p$, $\Psi_h(z, t) = Z(h(z), z, t)$ represents a C^k-flow of the differential equation

Invariant Manifolds

$$\dot{z} = A^0 z + R(h(z), z). \tag{4.21}$$

From the properties of the mappings $\tilde{P}$, $\tilde{Q}$ and Theorem 3.84 it follows that for arbitrary continuous mappings $u = u(t)$, $v = v(t)$, $w = w(t)$ on the interval $(-\infty, 0)((0, \infty))$ with images in R^m, R^n and R^p, respectively, there exists the integral

$$\int_{-\infty}^{0} e^{-A^{-}s} \tilde{P}(u(s), v(s), w(s)) \, ds \left(\int_{0}^{\infty} e^{-A^{+}s} \tilde{Q}(u(s), v(s), w(s)) \, ds \right),$$

and thus in the inequalities (4.19), (4.20) we can put $t = 0$ and pass to the limit for $t_0 \to -\infty$ in the equality (4.19) and for $t_0 \to \infty$ in the equality (4.20). Thus, we obtain that if there exists a centre C^k-manifold of the system (4.17) of the form $W^c = \{(x, y, z) : x = h_1(z), y = h_2(z)\}$, where $h_1 : R^p \to R^m$, $h_2 : R^p \to R^n$ are C^k-mappings then these mappings have to satisfy the system of equations as follows

$$h_1(z) = \int_{-\infty}^{0} e^{-A^{-}s} \tilde{P}(h(\Psi_h(z,s)), \Psi_h(z,s)) \, ds,$$

$$h_2(z) = -\int_{0}^{\infty} e^{-A^{+}s} \tilde{Q}(h(\Psi_h(z,s)), \Psi_h(z,s)) \, ds, \tag{4.22}$$

for all $z \in R^p$, where $h = (h_1, h_2)$, Ψ_h is a C^k-flow of the system (4.21). The system (4.21) can be viewed upon as a system of equations with the unknowns h_1, h_2. The equations (4.22) can be written as one equation

$$h(z) = \int_{-\infty}^{\infty} K(s) F(h(\Psi_h(z,s)), \Psi_h(z,s)) \, ds, \tag{4.23}$$

where $K : (-\infty, \infty) \to M(m+n)$, $K(t) = \text{diag}\{K_1(t), K_2(t)\}$, $K_1(t) = 0$, if $t > 0$, $K_1(t) = e^{-A^{-}t}$, if $t \leq 0$, $K_2(t) = -e^{-A^{+}t}$, if $t \geq 0$ and $K_2(t) = 0$, if $t < 0$, $F = (\tilde{P}, \tilde{Q})$. Let us define the mapping $\mathscr{F} : C_B^0 \to C_B^0$,

$$\mathscr{F}(h)(z) = \int_{-\infty}^{\infty} K(s) F(h(\Psi_h(z,s)), \Psi_h(z,s)) \, ds, \tag{4.24}$$

where $C_B^0 = C_B^0(R^p, R^{m+n})$ is a Banach space of continuous bounded mappings from R^p into R^{m+n} with the norm $\|f\|_0 = \sup_{z \in R^p} \|f(z)\|$. Let us denote as $X_L = X_L(R^p, R^{m+n})$, the set of all Lipschitz mappings from R^p into R^{m+n} with the Lipschitz constant L and such that $h(0) = 0$ for all $h \in X_L$. Obviously, the set X_L is a closed subset in C_B^0, and thus, it forms a complete metric space with the metrics $\varrho(f, g) = \|f - g\|_0$.

4.20 Lemma. *If the number Θ from definition of the system (4.17) is small enough, then the mapping $\mathscr{F}$ has a unique fixed point $h \in X_L(R^p, R^{m+n})$.*

Proof

First, we shall introduce some necessary inequalities. From the properties of the matrices A^-, A^+ and Lemma 3.84 it follows that there exist positive constants M, γ such that

$$\|K(t)\| \leq M \cdot e^{-\gamma|t|} \tag{4.25}$$

for all $t \in R$. Since the matrix A^0 has all its eigenvalues with zero real parts, from Lemma 3.84 it follows that for arbitrary $\beta > 0$ there exists a number $N(\beta)$, such that

$$\|e^{A^0 t}\| \leq N(\beta) \cdot e^{\beta|t|} \tag{4.26}$$

for all $t \in R$. Since $\tilde{P}$, $\tilde{Q}$ are globally Lipschitz with the Lipschitz constant $q \cdot \Theta$, then the mapping F is globally Lipschitz with the Lipschitz constant $\varrho(\Theta) = K \cdot q \cdot \Theta$, where $K > 0$. From this property of the mapping F, inequality (4.26) and equation (4.21) it follows that for arbitrary $(z,t) \in R^p \times R$, $h, h' \in C_B^0(R^p, R^{m+n})$ we have

$$\|\Psi_h(z,t) - \Psi_{h'}(z,t)\|$$
$$\leq \int_0^t \|e^{a^0(t-s)}\| \varrho(\Theta)[\|h - h'\|_0 + \|\Psi_h(z,s) - \Psi_{h'}(z,s)\|] \, ds$$
$$\leq \int_0^t N(\beta) e^{\beta|t-s|} \varrho(\Theta)[\|h - h'\|_0 + \|\Psi_h(z,s) - \Psi_{h'}(z,s)\|] \, ds$$

for all $t \in R$. From this inequality it follows that for all $t \in R$

$$e^{-\beta|t|}\|\Psi_h(z,t) - \Psi_{h'}(z,t)\| \leq (\varrho(\Theta)N(\beta)|\int_0^t e^{-\beta|s|} \, ds|)\|h - h'\|_0$$
$$+ N(\beta)\varrho(\Theta)|\int_0^t e^{-\beta|s|}\|\Psi_h(z,s) - \Psi_{h'}(z,s)\| \, ds|.$$

From the Gronwall inequality it follows that for all $t \in R$

$$e^{-\beta|t|}\|\Psi_h(z,t) - \Psi_{h'}(z,t)\| \leq \tilde{\Gamma}(\Theta,\beta,t)\|h - h'\|_0,$$

i.e. $\|\Psi_h(z,t) - \Psi_{h'}(z,t)\| \leq \Gamma(\Theta,\beta,t)\|h - h'\|_0$, where

$$\Gamma(\Theta,\beta,t) = \tilde{\Gamma}(\Theta,\beta,t)e^{\beta|t|},$$

$$\tilde{\Gamma}(\Theta,\beta,t) = \varrho(\Theta)N(\beta)\int_0^t e^{-\beta|s|} \, ds$$
$$+ \left|\int_0^t e^{N(\beta)\varrho(\Theta)|t-s|}\varrho(\Theta)N(\beta)\left(\int_0^s e^{-\beta|v|} \, dv\right) ds\right|.$$

With the use of this inequality and inequality (4.25) we get that for arbitrary $z \in R^p$, $h, h' \in X_L(R^p, R^{m+n})$

$$\|\mathscr{F}(h)(z) - \mathscr{F}(h')(z)\|$$
$$\leq \int_{-\infty}^{\infty} \|K(s)\|\varrho(\Theta)(\|h - h'\|_0 + \|\Psi_h(z,s) - \Psi_{h'}(z,s)\|)\,ds$$
$$\leq \sigma(\Theta)\|h - h'\|_0,$$

where

$$\sigma(\Theta) = M\varrho(\Theta)\omega(\beta,\gamma), \omega(\beta,\gamma) = \int_{-\infty}^{\infty} e^{-\gamma|s|}\Gamma(\Theta,\beta,s)\,ds.$$

The numbers β, Θ can be chosen arbitrarily small. Since γ is a fixed positive number, for small enough $\beta > 0$ it is $0 < \omega(\beta, \gamma) < \infty$. If $\Theta > 0$ is a number such that $\varrho(\Theta) < (M \cdot \omega(\beta, \gamma))^{-1}$, then $\varrho(\Theta) < 1$. This means that for the numbers β, Θ, chosen like this, $\|\mathscr{F}(h) - \mathscr{F}(h')\|_0 \leq \sigma(\Theta)\|h - h'\|_0$. Thus, the mapping $\mathscr{F}$ is contractive and for arbitrary $h \in C_B^0$ it is $\mathscr{F}(h) \in C_B^0$. Now we shall prove that $\mathscr{F}(X_L) \subset X_L$. Let $h \in X_L$. Then for arbitrary $x, y \in R^p$

$$\|\mathscr{F}(h)(x) - \mathscr{F}(h)(y)\| \leq \int_{-\infty}^{\infty} \|K(s)\| \cdot \|F(h(\Psi_h(x,s)), \Psi_h(x,s))$$
$$- F(h(\Psi_h(y,s)), \Psi_h(y,s))\|\,ds.$$

Since the mapping F is Lipschitz with the Lipschitz constant $\varrho(\Theta)$, when using the inequality (4.25) we get

$$\|\mathscr{F}(h)(x) - \mathscr{F}(h)(y)\|$$
$$\leq \tilde{M}(\Theta)\left|\int_{-\infty}^{\infty} e^{-\gamma|s|}\|\Psi_h(x,s) - \Psi_h(y,s)\|\,ds\right|, \quad (4.27)$$

where $\tilde{M}(\Theta) = M(L+1)\varrho(\Theta)$. From the equality (4.21) and inequality (4.26) the inequality

$$\|\Psi_h(x,t) - \Psi_h(y,t)\| \leq \|e^{A^0 t}\|\|x - y\|$$
$$+ \varrho(\Theta)\left|\int_0^t \|e^{A^0|t-\tau|}\|(M+1)\|\Psi_h(x,\tau) - \Psi_h(y,\tau)\|\,d\tau\right|$$
$$\leq N(\beta)e^{\beta|t|}\|x - y\| + r(\beta,\Theta)\left|\int_0^t e^{\beta|t-\tau|}\|\Psi_h(x,\tau) - \Psi_h(y,\tau)\|\,d\tau\right|$$

follows, where $r(\beta,\Theta) = \varrho(\Theta)N(\beta)(M+1)$. If $t \geq 0$, then for $u(t) = \|\Psi_h(x,t) - \Psi_h(y,t)\|$ we get the inequality

$$e^{-\beta t}u(t) \leq N(\beta)\|x - y\| + r(\beta,\Theta) \cdot \int_0^t e^{-\beta\tau}u(\tau)\,d\tau.$$

With the use of the Gronwall Lemma we get the inequality

$$u(t) \leq N(\beta)\|x - y\|e^{\tilde{r}(\beta,\Theta)t} \tag{4.28}$$

for all $t \geq 0$, where $\tilde{r}(\beta,\Theta) = r(\beta,\Theta) + \beta$. If $t < 0$, then

$$u(t) \leq N(\beta)e^{-\beta t}\|x - y\| + r(\beta,\Theta)\left|\int_0^t e^{-\beta t}e^{\beta\tau}u(\tau)\,d\tau\right|,$$

which yields the inequality

$$e^{\beta t}u(t) \leq N(\beta)\|x - y\| + r(\beta,\Theta)\left|\int_0^t e^{\beta\tau}u(\tau)\,d\tau\right|.$$

If, again, the Gronwall inequality is used, we obtain $u(t) \leq N(\beta)\|x - y\|e^{\tilde{r}(\beta,\Theta)(-t)}$ for all $t < 0$. From this inequality and the inequality (4.28) we get the inequality

$$\|\Psi_h(x,t) - \Psi_h(y,t)\| \leq N(\beta)\|x - y\|e^{\tilde{r}(\beta,\Theta)|t|} \tag{4.29}$$

for all $t \in R$. After substituting this inequality into the inequality (4.27) we get that $\|\mathscr{F}(h)(x) - \mathscr{F}(h)(y)\| \leq \sigma(\beta,\Theta,\gamma)\|x - y\|$, where

$$\sigma(\beta,\Theta,\gamma) = M(\Theta)N(\beta) \cdot \int_{-\infty}^{\infty} e^{-(\gamma-\tilde{r}(\beta,\Theta))|s|}\,ds.$$

Since for every number β with $\lim_{\Theta \to 0} \tilde{r}(\beta,\Theta) = 0$, there exists a positive number Θ such that $\gamma - \tilde{r}(\beta,\Theta) > 0$ and for such Θ it is $\sigma(\beta,\Theta,\gamma) < \infty$. Since both $\lim_{\Theta \to 0} M(\Theta) = 0$, and $\lim_{\Theta \to 0} \sigma(\beta,\Theta,\gamma) = 0$, thus, there exists a positive number Θ, such that $\sigma(\beta,\Theta,\gamma) < \max(1,L)$. Since the Lipschitz constant L is in common for all $h \in X_L$, in fact, we have proved that for a positive number Θ small enough $\mathscr{F}(X_L) \subset X_L$ and $\mathscr{F}/X_L$ is contractive. Now the proposition of the lemma immediately follows from the Banach Fixed Point Theorem. □

From Lemma 4.20 the next proposition follows.

4.21 Proposition.
1. *If Θ is a small enough positive number, then the system (4.17) has a unique global centre C^0-manifold $W^c = \{(x, y, z) : x = h_1(z), y = h_2(z)\}$, where $h = (h_1, h_2)$ is a bounded, global Lipschitz mapping;*
2. *The system (4.16) has at least one local centre C^0-manifold.*

4.22 Remark. The system (4.17) can possess several global centre manifolds but just one of them is defined by a bounded mapping. The system (4.13) from Example 4.16 is defined globally, it has an infinite number of global C^∞-manifolds but only one of them, namely $M(0) = \{(x, z) : x = h(z) \equiv 0\}$, is defined by a bounded function. Which of the local center manifolds of the system (4.16) is lying on a global centre manifold of the system (4.17) from Proposition 4.21 depends upon the choice of modification of the system (4.16).

In an analogous way to that used for proving the existence of the local centre C^0-manifold of the system (4.16) but technically much more complicated, the existence of the centre C^k-manifolds of this system is used to prove $1 \leq k \leq r - 1$. The complete proof of existence of a centre C^{r-1}-manifold for a system of the class C^r can be found in [2] or [79]. In the proof of the existence of a centre C^k-manifold for $k \geq 1$, the existence of a fixed point of the mapping $\mathscr{F} : C^k(R^p, R^{m+n})$ (see (4.24)) is to be proved, but in the class of C^k-mappings. In the proof of contractivity of the mapping $\mathscr{F}$ with respect to the metric in $C^k(R^p, R^{m+n})$ a suitable modification of the system is to be found, as well as suitable estimates for derivatives of the mapping $\mathscr{F}$ up to order k, which is technically rather exacting. For $k = n$, on searching for suitable estimates considerable complications arise, therefore it was only recently that the existence of a centre C^{k-1}-manifold was proved. In [158] the proof of existence of the centre C^r-manifold is presented. This proof is clever because it does not require estimations of norms of derivatives of the mapping $\mathscr{F}$ up to order. In spite of its relative simplicity it is quite extensive and for these reasons it is not presented. Thus, the following theorem holds.

4.23 Centre Manifold Theorem. *Let $1 \leq r < \infty$, $W \subset R^m \times R^n \times R^p$ is an open set containing the origin, $P \in C^r(W, R^m)$, $Q \in C^r(W, R^n)$, $R \in C^r(W, R^p)$, $P(0) = 0$, $d_0 P = 0$, $Q(0) = 0$, $d_0 Q = 0$, $R(0) = 0$, $d_0 R = 0$, the matrix $A^- \in M(m)$, ($A^+ \in M(n)$; $A^0 \in M(p)$) has all eigenvalues with negative (positive, zero) real part. Then the following assertions hold.*
1. *There exists the local stable (unstable) C^r-manifold $W^s_{loc} = \{(x, y, z) \in U : y = v^-(x), z = w^-(x)\}$ ($W^u_{loc} = \{(x, y, z) \in U : x = v^+(y), z = w^+(y)\}$) of the system (4.16), where $U \subset W$ is an open neighbourhood of the origin, $v^-(0) = 0$, $d_0 v^- = 0$, $w^-(0) = 0$, $d_0 w^- = 0$ ($v^+(0) = 0$, $d_0 v^+ = 0$, $w^+(0) = 0$, $d_0 w^+ = 0$).*
2. *There exists the local centre (centre stable, centre unstable)*

C^r-manifold $W^c_{loc} = \{(x, y, z) \in U : x = v^0(z), y = w^0(z)\}$ ($W^{cs}_{loc} = \{(x, y, z) \in U : x = h^{0-}(y, z)\}$; $W^{cu}_{loc} = \{(x, y, z) \in U : y = h^{0+}(x, z)\}$) of the system (4.16), where $U \subset W$ is an open neighbourhood of the origin, $v^0(0) = 0$, $d_0 v^0 = 0$, $w^0(0) = 0$, $d_0 w^0 = 0$ ($h^{0-}(0, 0) = 0$, $d_{(0,0)} h^{0-} = 0$, $h^{0+}(0, 0) = 0$, $d_{(0,0)} h^{0-} = 0$).

4.24 Corollary. *If E^s, E^c, E^u is a stable, centre and unstable invariant subspace, respectively, of linearization of the system (4.16), then $T_0(W^s_{loc}) = E^s$, $T_0(W^c_{loc}) = E^c$, $T_0(W^u_{loc}) = E^u$, i.e. $T_0(W^s_{loc}) \oplus T_0(W^c_{loc}) \oplus T_0(W^u_{loc}) = T_0(R^n)$ (Fig. 32).*

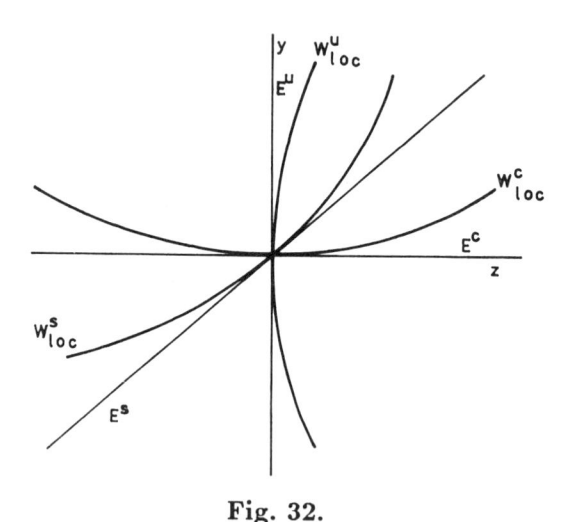

Fig. 32.

4.25 Definition. *Let $W^c_{loc} = \{(x, y, z) : x = h_1(z), y = h_2(z)\}$ be a local centre C^r-manifold of the system (4.16). The differential equation*

$$\dot{z} = A^0 z + R(h(z), z), \qquad (4.30)$$

where $h = (h_1, h_2)$, is called the reduction of the system (4.16) to the centre manifold W^c_{loc}.

If the system (4.16) is defined by analytical mappings, then it need not have a local centre manifold defined by analytical mappings, which is also the case in the following example.

4.26 Example. Let us consider the system of differential equations

$$\dot{x} = -x + z^2, \qquad (4.31)$$

$$\dot{z} = -z^3 \tag{4.32}$$

and let us suppose that it has the centre manifold $W^c = \{(x, z) : x = h(z)\}$, where $h(z) = \sum_{n=2}^{\infty} a_n z^n$ is an analytical function. The mapping h is defined by the equation (4.32) as follows

$$h(z) = \int_{-\infty}^{0} e^s (\Psi(s))^2 \, ds, \tag{4.33}$$

where $\Psi(t)$ is a solution of the differential equation (4.32) satisfying the initial condition $\Psi(0) = z$. By separating the variables in the equation (4.32) it can be shown that

$$(\Psi(s))^2 = z^2(2z^2 s + 1)^{-1} = \sum_{n=2}^{\infty} (-1)^n 2^n s^n z^{2n},$$

and thus,

$$h(z) = \left[\sum_{n=0}^{\infty} (-2)^n \int_{-\infty}^{0} e^s s^n \, ds\right] z^{2n+2}.$$

That is $a_2 = 1$, $a_{2n+1} = 0$ and $a_{2n+2} = n(n-1)a_{2n}$ for all $n \in N$, and thus, the function h cannot be analytical.

If the system (4.16) is defined by the mappings of class C^{∞}, then it need not have a local centre C^{∞}-manifold, the following example of the system of differential equations demonstrates this

$$\dot{x} = -x + z_2^2, \dot{z}_1 = 0, \dot{z}_2 = -z_1 z_2 - z_2^3. \tag{4.34}$$

The proof of non-existence of the local centre C^{∞}-manifold for this system can be found in [146] and it will not be presented here.

In the bifurcation theory of vector fields a certain modification of the Centre Manifold Theorem for parametrized differential equations is needed that will now be formulated here.

Let us consider the parametrized system of differential equations

$$\begin{aligned} \dot{x} &= A^- x + P(x, y, z, \varepsilon), \\ \dot{y} &= A^+ y + Q(x, y, z, \varepsilon), \\ \dot{z} &= A^0 z + R(x, y, z, \varepsilon), \end{aligned} \tag{4.35}$$

where the matrices A^-, A^+, A^0 have the same properties as in Theorem 4.23, $x \in R^m$, $y \in R^n$, $z \in R^p$, $\varepsilon \in R^k$ is the parameter, $P, Q, R \in C^r$. If to the system (4.35) the equation

$$\dot{\varepsilon} = 0, \tag{4.36}$$

is added, we arrive at the system of the form (4.16) in which instead of the state variable z there is the state variable (z, ε) and in place of the matrix A^0 the diag $\{A^0, 0\}$ is standing, where $0 = 0_k \in M(k)$.

4.27 Definition. *Let there exist C^k-mappings $(0 \leq k \leq r)$ $x = v^0(z, \varepsilon)$, $y = w^0(z, \varepsilon)$ defined on some neighbourhood of the origin, which have at the origin both zero values and zero derivatives and such that $V_{loc}^c = \{(x, y, z, \varepsilon) : x = v^0(z, \varepsilon), y = w^0(z, \varepsilon)\}$ is an invariant set of the system (4.35), (4.36). Then the set V_{loc}^c is called the local centre C^k-manifold of the parametrized system (4.35).*

Now, as a direct consequence of Theorem 4.23 the following theorem can be formulated.

4.28 Parametrized Centre Manifold Theorem. *Let $1 \leq r < \infty$, $W \subset R^m \times R^n \times R^p \times R^k$ be an open set containing the origin, $P \in C^r(W, R^m)$, $Q \in C^r(W, R^n)$, $R \in C^r(W, R^p)$, $P(0) = 0$, $d_0 P = 0$, $Q(0) = 0$, $d_0 Q = 0$, $R(0) = 0$, $d_0 R = 0$ and the matrix $A^-(A^+; a^0)$ has all its eigenvalues with the negative (positive, zero) real part. Then there exists the local center C^r-manifod $V_{loc}^c = \{(x, y, z, \varepsilon) : x = v^0(z, \varepsilon), y = w^0(z, \varepsilon)\}$ of the parametrized system (4.35), where $v^0, w^0 \in C^r$, $v^0(0, 0) = 0$, $d_{(0,0)} v^0 = 0$, $w^0(0, 0) = 0$, $d_{(0,0)} w^0 = 0$.*

4.29 Definition. *Let $V_{loc}^c = \{(x, y, z, \varepsilon) : x = h_1(z, \varepsilon), y = h_2(z, \varepsilon)\}$ be a local centre C^r-manifold of the parametrized system (4.35). Then the parametrized differential equation.*

$$\dot{z} = A^0 z + R(h(z, \varepsilon), z, \varepsilon), \qquad (4.37)$$

where $h = (h_1, h_2)$, is called the reduction of the system (4.35) to the centre manifold V_{loc}^c or also the bifurcation equation of the system (4.35).

The next theorem that was proved by A. N. Šošitaišvili in [150] can be taken for a generalization of the Grobman–Hartmann Theorem. The proof of this theorem is quite complicated, and thus, it is not presented here.

4.30 Šošitaišvili Theorem. *Let the assumptions of Theorem 4.28 be fulfilled. Then there exists a centre C^r-manifold $V_{loc}^c = \{(x, y, z, \varepsilon) : x = h_1(z, \varepsilon), y = h_2(z, \varepsilon)\}$ of the parametrized system (4.35) such that the germ at the point 0, represented by this system is orbitally topologically equivalent to the germ at the point 0 represented by the following parametrized system of differential equations*

$$\dot{x} = -x, \quad \dot{y} = y, \qquad (4.38)$$

$$\dot{z} = A^0 z + R(h(z, \varepsilon), z, \varepsilon), \qquad (4.39)$$

where $h = (h_1, h_2)$, $x \in R^m, y \in R^n$.

Let us note that according to the original Šošitaišvili Theorem from [150] the local centre manifold V_{loc}^c is just of the class C^{r-1}. According to

[158] this centre manifold is of the class C^r, i.e. of the same class as the parametrized system (4.35).

The linear system (4.38) is independent both of z and ε, it is hyperbolic and Theorems 3.84, 3.87 give a complete image of the topological structure of its trajectories. This means that bifurcations of the parametrized system (4.35) in a neighbourhood of the origin small enough and for ε small enough are uniquely defined by bifurcations of the parametrized differential equation (4.39), and thus, its name the bifurcation equation is really feasible. The Šošitaišvilli Theorem in the literature is also called the principle of reduction onto centre manifold.

4.31 Remark. The parametrized system (4.38), (4.39) has the local centre manifold $V^c_{\text{loc}} = \{(x, y, z, \varepsilon) : x = 0, y = 0\}$. From hyperbolicity of the linear system (4.38) it follows that if for some ε from a neighbourhood of the origin small enough there exists the periodic solution $(x(t), y(t), z(t))$ of the system (4.38), (4.39) such that its corresponding trajectory lies in a sufficiently small neighbourhood of the origin of the state space, then $x(t) \equiv 0, y(t) \equiv 0$ and $(0, 0, z(t), \varepsilon) \in V^c_{\text{loc}}$ for all $t \in R$. This means that periodic trajectories of the system (4.38), (4.39) for ε from a sufficiently small neighbourhood of the origin lying in a sufficiently small neighbourhood of the origin of the state space must lie on a submanifold of the state space defined by the equations $x = 0, y = 0$ and are uniquely determined by periodic trajectories of the bifurcation equation (4.39). When trying to find periodic trajectories of the bifurcation equation, for example using the method of normal forms, it is important that its right side had a high enough order of differentiability.

If the mapping $h = (h_1, h_2)$ is known and by defining the centre manifold V^c_{loc} from the Šošitaišvilli Theorem, then the problem of bifurcations of the system (4.35) is definitely reduced to the problem of bifurcations of the bifurcation equation (4.39). In general, however, this mapping cannot be found explicitly, so that the right side of the bifurcation equation is not expressed explicitly. However, bifurcations very often depend upon a finite number of the terms of the Taylor expansion of the right side of the bifurcation equation and for their determination it is sufficient to know a certain finite number of the Taylor expansion of the mapping h. Therefore the theorem on approximation of centre manifold, which will be formulated and proved below, is very important. The Theorem on Approximation of Centre Manifold for autonomous differential equations whose state space is a special infinite dimensional Banach space, is formulated and proved in the book [62] by D. Henry. With the use of such equations in Henry's book problems of geometrical theory of the so-called semi-linear parabolic differential equations and using the reduction on the centre manifold the bifurcation problems for such equations can be reduced to the bifurcation problems of one or just several scalar ordinary differential equations. The

Chapter 4

Theorem on Approximation of Centre Manifold was reformulated in the paper [28] by J. Carr for the systems of the form (4.16) but both the Henry's and the Carr's formulation of the theorem contains the assumption that linearization of the system does not have at the same time the eigenvalues with both the negative and the positive real part. Estimates from the proof of Lemma 4.20 enable us to prove the Theorem on Approximation of Centre Manifold for general form of the system (4.16).

Let $\Phi_1 : R^p \to R^m, \Phi_2 : R^p \to R^n$ be C^1-mappings and $\Phi = (\Phi_1, \Phi_2)$. Let us define the mapping $M\Phi : R^p \to R^{m+n}$,

$$(M\Phi)(z) = d_z\Phi(A^0 z + R(\Phi(z), z)) - B\Phi(z) - G(\Phi(z), z), \quad (4.40)$$

where $B = \text{diag}\{A^-, A^+\}$, $G = (P, Q)$, the matrices A^-, A^+, A^0 and the mappings P, Q, R are the mappings from the right side of the system (4.16) and have the properties as in Theorem 4.23.

4.32 Theorem on Approximation of Centre Manifold. *Let $W^c_{\text{loc}} = \{(x, y, z) : x = h_1(z), y = h_2(z)\}$ is a local centre C^r-manifold of the system (4.16), where $r \geq 2$, $h = (h_1, h_2)$. Let us assume that there exists a natural number $j > 1$ and C^1-mappings $\Phi_1 : R^p \to R^m, \Phi_2 : R^p \to R^n$, such that for the mapping $\Phi_1 : R^p \to R^m, \Phi_2 : R^p \to R^n$, such that for the mapping $\Phi = (\Phi_1, \Phi_2)$ it holds that*
1. *$\Phi(0) = 0, d_0\Phi = 0$;*
2. *$(M\Phi)(z) = 0(\|z\|^j)$ for $z \to 0$. Then*

$$\|h(z) - \Phi(z)\| = 0(\|z\|^j) \text{ for } z \to 0. \quad (4.41)$$

Proof

The theorem is of local character, and thus, it will suffice to prove it for the modification (4.17) of the system (4.16). Instead of the mapping Φ it suffices to take a C^1-mapping $\tilde{\Phi} : R^p \to R^{m+n}$, such that $\tilde{\Phi}(z) = \Phi(z)$ for all $z \in U_1$, $\tilde{\Phi}(z) = 0$ for all $z \in R^p \setminus U_2$ and $\|d_z\tilde{\Phi}\| \leq q\Theta$ for all $z \in R^p$, where $U_1, U_2 \subset R^p$ are open neighbourhoods of the origin, $\overline{U}_1 \subset U_2, q > 0$ and Θ is the positive number from the definition of modification of the system (4.16). According to Lemma 3.128 such a C^1-mapping exists and from its above mentioned properties it follows that it is globally Lipschitz with the Lipschitz constant $\varrho(\Theta) = q\Theta$. Thus instead of the mapping $M\Phi$ it will suffice to take the mapping $\tilde{M}\tilde{\Phi} : R^p \to R^{m+n}$,

$$(\tilde{M}\tilde{\Phi})(z) = d_z\tilde{\Phi}(A^0 z + R(\tilde{\Phi}(z), z)) - B\tilde{\Phi}(z) - F(\tilde{\Phi}(z), z),$$

where $F = (\tilde{P}, \tilde{Q})$. From assumption 2 of the theorem it obviously follows that $(\tilde{M}\tilde{\Phi})(z) = O(\|z\|^j)$ for $z \to 0$. According to Lemma 4.20 the mapping

h is the fixed point of the mapping $\mathscr{F}$ defined by the formula (4.24). If denoting $\eta = h - \tilde{\Phi}$, then $h = \tilde{\Phi} + \eta$ and thus, the equality $\mathscr{F}(h) = h$ can be written in the form

$$\mathscr{F}(\tilde{\Phi} + \eta) - \tilde{\Phi} = \eta. \tag{4.42}$$

In the proof of Lemma 4.20 it was shown that there exist positive numbers L, Θ, such that the mapping $\mathscr{F}/X_L$ is contractive. Therefore the mapping $H : X_L \to X_L$, $H(u) = \mathscr{F}(\tilde{\Phi} + u) - \tilde{\Phi}$ is contractive and according to the Banach Fixed Point Theorem it has a unique fixed point $u_0 \in X_L$. Since $h = \tilde{\Phi} + \eta$ is the only fixed point of the mapping $\mathscr{F}$, from the equality (4.42) we get that $u_0 = \eta$. For the positive number k we define the set

$$Y(k) = \left\{ g \in X_L : \|g(z)\| \leq k\|z\|^j \text{ for all } z \in R^p \right\}.$$

This set is obviously a closed subset in X_L, and thus, it forms a complete metric system. From the contractivity of the mapping H, the contractivity of the mapping $\tilde{H} = H/Y(k)$ follows. Now it will suffice to prove that there exists a positive number k such that $\tilde{H}(Y(k)) \subset Y(k)$, since in case of validity of this inclusion the mapping $\tilde{H}$ must have a unique fixed point $v_0 \in Y(k)$ that must be also the only fixed point of the mapping H, and thus, $v_0 = \eta$. From the definition of the set $Y(k)$ it follows that $\|h(z) - \tilde{\Phi}(z)\| = \|\eta(z)\| \leq k\|z\|^j$, and thus, the equality (4.41) holds. Let $v \in Y(k)$. Then

$$\mathscr{F}(v + \tilde{\Phi})(z) = \int_{-\infty}^{\infty} K(s) F((v + \tilde{\Phi})(\Psi(z,s)), \Psi(z,s))\,ds, \tag{4.43}$$

where Ψ is the flow of the differential equation

$$\dot{z} = A^0 z + R(v(z) + \tilde{\Phi}(z), z). \tag{4.44}$$

From the properties of the matrices A^-, A^+ and from the definition of the mapping K it follows that

$$-\Phi(z) = -\int_{-\infty}^{\infty} \frac{d}{ds}(K(s)\tilde{\Phi}(\Psi(z,s)))\,ds$$
$$= \int_{-\infty}^{\infty} K(s)(B\tilde{\Phi}(z,s)) - \frac{d}{ds}\tilde{\Phi}(\Psi(z,s)))\,ds, \tag{4.45}$$

where $B = \text{diag}\{A^-, A^+\}$. If instead of $\Psi(z,s)$ and $\tilde{\Phi}(z)$ we write just Ψ and Φ, respectively, then from definition of the mapping $\tilde{M}\tilde{\Phi}$ and from (4.44) we get

$$B\tilde{\Phi}(\Psi) - \frac{d}{ds}\tilde{\Phi}(\Psi) = B\tilde{\Phi}(\Psi) - d_\Psi \tilde{\Phi}(A^0 \Psi + \tilde{R}((v + \tilde{\Phi})(\Psi), \Psi))$$
$$= -(\tilde{M}\tilde{\Phi})(\Psi) - F(\tilde{\Phi}(\Psi), \Psi) + d_\Psi \Phi(\tilde{R}(\tilde{\Phi}(\Psi), \Psi)$$
$$- \tilde{R}((v + \Phi)(\Psi), \Psi)).$$

After substituting this equality into (4.45) we obtain that

$$\tilde{H}(v)(z) = \mathcal{F}((v + \tilde{\Phi})(z)) - \tilde{\Phi}(z)$$

$$= \int_{-\infty}^{\infty} K(s) S(v(\Psi(z,s)), \Psi(z,s)) \, ds, \tag{4.46}$$

$$S(v(\Psi), \Psi) = F((v + \tilde{\Phi})(\Psi), \Psi) - F(\tilde{\Phi}(\Psi), \Psi)$$
$$- (\tilde{M}\tilde{\Phi})(\Psi) + (d_\Psi \tilde{\Phi})(\tilde{R}(\tilde{\Phi}(\Psi), \Psi) - \tilde{R}((v + \tilde{\Phi})(\Psi), \Psi). \tag{4.47}$$

Since $\text{Lip}(\tilde{\Phi}) = \varrho(\Theta)$, so $\|\tilde{\Phi}(z)\| \leq \varrho(\Theta)\|z\|$ for all $z \in R^p$ and since the mapping $\tilde{\Phi}$ has zero values on the set $R^p \setminus U_2$ only, then

$$\|\tilde{\Phi}(z)\| \leq \varrho_1(\Theta) \tag{4.48}$$

where ϱ_1 is a positive function of the variable Θ, such that $\lim_{\Theta \to 0} \varrho_1(\Theta) = 0$. Since $(\tilde{M}\tilde{\Phi})(z) = O(\|z\|^j)$ for $z \to 0$, from global properties of the mapping $\tilde{\Phi}$ we find that there exists a positive number K_1 such that

$$\|(\tilde{M}\tilde{\Phi})(z)\| \leq K_1 \|z\|^j \tag{4.49}$$

for all $z \in R^p$. From (4.16), (4.47), (4.48) and the Lipschitz property of the mapping F and R we get that

$$\|S(v(\Psi), \Psi)\| \leq \varrho(\Theta)\|v(\Psi)\| + \|(\tilde{M}\tilde{\Phi})(\Psi)\| + (\varrho(\Theta))^2 \|v(\Psi)\|$$
$$\leq \delta(\Theta, k) \|\Psi\|^j, \tag{4.50}$$

where $\delta(\Theta, k) = \varrho(\Theta)k + (\varrho(\Theta))^2 k + K_1$. In the same way as in the proof of Lemma 4.20 the inequality (4.29) was proved, it can be shown that for arbitrary positive number β there exist positive continuous functions $N(\beta, \Theta), r(\beta, \Theta)$, such that $\lim_{\Theta \to 0} N(\beta, \Theta) = 0, \lim_{\Theta \to 0} r(\beta, \Theta) = 0$ and

$$\|\Psi(z, t)\| \leq N(\beta, \Theta) e^{r(\beta, \Theta)|t|} \|z\| \tag{4.51}$$

for all $z \in R^p, t \in R$. From (4.25), (4.46), (4.50) and (4.51) it follows that

$$\|\tilde{H}(v)(z)\| \leq \int_{-\infty}^{\infty} \|K(s)\| \|S(v(\Psi(z,s)), \Psi(z,s))\| \, ds$$
$$\leq M \int_{-\infty}^{\infty} e^{-\gamma|s|} \delta(\Theta, k) N(\beta, \Theta) e^{r(\beta, \Theta)|s|} \, ds \|z\|^j = \alpha(\beta, \Theta) \|z\|^j,$$

where $\alpha(\beta, \Theta) = M\delta(\Theta, k) N(\beta, \Theta)(\gamma - q \cdot r(\beta, \Theta))^{-1}$. If we choose

$$k > K_1 M \cdot N(\beta, \Theta)(\gamma - q \cdot r(\beta, \Theta) - M(\varrho(\Theta) + (\varrho(\Theta))^2)^{-1},$$

where Θ is a positive number such that the right side of this inequality is positive, then $\tilde{H}(v) \in Y(k)$. □

4.33 Corollary. *Let the assumptions of Theorem 4.32 be fulfilled and let* $W^c_{\text{loc}} = \{(x,y,z) : x = h_1(z), y = h_2(z)\}$, $\tilde{W}^c_{\text{loc}} = \{(x,y,z) : x = \tilde{h}_1(z), y = \tilde{h}_2(z)\}$ *be local center C^1-manifolds of the system (4.16). Then for arbitrary $j \in N$*

$$\|h(z) - \tilde{h}(z)\| = O(\|z\|^j) \quad \text{for} \quad z \to 0, \tag{4.52}$$

where $h = (h_1, h_2), \tilde{h} = (\tilde{h}_1, \tilde{h}_2)$.

Proof

According to Theorem 4.32 there exists a neighbourhood U of the origin and constants K_1, K_2 such that the mappings $\Phi, h, \tilde{h}$ are defined on U and for arbitrary $z \in U$ we have $\|h(z) - \Phi(z)\| \leq K_1\|z\|^j, \|\tilde{h}(z) - \Phi(z)\| \leq K_2\|z\|^j$, and thus,

$$\|h(z) - \tilde{h}(z)\| \leq \|h(z) - \Phi(z)\| + \|\tilde{h}(z) - \Phi(z)\| \leq (K_1 + K_2)\|z\|^j,$$

i.e. $\|h(z) - \tilde{h}(z)\| = O(\|z\|^j)$. □

4.34 Example. Let us consider the system of differential equations

$$\dot{y} = -y + az^2, \quad \dot{z} = byz, \tag{4.53}$$

where $a, b \in R$. If $y = \Phi(z)$ is a C^1-function, then $(M\Phi)(z) = b\dot{\Phi}(z)\Phi(z)z + \Phi(z) - az^2$. Let $\Phi(z) = \alpha z^2 + \beta z^3 + \gamma z^4 + O(|z|^5)$. Then

$$(M\Phi)(z) = (\alpha - a)z^2 + \beta z^3 + (\gamma - 2\alpha\beta)z^4 + O(|z|^5).$$

If $\alpha = a, \beta = 0, \gamma = 0$, then $(M\Phi)(z) = O(|z|^5)$. If $W^c_{\text{loc}} = \{(y,z) : y = h(z)\}$ is a local centre manifold of the system (4.53), where $h \in C^1$, then according to Theorem 4.32 it is $|h(z) - \Phi(z)| = O(|z|^5)$, and thus, $h(z) = az^2 + O(|z|^5)$. Reduction of the system (4.53) to the centre manifold W^c_{loc} is of the form

$$\dot{z} = h(z)z = az^3 + O(|z|^6). \tag{4.54}$$

According to Corollary 4.33 the reduction of the system (4.53) to arbitrary centre manifold of this system is of the form (4.54). According to Šošitaišvili Theorem the system (4.53) is orbitally topologically equivalent in a neighbourhood of the origin with the system $\dot{y} = -y, \dot{z} = az^3 + O(|z|^6)$.

Let us now consider the diffeomorphism $f = (f_1, f_2, f_3)$, where

$$\begin{aligned} f_1(x,y,z) &= A^s x + P(x,y,z), \\ f_2(x,y,z) &= A^u y + Q(x,y,z), \\ f_3(x,y,z) &= A^c z + R(x,y,z), \end{aligned} \tag{4.55}$$

where the C^r-mappings P, Q, R are defined on an open subset $W \subset R^N = R^m \times R^n \times R^p$ containing the origin, and they have at the origin both zero values and zero first order derivatives and all eigenvalues of the matrix $A^s(A^u; A^c)$ have their absolute values less than (greater than, equal to) 1.

By analogy, as defined for the invariant sets W^s, W^u, W^{cs}, W^{cu}, W^c for the system of differential equations (4.16) the similar sets can be defined also for the diffeomorphism f and they will be denoted as $W^s(f), W^u(f)$, etc. Let us define at least the set $W^c(f)$.

4.35 Definition. *If there exist C^k-mappings $x = h_1(z), y = h_2(z)$, having at the origin both zero values and zero first derivatives such that the set*

$$W^c(f) = \{(x, y, z) : x = h_1(z), y = h_2(z)\}$$

is an invariant set of the diffeomorphism f, then this set is called the centre C^k-manifold of the diffeomorphism f. If the mapping $h = (h_1, h_2)$ is defined only in a neighbourhood of the origin (not on the whole set W), then this set is called a local centre C^k-manifold of the diffeomorphism f denoted as $W^c_{loc}(f)$.

Let us now consider the parametrized C^r-diffeomorphism $g = (g_1, g_2, g_3)$, where

$$\begin{aligned} g_1(x, y, z, \varepsilon) &= A^s x + F(x, y, z, \varepsilon), \\ g_2(x, y, z, \varepsilon) &= A^u y + G(x, y, z, \varepsilon), \\ g_3(x, y, z, \varepsilon) &= A^c z + H(x, y, z, \varepsilon), \end{aligned} \quad (4.56)$$

where the C^r-mappings F, G, H are defined on an open set $W \times V$ containing the origin, where $W \subset R^N = R^m \times R^n \times R^p$, $V \subset R^k$, ε being the parameter and the matrices A^s, A^u, A^c have the same properties as in the definition of the diffeomorphism (4.55). Let us suppose that the mappings F, G, H have at the origin both zero values and zero first-order derivatives.

4.36 Definition. *Let there exist C^k-mappings $x = h_1(z, \varepsilon), y = h_2(z, \varepsilon)$, having at the origin both zero values and zero first derivatives such that the set*

$$V^c(g) = \{(x, y, z, \varepsilon) : x = h_1(z, \varepsilon), y = h_2(z, \varepsilon)\}$$

is an invariant set of the diffeomorphism $\tilde{g}_\bullet : W \times V \to R^N \times R^k$, $\tilde{g}(x, y, z, \varepsilon) = (g(x, y, z, \varepsilon), \varepsilon)$, where g is the parametrized diffeomorphism (4.56). Then the set $V^c(g)$ is called the centre C^k-manifold of the parametrized diffeomorphism g. If the mapping $h = (h_1, h_2)$ is defined only locally in a certain neighbourhood of the origin (not on the whole set

$W \times V$), *then this set is called the local centre C^k-manifold for g denoting it as $V^c_{\text{loc}}(g)$.*

The following theorems on the existence of centre manifolds of diffeomorphism, or on the parametrized diffeomorphisms follow from much more general theory of invariant manifolds of diffeomorphisms, presented in [67 and 130]. This theory is more general in the respect that the invariant manifolds are defined not only for the fixed, or periodic points of the diffeomorphisms, but also for more complicated invariant sets of diffeomorphisms and vector fields. We shall not treat this general rather exacting theory in more detail. The reader who might be interested in deeper study of this theory is referred to the simpler case of invariant sets of diffeomorphisms, so-called hyperbolic sets of diffeomorphisms, such as presented in [66]. The following theorems will be presented without proof.

4.37 Theorem. *Let $1 \leq r < \infty$, $W \subset R^N = R^m \times R^n \times R^p$ be an open set containing the origin, $P \in C^r(W, R^m), Q \in C^r(W, R^n), R \in C^r(W, R^p), P(0) = 0, d_0 P = 0, Q(0) = 0, d_0 Q = 0, R(0) = 0, d_0 R = 0$, for all eigenvalues of the matrix $A^s(A^u; A^c)$ having their absolute values less than (greater than, equal to) 1. Then there exists the local centre C^r-manifold $W^c_{\text{loc}}(f)$ of the diffeomorphism (4.55).*

4.38 Theorem. *Let $1 \leq r < \infty$, $W \subset R^N = R^m \times R^n \times R^p$, $V \subset R^k$ be open sets containing the origin, $F \in C^r(W \times V, R^m)$, $G \in C^r(W \times V, R^n)$, $H \in C^r(W \times V, R^p)$, $F(0,0) = 0$, $d_{(0,0)}F = 0$, $G(0,0) = 0$, $d_{(0,0)}G = 0$, $H(0,0) = 0$, $d_{(0,0)}H = 0$ and the matrices A^s, A^u, A^c have the same properties as in Theorem 4.37. Then there exists the local centre C^r-manifold $V^c_{\text{loc}}(g)$ of the parametrized diffeomorphism (4.56).*

4.39 Theorem. *Let the assumptions of Theorem 4.38 be fulfilled. Then there exists the local centre C^r-manifold $V^c_{\text{loc}} = \{(x, y, z, \varepsilon) : x = h_1(z, \varepsilon), y = h_2(z, \varepsilon)\}$ of the parametrized diffeomorphism (4.56) such that the germ at the point 0 represented by this diffeomorphism is orbitally topologically equivalent to that at 0 point, that is represented by the parametrized diffeomorphism $\tilde{g} = (\tilde{g}_1, \tilde{g}_2, \tilde{g}_3)$, where*

$$\tilde{g}_1(x, y, z, \varepsilon) = A^s x, \quad \tilde{g}_2(x, y, z, \varepsilon) = A^u y, \tag{4.57}$$

$$\tilde{g}_3(x, y, z, \varepsilon) = A^c z + H(h(z, \varepsilon), z, \varepsilon), \tag{4.58}$$

where $h = (h_1, h_2)$.

4.40 Definition. *Let $V^c_{\text{loc}}(g) = \{(x, y, z, \varepsilon) : x = h_1(z, \varepsilon), y = h_2(z, \varepsilon)\}$ be a local centre C^r-manifold of the parametrized diffeomorphism (4.46).*

The parametrized diffeomorphism $g_c(z,\varepsilon) = A^c z + H(h(z,\varepsilon), z, \varepsilon)$, where $h = (h_1, h_2)$, is called the reduction of the parametrized diffeomorphism g to the centre manifold $V^c_{loc}(g)$.

To conclude the present chapter let us mention the problem of versality of the germs of parametrized vector fields that will be studied in more detail in the next chapter. A. N. Šošitaišvili in [150] proved the following theorem.

4.41 Theorem. *The germ $[F]_0$ represented by the parametrized system of differential equations (4.35) is versal iff the germ $[G]_0$ represented by the reduction (4.39) of the system (4.35) to its centre manifold V^c_{loc} from the Šošitaišvili Theorem is versal.*

Proof of this theorem is directly connected with the proof of the Šošitaišvili Theorem and since it is also quite exacting, we shall not introduce it here.

5 GENERIC BIFURCATIONS OF VECTOR FIELDS AND DIFFEOMORPHISMS

The aim of this chapter is to describe some generic properties or generic bifurcations of parametrized vector fields and therefore we shall deal only with those parametrized differential equations whose linearizations have the generic properties in the space of parametrized differential equations in the sense that they were dealt with in Chapter 2.

5.1 Ljapunov–Schmidt Method

In the preceding chapter we mentioned the method of the reduction of a parametrized system of differential equations to the centre manifold and we derived the bifurcation equation containing the minimum number of parametrized differential equations necessary for complete determination of bifurcations of the original system. Now, we mention a functional method of solving equations in Banach spaces which is often used in the bifurcation theory of differential equations. In many cases using this method the problem of bifurcations of equations on infinite-dimensional Banach spaces can even be reduced to the problem of bifurcations of the so-called bifurcation equation containing just a finite number of equations and the number of these equations is minimal. In this paragraph we use this method for the derivation of the bifurcation equation for non-autonomous differential equation in R^n periodic in time and we shall use this result in the following paragraph in the proof of the existence of the so-called Hopf bifurcation for parametrized autonomous differential equations.

Let X, Y be Banach spaces and $F \in L(X, Y)$. Let us remember that we denote the kernel by the sign $\operatorname{Ker} F$ and the image of the mapping F by the sign $\operatorname{Image}(F)$, i.e. $\operatorname{Ker} F = \{x \in X : F(x) = 0\}$, $\operatorname{Image}(F) = \{F(x) : x \in X\}$.

5.1 Definition. *The mapping $P \in L(X)$ is called the projector in the Banach space X if $P^2 = P$, i.e. $P(Px) = Px$ for all $x \in X$. Let us denote the image* $\operatorname{Image}(P)$ *of the projector P as X_P.*

5.2 Example. Let $X = R^3$ and $P \in L(X)$ be the mapping represented by the matrix

$$[P] = \begin{bmatrix} 0 & -1 & 0 \\ 0 & 1 & 0 \\ 0 & 0 & 1 \end{bmatrix}.$$

Then $[P]^2 = [P]$ and therefore $P^2 = P$, i.e. P is a projector, $\operatorname{Ker} P = \{(x, y, z) : y = 0, z = 0\}$, $\operatorname{Image}(P) = X_P = \{(x, y, z) : x = -y\}$, i.e. $\operatorname{Ker} P$ is a linear one-dimensional subspace in X and X_P is a linear two-dimensional subspace in X.

5.3 Lemma. *Let $P \in L(X)$ be a projector in the Banach space X. Then these equalities hold:*
  1. $X_P = \operatorname{Ker}(\operatorname{id} - P)$,
  2. $X = (\operatorname{Ker} P) \oplus X_P$,
  3. $X = X_{\operatorname{id} - P} \oplus X_P$,
where $\operatorname{id} = \operatorname{id}_X$.

Proof

As P is a projector, thus $(\operatorname{id} - P)(Px) = 0$ for all $x \in X$, and thus $X_P \subset \operatorname{Ker}(\operatorname{id} - P)$. If $x \in \operatorname{Ker}(\operatorname{id} - P)$, then $x = Px \in X_P$ and thereby assertion 1 is proved. From assertion 1 it follows that $X_{\operatorname{id} - P} = \operatorname{Ker}(\operatorname{id} - (\operatorname{id} - P)) = \operatorname{Ker} P$. Therefore if $x \in (\operatorname{Ker} P) \cap X_P$ thus $Px = 0$, $x - Px = 0$ and thus $x = 0$. Each element $x \in X$ can be written in the form $x = (x - Px) + Px$ and as $x - Px \in \operatorname{Ker} P$, thus $X = (\operatorname{Ker} P) + X_P$. The mapping P is continuous and therefore the spaces $\operatorname{Ker} P$ and $X = \operatorname{Ker}(\operatorname{id} - P)$ are closed. Thereby assertion 2 is proved. Assertion 3 follows from assertions 1 and 2. □

5.4 Lemma. *Let X, Y be Banach spaces, $A \in L(X, Y)$ and let us assume that there exist mappings $B \in L(X)$, $C \in L(Y)$ such that B is a projector on X, C is a projector on Y, $\operatorname{Ker} A = X_B$ and $\operatorname{Image}(A) = Y_C$. Then there exists a mapping $D \in L(Y_C, X_{\operatorname{id} - B})$, such that $AD = \operatorname{id}$ on Y_C and $DA = \operatorname{id} - B$ on X.*

5.5 Remark. As B and C are projectors so $Y_C, X_{\operatorname{id} - B}$ are closed subspaces, and thus they form the Banach spaces. According to Lemma 5.3 $X = X_B \oplus X_{\operatorname{id} - B}$ and $Y = Y_{\operatorname{id} - C} \oplus Y_C$. The assertion of Lemma 5.4 can be graphed as in Fig. 33.

Proof of Lemma 5.4

According to assertion 5 of Lemma 5.3 $X = X_B \oplus X_{\operatorname{id} - B}$. As $\operatorname{Ker} A = X_B$, the mapping $\tilde{A} = A/X_{\operatorname{id} - B}$ is surjective and $\operatorname{Image}(\tilde{A}) = \operatorname{Image}(A) = Y_C$. Therefore there exists a linear mapping $D : Y_C \to X_{\operatorname{id} - B}$ such that

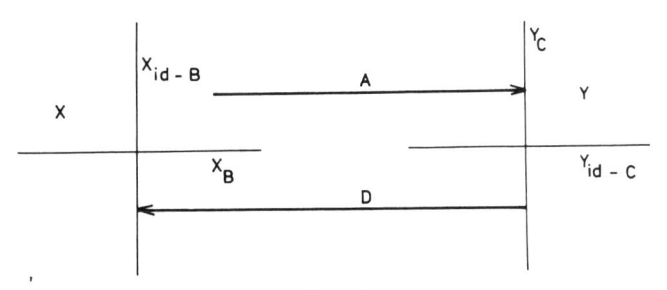

Fig. 33.

$\tilde{A}D = \text{id}$ on Y_C and $D\tilde{A} = \text{id} - B$ on X. The continuity of the mapping D follows from the Closed Graph Theorem. □

Let us assume now that the assumptions of Lemma 5.4 are fulfilled. Let a C^r-mapping be given $G : X \times \Lambda \to Y$, where $1 \leq r < \infty$, such that

$$G(0,0) = 0, d_{(0,0)}G = 0, \tag{5.1}$$

where Λ is the Banach space which we shall consider as a space of parameters. We shall solve the equation

$$Ax = G(x, \lambda) \tag{5.2}$$

in a neighbourhood of the point $(x, \lambda) = (0, 0)$.

If $X = Y$ and $\text{Ker } A = \{0\}$, then A is a surjective mapping and it follows from the Implicit Function Theorem that there exists a C^r- mapping $x = x(\lambda)$ defined in a neighbourhood V of the point $\lambda = 0$ that $x(0) = 0$ and $Ax(\lambda) = G(x(\lambda), \lambda)$ for all $\lambda \in V$. However, in general equation (5.2) cannot be solved by the Implicit Function Theorem. Now we show that under the assumptions of Lemma 5.4 equation (5.2) can be reduced with the help of the Implicit Function Theorem to the so-called bifurcation equation.

As $Y = Y_C \oplus Y_{\text{id}-C}$ we can write equation (5.2) as the system of equations

$$CAx = CG(x, \lambda), \tag{5.3}$$
$$(\text{id} - C)Ax = (\text{id} - C)G(x, \lambda). \tag{5.4}$$

As $X = X_B \oplus X_{\text{id}-B}$ thus $x = y + z$, where $y \in X_B, z \in X_{\text{id}-B}$ and thus we can write $y + z$ instead of x in the system (5.3), (5.4) where y, z can be considered as new variables. From the assumption $\text{Image}(A) = Y_C$ we obtain that $CA = A$, i.e.

$$(\text{id} - C)A = 0. \tag{5.5}$$

According to Lemma 5.4

$$DA = \text{id} - B \quad \text{on} \quad X. \tag{5.6}$$

When using the equality (5.5) we can rewrite the equation (5.3) in the form

$$Ax = CG(x, \lambda). \tag{5.7}$$

As $x = y + z$ where $y \in X_B$, $z \in X_{\text{id}-B}$ and $\operatorname{Ker} A = X_B$, then $Bz = 0$, $Ay = 0$, and thus we obtain from the equality (5.6) that $DAx = DA(y + z) = DAy + DAz = DAz = (\text{id} - B)z = z$, i.e.

$$DAx = z. \tag{5.8}$$

On the basis of this equality we can rewrite the equation (5.7) in the form $z = DCG(y+z, \lambda)$. Thus we obtain that $x = y + z$ is the solution of the system of equations

$$z = DCG(y + z, \lambda), \tag{5.9}$$

$$(\text{id} - C)G(y + z, \lambda) = 0. \tag{5.10}$$

From the Implicit Function Theorem it follows that there exists an open neighbourhood $W \times V$ of the point $(y, \lambda) = (0, 0)$ in $X_B \times \Lambda$ and just one C^r-mapping $u : W \times V \to Y$ such that $u(0,0) = 0, (\partial_1)_{(0,0)} u = 0$ and $u(y, \lambda) = DCG(y + u(y, \lambda), \lambda)$ for all $(y, \lambda) \in W \times V$. If $x = y + u(y, \lambda)$ is a solution of the equation (5.2) then the pair (y, λ) must satisfy the equation

$$H(y, \lambda) \stackrel{\text{def}}{=} (\text{id} - C)G(y + u(y, \lambda), \lambda) = 0. \tag{5.11}$$

Equation (5.11) is called bifurcation equation of the equation (5.2) and the method we have used for its derivation is called the Ljapunov–Schmidt method or alternative method. We have proved the following theorem.

5.6 Theorem. *Let the assumptions of Lemma 5.4 be fulfilled. Let us assume that Λ is a Banach space and $G : X \times \Lambda \to Y$ is a C^r-mapping, where $1 \leq r < \infty$ and fulfills condition (5.1). Then there exists an open neighbourhood U of the point $x = 0$ in X, an open neighbourhood $W \times V$ of the point $(y, \lambda) = (0, 0)$ in $X_B \times \Lambda$ and a C^r-mapping $u : W \times V \to Y$ such that $u(0,0) = 0, (\partial_1)_{(0,0)} u = 0, z = u(y, \lambda)$ is the unique solution of equation (5.9) for all $(y, \lambda) \in W \times V$ and each solution $x = x(\lambda)$, where $\lambda \in V$, of equation (5.2) with values in U has the form $x(\lambda) = y(\lambda) + u(y(\lambda), \lambda)$, where $y = y(\lambda)$ is the solution of the bifurcation equation (5.11).*

5.7 Remark. The Ljapunov–Schmidt method has wide possibilities of application. Some of its applications in the theory of bifurcations of ordinary and partial differential equation can be found for example in [74]. In [53] there is a brief but understandable survey of basic results from the theory of bifurcations of differential equations obtained by this method. The Ljapunov–Schmidt method as well as more general alternative methods are used effectively in [157] in searching for bifurcations of parametrized differential equations fulfilling various symmetry conditions.

Now we use the Ljapunov–Schmidt method in searching for a periodic solution of the differential equation

$$\dot{x} = \frac{dx}{d\Theta} = f(x, \lambda, \Theta), \tag{5.12}$$

where $x \in R, \lambda \in R$ is a parameter, $f \in C^r(R^3, R)$, where $1 \leq r < \infty$, while the function $f(x, \lambda, \Theta)$ is periodic in the variable Θ with the period 2π, $f(0, \lambda, \Theta) = 0$ for all $(\lambda, \Theta) \in R \times R$ and

$$\frac{\partial f(0, 0, \Theta)}{\partial x} = 0$$

for all $\Theta \in R$. Let

$$C_{2\pi}^m = \{x \in C^m(R, R) : x(\Theta + 2\pi) = x(\Theta) \quad \text{for all} \quad \Theta \in R\},$$

where $m \geq 0$. The sets $X = C_{2\pi}^1$, $Y = C_{2\pi}^0$ with the norm $\|x\| = \max_{\Theta \in R}(|x(\Theta)|$ respectively $\|x\| = \max_{\Theta \in R}|x(\Theta)| + |\dot{x}(\Theta)|)$ are Banach spaces and the mapping

$$A : X \to Y, (Ax)(\Theta) = \frac{dx(\Theta)}{d\Theta} \tag{5.13}$$

is obviously continuous and linear, i.e. $A \in L(X, Y)$. Let us define the mappings $G \in C^r(X \times R, Y)$, $G(x, \lambda)(\Theta) = f(x(\Theta), \lambda, \Theta)$. Equation (5.12) has a periodic solution with period 2π iff the equation

$$Ax = G(x, \lambda) \tag{5.14}$$

has the solution $x = x(\lambda) \in X = C_{2\pi}^1$. For the mapping A

$$\text{Image}\,(A) = \left\{z \in C_{2\pi}^0 : z(\Theta) = \frac{dx(\Theta)}{d\Theta} \text{ for all } \Theta \in R, \text{ where } x \in C_{2\pi}^1\right\}$$

and $\quad \text{Ker}\, A = \left\{x \in C_{2\pi}^1 : \frac{dx(\Theta)}{d\Theta} = 0 \text{ for all } \Theta \in R\right\}.$

On the set $X = C_{2\pi}^1$ we can define the scalar product

$$\langle x, y \rangle = \frac{1}{2\pi} \int_0^{2\pi} x(s)y(s)\,\mathrm{d}s.$$

5.8 Lemma. *Let $A \in L(C_{2\pi}^1, C_{2\pi}^0)$, be the mapping defined by the relation (5.13). Then $z \in \mathrm{Image}\,(A)$, iff $\langle y, z \rangle = 0$ for all $y \in \mathrm{Ker}\,A$.*

Proof

If $z \in \mathrm{Image}\,(A)$ and $y \in \mathrm{Ker}\,A$ then

$$\frac{\mathrm{d}y(\Theta)}{\mathrm{d}\Theta} = 0 \quad \text{for all } \Theta \in R$$

and there exists such a function $x \in C_{2\pi}^1$ such that

$$z(\Theta) = \frac{\mathrm{d}x(\Theta)}{\mathrm{d}\Theta} \quad \text{for all } \Theta.$$

Therefore

$$\langle x, y \rangle = \frac{1}{2\pi} \int_0^{2\pi} y(t)z(t)\,\mathrm{d}t = \frac{1}{2\pi} \int_0^{2\pi} y(t)\frac{\mathrm{d}x(t)}{\mathrm{d}t}\,\mathrm{d}t$$

$$= -\frac{1}{2\pi} \int_0^{2\pi} \frac{\mathrm{d}y(t)}{\mathrm{d}t} x(t)\,\mathrm{d}t = 0.$$

Let $z \in C_{2\pi}^0$ and $\langle y, z \rangle = 0$ for all $y \in \mathrm{Ker}\,A$. As

$$\frac{\mathrm{d}y(\Theta)}{\mathrm{d}\Theta} = 0$$

thus $y(\Theta) \equiv c \in R$. If we choose $y \in \mathrm{Ker}\,A$ so that $y(\Theta) \equiv 1$ then we obtain the equality

$$\int_0^{2\pi} z(t)\,\mathrm{d}t = 0.$$

If we define the function

$$x(\Theta) = \int_0^{\Theta} z(s)\,\mathrm{d}s,$$

then from the preceding equality it follows that $x \in C_{2\pi}^1$ and obviously

$$z(\Theta) = \frac{\mathrm{d}x(\Theta)}{\mathrm{d}\Theta} \quad \text{for all } \Theta \in R. \quad \square$$

5.9 Definition. *Let us define the mappings:*

$$B : X \to X, \quad (Bx)(\Theta) = \int_0^{2\pi} x(s)\,ds,$$

$$C : Y \to Y, \quad (Cy)(\Theta) = y(\Theta) - \frac{1}{2\pi}\int_0^{2\pi} y(s)\,ds.$$

5.10 Lemma. *If B, C are the mappings from the Definition 5.9 then these assertions hold.*
  1. *The mapping B is a continuous projector in X and $X_B = \mathrm{Ker}\,A$.*
  2. *The mapping C is a continuous projector in Y and $Y = \mathrm{Image}\,A$, where A is the mapping defined by the relation (5.13).*

Proof

The mapping B is obviously a continuous projector in X and obviously both X and $\mathrm{Ker}\,A$ consist of constant functions on X so $X_B = \mathrm{Ker}\,A$, i.e. assertion 1 holds. The mapping C is obviously continuous and linear. For $y \in C^1_{2\pi}$

$$(C(Cy))(\Theta) = (Cy)(\Theta) - \frac{1}{2\pi}\int_0^{2\pi}(Cy)(s)\,ds$$
$$= (Cy)(\Theta) - \frac{1}{2\pi}\int_0^{2\pi}\left(y(s) - \frac{1}{2\pi}\int_0^{2\pi}y(\sigma)d\sigma\right)ds = (Cy)(\Theta),$$

i.e. C is a continuous projector on Y. From Lemma 5.8 it follows that

$$\mathrm{Image}\,(A) = \left\{ z \in Y : \int_0^{2\pi} z(s)\,ds = 0 \right\}$$

thus we obtain that

$$Y_C = \mathrm{Image}\,(C)\left\{ z \in Y : z(\Theta) = y(\Theta) - \frac{1}{2\pi}\int_0^{2\pi}y(s)\,ds, y \in Y \right\}$$
$$\subset \mathrm{Image}\,(A).$$

If $z \in \mathrm{Image}\,(A)$, then $\langle 1, z \rangle = 0$ and therefore we can write z in the form

$$z(\Theta) = z(\Theta) - \frac{1}{2\pi}\int_0^{2\pi} z(s)\,ds$$

thus $z \in Y_C$. It means that $\mathrm{Image}\,(A) \subset Y_C$. □

Lemma 5.10 guarantees that for equation (5.14) the assumptions of the Lemma 5.4 are fulfilled and thus we can use the Ljapunov–Schmidt method.

According to Theorem 5.6 there exists an open neighbourhood U of the point $x = 0$ in $C_{2\pi}^1$, an open neighbourhood $W \times R > V$ of the point $(y, \lambda) = (0, 0)$ in $X_B \times R$ and just one C^r-mapping $u : W \times V \to C_{2\pi}^0$ such that $u(0, 0) = 0$, $(\partial_1)_{(0,0)} u = 0$,

$$u(y, \lambda) = DCG(y + u(y, \lambda), \lambda) \tag{5.15}$$

for all $(y, \lambda) \in W \times V$ and each solution of the equation (5.14) with values in U is of the form $x = y + u(y, \lambda)$, where y satisfies the bifurcation equation

$$QG(y + u(y, \lambda), \lambda) = 0, \tag{5.16}$$

$Q = \mathrm{id} - C$, where $y \in \mathrm{Ker}\, A$, i.e. y is a real number and $u(y, \lambda) \in \mathrm{Image}\,(A)$ and thus

$$\int_0^{2\pi} u(y, \lambda)(s)\, ds = 0.$$

The last equality means that the function $u(y, \lambda)$ has the zero mean value. In accordance with the Lemma 5.4 $AD = \mathrm{id}$ on Y_C, from the equation (5.15) we obtain that

$$Au(y, \lambda) = CG(y + u(y, \lambda), \lambda). \tag{5.17}$$

As

$$(Ax)(\Theta) = \frac{dx(\Theta)}{d\Theta},$$

$$(Cy)(\Theta) = y(\Theta) - \frac{1}{2\pi}\int_0^{2\pi} y(s)\, ds,$$

$$(Qy)(\Theta) = \frac{1}{2\pi}\int_0^{2\pi} y(s)\, ds \text{ and }$$

$$G(y + u(y, \lambda), \lambda)(s) = f(y + u(y, \lambda)(s), \lambda, s)$$

we can write the system of equations (5.16), (5.17) in the form

$$H(y) \stackrel{\mathrm{def}}{=} \int_0^{2\pi} f(y + u(y, \lambda)(s), \lambda, s)\, ds = 0, \tag{5.18}$$

$$\frac{du(y, \lambda)(\Theta)}{d\Theta} = f(y + u(y, \lambda)(\Theta), \lambda, \Theta) \\ - \frac{1}{2\pi}\int_0^{2\pi} f(y + u(y, \lambda)(s), \lambda, s)\, ds, \tag{5.19}$$

$$\int_0^{2\pi} u(y,\lambda)(s)\,ds = 0. \qquad (5.20)$$

The existence of the function $u(y,\lambda)$ is guaranteed by Theorem 5.6 and thus the number of periodic solutions with period 2π of the differential equation (5.12) equals the number of solutions of the bifurcation equation (5.18). Each such solution of the equation (5.12) is of the form $x = y + u(y,\lambda)$, where y is a solution of the bifurcation equation (5.18).

Now we show how, relatively simply, the stability properties of periodic solutions of the bifurcation equation (5.12) can be found. However, first we prove the lemma whose first part will also be important in further parts.

5.11 Lemma. *Let $U \subset R^n$ be an open set, $g \in C^r(U, R^n)$, where $1 \leq r \leq \infty$. Then these assertions hold.*

1. If $\varrho \in C^r(U,R)$ and $\varrho(x) \neq 0$ for all $x \in U$ then the differential equations

$$\dot{x} = \varrho(x)g(x), \qquad (5.21)$$
$$\dot{y} = (\operatorname{sign} \varrho(y))g(y) \qquad (5.22)$$

are orbitally C^r-equivalent.

2. If $\sigma \in C^r(U \times R, R)$, $0 < a < |\sigma(x,t)| < b < \infty$ for all $(x,t) \in U \times R$ and $g(x_0) = 0$ for a $x_0 \in U$ then $x(t) \equiv x_0$ is a stable (unstable) solution of the differential equation

$$\dot{x} = \sigma(x,t)g(x) \qquad (5.23)$$

iff it is also a stable (unstable) solution of the differential equation

$$\dot{y} = (\operatorname{sign} \sigma(y,t))g(y). \qquad (5.24)$$

Proof

First, let us assume that the function ϱ is positive. Let γ_x be a trajectory of the differential equation (5.22) defined by the solution $y : (\alpha, \beta) \to U$ fulfilling the initial condition $y(0) = x$, where $\alpha < 0 < \beta$. Let us define the function

$$\tau : (\alpha, \beta) \to R, \quad \tau(t) = \int_0^t (\varrho(y(s)))^{-1}\,ds.$$

As $\varrho(x) \neq 0$ for all $x \in U$,

$$\frac{d\tau(t)}{dt} = (\varrho(y(t)))^{-1} \neq 0 \qquad \text{for all } t \in (\alpha, \beta).$$

Therefore there exists an interval $(\alpha', \beta'), \alpha' < 0 < \beta'$ such that the mapping τ has inverse $\omega : (\alpha', \beta') \to (\alpha, \beta)$, for which it holds that

$$\frac{d\omega(s)}{ds} = \left(\frac{d\tau(\omega(s))}{dt}\right)^{-1} = \varrho(y(\omega(s))).$$

From this equality it follows that the mapping $x : (\alpha', \beta') \to U$, $x(t) = y(\omega(t))$ has the derivative

$$\frac{dx(t)}{dt} = \frac{d\omega(t)}{dt} \cdot \frac{dy(\omega(t))}{ds} = \varrho(y(\omega(t))) \cdot g(y(\omega(t))) = \varrho(x(t))g(x(t))$$

for all $t \in (\alpha', \beta')$. As $x(0) = y(\omega(0)) = y(0) = x$ we obtain that $\tilde{\gamma}_x = \{x(t) : t \in (\alpha', \beta')\}$ is a trajectory of the differential equation (5.21) and obviously $\gamma_x = \tilde{\gamma}_x$, while the trajectories $\gamma_x, \tilde{\gamma}_x$ have the same orientations. As we can write equation (5.22) in the form $\dot{y} = (\varrho(y))^{-1}\varrho(y)g(y)$ we can make the same consideration as in the preceding part of the proof when changing the tasks of differential equations and we obtain that each trajectory of differential equation (5.21) is also the trajectory of differential equation (5.22) with the same orientation. If the function ϱ is negative then we put $-g$ instead of g and $-\varrho$ instead of ϱ in the differential equations (5.21), (5.22). From the preceding part of the proof we obtain that the differential equation $\dot{x} = -(\varrho(x))(-g(x))$ is orbitally C^r-equivalent to the differential equation $\dot{y} = -g(y)$ and thereby assertion 1 is proved. Now we prove assertion 2. If $y : (\alpha, \beta) \to U$ is the solution of differential equation (5.24) fulfilling the initial condition $x(0) = x$ then $x(t) = y(\omega(t))$ is the solution of differential equation (5.23) fulfilling the initial condition $x(0) = x$, where ω is the inverse function of the function $\tau : (\alpha, \beta) \to R$ defined as the solution of the differential equation

$$\frac{d\tau}{dt} = \sigma(y(t), \tau)$$

fulfilling the initial condition $\tau(0) = 0$. From Theorem 1.41 the local existence of the function τ in a neighbourhood of the point $t = 0$ follows and from Theorem 1.42 its existence on the whole R also follows. As $\varrho(y, t) \neq 0$ for all $(y, t) \in U \times R$ thus

$$\frac{d\tau(t)}{dt} \neq 0 \quad \text{for all } t \in R$$

and therefore the function τ has its inverse ω. If $x(t)$ is a solution of the differential equation then $y(t) = x(\tau(t))$ is a solution of differential equation (5.24). From the relation between the differential equations (5.23), (5.24) which we have derived now, assertion 2 follows. Now let us return to the problem of searching stability properties of solutions of the differential equation (5.12).

Let $x = \Psi(z) = z + u(z,\lambda)$, where u is the mapping defined by the equation (5.15). When using the form of the equations (5.12) and (5.19) we obtain for $x(\Theta) = z(\Theta) + u(z,\lambda)(\Theta)$ the equality

$$f(z(\Theta) + u(z,\lambda)(\Theta), \lambda, \Theta) = \frac{dx(\Theta)}{d\Theta}$$

$$= \frac{dz(\Theta)}{d\Theta} + \frac{\partial u(z,\lambda)(\Theta)}{\partial y} \cdot \frac{dz(\Theta)}{d\Theta} + \frac{du(z,\lambda)(\Theta)}{d\Theta}$$

$$= \frac{dz(\Theta)}{d\Theta}\left(1 + \frac{du(z,\lambda)(\Theta)}{dy}\right) + \frac{1}{2\pi}\int_0^{2\pi} f(z(\Theta) + u(z,\lambda)(\Theta), \lambda, \Theta)\,d\Theta.$$

As $u(0,0) = 0$, $(\partial_1)_{(0,0)}u = 0$, for $|\lambda|$ sufficiently small we obtain that $z = z(\Theta)$ is a solution of the differential equation

$$\dot{z} = \left(1 + \frac{\partial u(z,\lambda)(\Theta)}{\partial y}\right)^{-1} H(z), \tag{5.25}$$

where $H(z)$ is the left side of the bifurcation equation (5.18). The function

$$\sigma(z,\lambda,\Theta) = \left(1 + \frac{\partial u(z,\lambda)(\Theta)}{\partial y}\right)^{-1}$$

is a periodic function of the variable Θ and there exist such positive constants a,b that for all $|\lambda|$ sufficiently small $a < \sigma(z,\lambda,\Theta) < b$ for all $(z,\Theta) \in U \times R$, where $U \subset R$ is a neighbourhood of the origin. We have proved that each 2π-periodic solution of the differential equation (5.12) is of the form $x(\Theta) = y + u(y,\lambda)(\Theta)$, where y is a solution of the bifurcation equation (5.18), i.e. $H(y) = 0$. From this form of the solution $x(\Theta)$ it follows that $x(\Theta)$ is stable (unstable) iff $z(t) \equiv y$ is a stable (unstable) solution of differential equation (5.25). According to the Lemma 5.11 this solution is a stable (unstable) iff the point y is a stable (unstable) singular point of the autonomous differential equation

$$\dot{z} = H(z). \tag{5.26}$$

We have proved the following theorem.

5.12 Theorem. *Let $r \in N, 1 \leq r < \infty$ and $f \in C^r(R^3, R)$ be a function of the variables $x, \lambda, \Theta, 2\pi$-periodic in the variable Θ, where $f(0,\lambda,\Theta) = 0$ for all $(\lambda,\Theta) \in R \times R$ and*

$$\frac{\partial f(0,0,\Theta)}{\partial x} = 0 \quad \text{for all } \Theta \in R.$$

Then there exists an open neighbourhood V of the point $\lambda = 0$ such that for $\lambda \in V$ the differential equation (5.12) has k periodic solutions with the period 2π iff the bifurcation equation (5.18) has k real solutions. Furthermore for $\lambda \in V$ these assertions hold.

1. Each 2π-periodic solution $x = x(\Theta)$ of the differential equation (5.12) is of the form $x = y + u(y, \lambda)$, where y is a real solution of the bifurcation equation (5.18) and $u(y, \lambda)$ is a 2π-periodic solution of the differential equation (5.19) fulfilling the condition (5.20).

2. The 2π-periodic solution $x = y + u(y, \lambda)$ of the differential equation (5.12) is stable (unstable) iff the singular point y of the differential equation (5.26) is stable (unstable), where $H(z)$ is the left side of the bifurcation equation (5.18).

5.2 Generic Bifurcations of 1-parameter Systems of Vector Fields in Neighbourhoods of Singular Points

Let us consider a germ $[f]_0 \in V_0^r (U \times V, R^N)$, i.e. the germ at the point $(u, \varepsilon) = (0, 0) \in R^N \times R$ of the parametrized vector field f, represented by the differential equation

$$\dot{u} = f(u, \varepsilon), \qquad (5.27)$$

where $f \in C^r(U \times V, R^N)$, $2 \leqq r \leqq \infty$, $U \times V \subset R^N \times R$ is an open neighbourhood of the origin and $f(0, 0) = 0$. Let us suppose that the linear differential equation

$$\dot{u} = Au, \qquad (5.28)$$

where $A = [d_0 f_0]$ ($f_0(u) = f(u, 0)$), has a quasi-hyperbolic singular point of degree 1. Only two cases can occur, namely: Either the matrix A has the eigenvalue $\lambda = 0$ of multiplicity 1 or it has a pair of complex conjugated pure imaginary eigenvalues of multiplicity 1, whereby in both cases it has no other eigenvalues with zero real part. Without loss of generality we can assume that $A = \text{diag}\{A^-, A^+, A^0\}$, where the matrix $A^-(A^+; A^0)$ has all its eigenvalues with negative (positive; zero) real part. According to the Šošitaišvili Theorem, there exists such a local centre C^r-manifold V_{loc}^c of the parametrized differential equation (5.27) that the germ $[f]_0$ is orbitally topologically equivalent to the germ at the point 0, represented by the parametrized system of differential equations of the form

$$\dot{x} = -x, \ \dot{y} = y, \qquad (5.29)$$

$$\dot{z} = g(z, \varepsilon), \tag{5.30}$$

where $x \in R^m, y \in R^n$ and differential equation (5.30) represents the reduction of parametrized differential equation (5.27) (bifurcation equation) to the centre manifold V_{loc}^c.

5.13 Definition. *If $f(u, \varepsilon) = 0$, then the point (u, ε) is called the singular point of parametrized differential equation (5.27). The set of all the singular points of this equation will be referred to as $K(f)$.*

5.14 Definition. *Let $[f]_0$ be the germ represented by the parametrized differential equation (5.27). We say that this germ is of the type $m(0)$ $(m(I))$, if the linear differential equation (5.28) has such a quasihyperbolic singular point of degree 1 that the matrix A has the eigenvalue $\lambda = 0$ (a pair of complex conjugated pure imaginary eigenvalues).*

5.15 Theorem on the Saddle-Node Type Bifurcation. *Let $[f]_0$ be a germ of the type $m(0)$, represented by the parametrized differential equation (5.27). Let us assume that*

$$\alpha = \frac{\partial g(0,0)}{\partial \varepsilon} \neq 0, \quad \text{and} \quad \beta = \frac{1}{2}\frac{\partial^2 g(0,0)}{\partial z^2} \neq 0$$

where g is the right side of the bifurcation equation (5.30) corresponding to the equation (5.27). Then the following assertions hold.

1. There exists an open neighbourhood W of the origin in R^2 such that the set $K(g) \cap W$ is a one-dimensional C^r-submanifold in R^2 ($K(g)$ is the set of singular points of the bifurcation equation (5.30)).

2. If $\alpha\beta < 0$ $(\alpha\beta > 0)$, then for $\varepsilon > 0$ $(\varepsilon < 0)$ the differential equation (5.27) has two singular points $K_{1\varepsilon}$, $K_{2\varepsilon}$, for $\varepsilon = 0$ it has one singular point $K_0 = (0, 0)$ and for $\varepsilon < 0$ $(\varepsilon > 0)$ it has no singular point.

3. Let $L(K_{1\varepsilon})$ $(L(K_{2\varepsilon}))$ be the matrix of linearization of the differential equation (5.27) at the point $K_{1\varepsilon}(K_{2\varepsilon})$ and let s_1, s_2 be the number of positive eigenvalues of the matrix $L(K_{1\varepsilon})$ and $L(K_{2\varepsilon})$, respectively. Then $|s_1 - s_2| = 1$.

Proof

Since $g(0, 0) = 0$ and

$$\alpha = \frac{\partial g(0,0)}{\partial \varepsilon} \neq 0,$$

from the Implicit Function Theorem it follows that there exists an open neighbourhood $W = \tilde{U} \times \tilde{V}$ of the origin in R^2 and a function $\varphi \in C^r(\tilde{U}, R)$, $\varepsilon = \varphi(z)$, such that $\varphi(0) = 0$ and

Chapter 5

$$g(z, \varphi(z)) = 0 \qquad (5.31)$$

for all $z \in \tilde{U}$. As the germ $[f]_0$ is of the type $m(0)$, then $\partial g(0,0)/\partial z = 0$ and the derivation of the equality (5.31) with respect to z yields

$$\frac{\mathrm{d}\varphi(0)}{\mathrm{d}z} = 0 \quad \text{and} \quad \frac{\mathrm{d}^2\varphi(0)}{\mathrm{d}z^2} = -\frac{\beta}{\alpha}.$$

Let us assume that $\alpha\beta < 0$ (the case $\alpha\beta > 0$ is analogous). Then $K(g) \cap W = \text{graph}(\varphi)$. According to the Šošitaišvili Theorem the parametrized differential equation (5.27) is orbitally topologically equivalent the parametrized system (5.29), (5.30), and thus, assertions 1 and 2 follow from the properties of the sets $K(g)$ and graph (φ), introduced above. In the proof of assertion 3 it suffices to assume that the equation (5.27) is in the form of the system (5.29), (5.30). Then $K_i = (0, 0, z_i)$, where $g(z_i, \varepsilon) = 0$, and $L(K_{i\varepsilon}) = \text{diag}\{-I_m, I_n, h(z_i)\}$, $i = 1, 2$,

$$h : \tilde{U} \to R, \quad h(z) = \frac{\partial g(z, \varphi(z))}{\partial z}.$$

Since

$$\frac{\mathrm{d}h(z)}{\mathrm{d}z} = \frac{\partial^2 g(z, \varphi(z))}{\partial z^2} + \frac{\partial^2 g(z, \varphi(z))}{\partial \varepsilon \partial z} \frac{\mathrm{d}\varphi(z)}{\mathrm{d}z},$$

we obtain

$$\frac{\mathrm{d}h(0)}{\mathrm{d}z} \neq 0.$$

This means that the function h changes its sign at the point 0. As the points $(z_1, \varphi(z_1)), (z_2, \varphi(z_2))$ lie in different connected components of the set $K(g) \cap W \setminus \{0\}$, it must hold that $|s_1 - s_2| = 1$. □

On analysing the qualitative properties of singular points the following lemma which is the direct consequence of the Malgrange–Weierstrass Theorem will be used to great advantage.

5.16 Lemma. *Let $g \in C^r(U, R)$, where $1 \leq k \leq r \leq \infty$, $k < \infty$, $U \subset R$ is an open neighbourhood of the origin and*

$$g(0) = 0, \quad \frac{\mathrm{d}^j g(0)}{\mathrm{d}z^j} = 0, \quad j = 1, 2, \ldots, k-1.$$

Therefore exists an open neighbourhood $V \subset U$ of the origin and such a function $h \in C^r(V, R)$ that $g(z) = z^k h(z)$ for all $z \in V$.

Now let us come back to the bifurcation equation (5.30). Let us assume that the assumptions of Theorem 5.15 on the bifurcation equation (5.30) are fulfilled. According to Lemma 5.16 there exists an open neighourhood V of the origin and a function $h \in C^r(V, R)$ such that $g(z, 0) = z^2 h(z)$ for all $z \in V$, whereby from the assumptions of Theorem 5.15 it follows that $h(0) = 1/2\beta \neq 0$. Therefore from Lemma 5.11 it follows that the differential equation (5.30) for $\varepsilon = 0$ is orbitally C^r-equivalent in a neighbourhood of the point 0 to the differential equation

$$\dot{z} = \sigma z^2, \qquad (5.32)$$

where $\sigma = \text{sign}\,\beta$. The differential equation for $\sigma = -1$ can be obtained from that for $\sigma = 1$ by transforming the time $t \to -t$, and so without loss of generality one can assume that $\sigma = 1$. The differential equation

$$\dot{z} = z^2 \qquad (5.33)$$

has the flow of the form $\varphi(z,t) = -z(tz-1)^{-1}$. If $z < 0$ ($z > 0$) then $\lim_{t \to \infty} \varphi(z,t) = 0$ ($\lim_{t \to -\infty} \varphi(z,t) = 0$), and thus, from Theorem 5.15 it follows that bifurcations of the bifurcation equation (5.30) are such as shown in Fig. 34. The topological structure of trajectories of equation (5.30) for $\varepsilon = 0$ will originate in the case of $n = 2$ by merging the saddle with the node and that is why the singular point of differential equation (5.32) is called the bifurcation of the saddle-node type and the corresponding bifurcation is called the saddle-node bifurcation. The above bifurcation of the bifurcation equation (5.30) defines the bifurcation of the whole system (5.29) and (5.30) and thus, of the system (5.27), as well. For example, the system

$$\dot{x} = -x, \ \dot{z} = z^2, \ \dim x = 1$$

has the bifurcations as shown in Fig. 35.

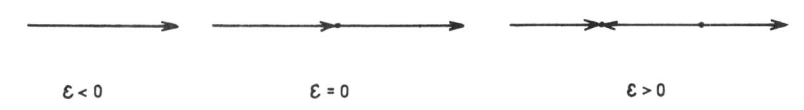

Fig. 34. Bifurcation of the saddle-node type.

The set of singular points $K(g)$ of the bifurcation equation (5.30) in a neighbourhood of the origin was successfully expressed using the Implicit Function Theorem and this set does not differ there from the parabola $S = \{(z,\varepsilon) : \varepsilon = z^2\}$. We shall prove that there exist coordinates in a neighbourhood of the origin of the state space and of the space of parameters in which the set of singular points of the bifurcation equation (5.30) coincides with parabola S. More precisely, it is expressed by the following theorem.

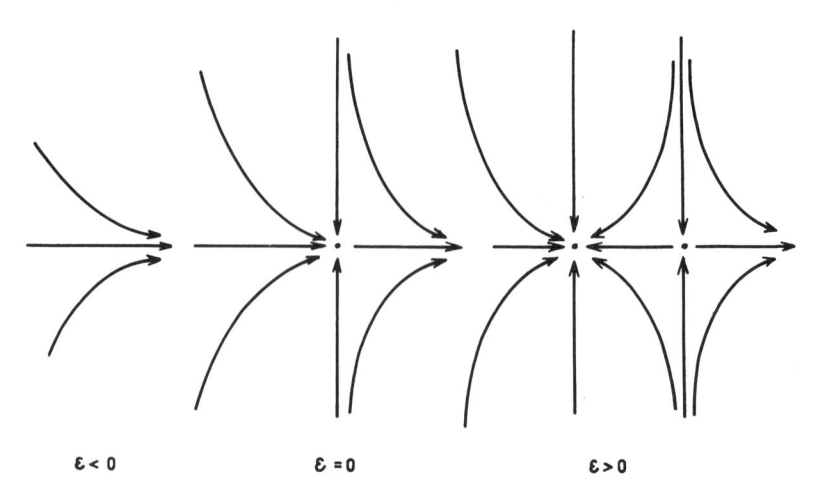

Fig. 35. Bifurcation of the saddle-note type.

5.17 Theorem. *Let the assumptions of Theorem 5.15 be satisfied and let $\sigma = \operatorname{sign}\beta$. Then the germ $[G]_0$ represented by the system (5.29), (5.30) is orbitally equivalent to the germ $[H]_0$ which is represented by the system*

$$\dot{x} = -x, \quad \dot{y} = y \tag{5.34}$$
$$\dot{v} = \sigma v^2 - \mu \tag{5.35}$$

Proof

Without loss of generality we can assume that $\sigma = 1$. From the Malgrange–Weierstrass Theorem it follows that there exist C^r-functions $\Theta(z,\varepsilon), a(\varepsilon), b(\varepsilon)$ such that $\Theta(0,0) = 1$, $a(0) = 0$, $b(0) = 0$,

$$\frac{da(0)}{d\varepsilon} = \alpha \neq 0$$

and $g(z,\varepsilon) = \Theta(z,\varepsilon)\,(z^2 + b(\varepsilon)z + a(\varepsilon))$ for (z,ε) from a sufficiently small neighbourhood of the origin. From Lemma 5.11 it follows that the germ at point 0, represented by the differential equation (5.30) is orbitally C^r-equivalent to the germ at point 0, which is represented by the differential equation

$$\dot{z} = z^2 + b(\varepsilon)z + a(\varepsilon). \tag{5.36}$$

Since $z^2 + b(\varepsilon)z + a(\varepsilon) = (z + (1/2)b(\varepsilon))^2 + a(\varepsilon) - (1/4)(b(\varepsilon))^2$, the mappings $v = z + (1/2)b(\varepsilon)$, $\mu = a(\varepsilon) - (1/4)(b(\varepsilon))^2$ define the germ of C^r-diffeomorphisms which is the conjugating germ of the germs at point 0, represented by differential equations (5.35) and (5.36). Since the systems (5.29), (5.34) are identical, the theorem is thus proved. □

5.18 Theorem. *The germ $[G]_0$, represented by the parametrized system of differential equations (5.34), (5.35) is versal.*

Proof

According to Theorem 4.40 it suffices to prove that the germ $[\tilde{H}]_0$, represented by the bifurcation equation (5.35) is versal. Let the deformation $[K]_0$ of the germ at point 0 be given. It is represented by differential equation (5.35) for $\mu = 0$ which is represented by the parametrized differential equation

$$\dot{v} = k(v, \varepsilon), \tag{5.37}$$

where $\varepsilon \in R^q$, $k \in C^r$. Since $k(v,0) = v^2$, by analogous use of the Malgrange–Weierstrass Theorem and Lemma 5.11 as that used on proving Theorem 5.17 it can be shown that the germ $[K]_0$ is orbitally C^r-equivalent to a germ $[L]_0$ which is represented by the parametrized differential equation

$$\dot{z} = z^2 + B(\varepsilon)z + A(\varepsilon), \tag{5.38}$$

where $A, B \in C^r$, $A(0) = 0$, $B(0) = 0$. If we choose $\mu = \Psi(\varepsilon) = A(\varepsilon) - (1/4)(B(\varepsilon))^2$ then the germ $[M]_0$ represented by the parametrized differential equation

$$\dot{v} = v^2 - \Psi(\varepsilon) \tag{5.39}$$

is induced from germ $[\tilde{H}]_0$ and that is orbitally C^r-equivalent to the germ $[L]_0$, where $z = v - (1/2)B(\varepsilon)$ is the conjugating diffeomorphism of the equations (5.38), (5.39). $\square$

We shall now look for local bifurcations in a neighbourhood of point 0 of vector fields of the form (5.27), which have at point 0 the germ of the type $m(I)$.

5.19 Lemma. *The germ $[f]_0$ of the type $m(I)$, represented by the parametrized differential equation (5.27) is orbitally topologically equivalent to the germ at point 0, which is represented by the parametrized system of differential equations (5.29), (5.30), where the bifurcation equation (5.30) is of the form*

$$\dot{z} = B(\varepsilon)z + R(z, \varepsilon), \tag{5.40}$$

$B \in C^r(V, M(2))$, $R \in C^r(U \times V, R^2)$, $U \times V$ *is an open neighbourhood of the origin in* $R^2 \times R$, $R(0, \varepsilon) = 0$ *for all* $\varepsilon \in V$, $d_{(0,0)}R = 0$, *where*

$$B(\varepsilon) = \begin{bmatrix} \lambda_1(\varepsilon) & \lambda_2(\varepsilon) \\ -\lambda_2(\varepsilon) & \lambda_1(\varepsilon) \end{bmatrix}, \qquad (5.41)$$

$\lambda_j \in C^r(V, R)$, $j = 1, 2$, $\lambda_1(0) = 0, \lambda_2(0) = 1$.

Proof

Since the germ $[f]_0$ is of the type $m(I)$ and the matrix $A = [d_0 f_0]$ has pure imaginary eigenvalues $\lambda_0 = i\omega, \overline{\lambda}_0$ of multiplicity 1 and has no other eigenvalues with zero real part. From the Šošitaišvili Theorem it follows that the germ $[f]_0$ is orbitally topologically equivalent to the germ at point 0 which is represented by a parametrized system of differential equations of the form (5.29), (5.30). From Lemma 5.11 it follows that the germ at point 0 represented by bifurcation equation (5.30) is orbitally equivalent to the germ at point 0 which is represented by the system of parametrized differential equations (5.29), (5.30). From Lemma 5.11 it follows that the germ at point 0 represented by the bifurcation equation (5.30) is orbitally equivalent to the germ at point 0 represented by the parametrized differential equation $\dot{z} = \tilde{g}(z, \varepsilon) = \omega^{-1} g(z, \varepsilon)$. The matrix $\tilde{A} = [d_0 \tilde{g}_0]$ obviously has the eigenvalues $\nu_0 = i, \overline{\nu}_0$ and has no other eigenvalues with zero real part. Since $\tilde{g}(0, 0) = 0$ and $[d_0 \tilde{g}_0]$ is a regular matrix, from the Implicit Function Theorem it follows that there exists an open neighbourhood $U \times V$ of the origin in $R^2 \times R$ and a C^r-function $\varphi : V \to U, x = \varphi(\varepsilon)$ such that $\varphi(0) = 0$ and $\tilde{g}(\varphi(\varepsilon), \varepsilon) = 0$ with $\varepsilon(0) = 0$ and $\tilde{g}(\varphi(\varepsilon), \varepsilon) = 0$ for all $\varepsilon \in V$. If we introduce a new variable $y = z - \varphi(\varepsilon)$, we obtain the equation $\dot{y} = \hat{g}(y, \varepsilon) = \tilde{g}(y + \varphi(\varepsilon), \varepsilon)$ for which it is obviously true that $\hat{g}(0, \varepsilon) = 0$ for all $\varepsilon \in V$. From Lemma 3.65 it follows that the germ of this parametrized equation at point 0 is orbitally C^r-equivalent with that at point 0 which is represented by a parametrized differential equation (5.40), where B and R have the properties as specified in Lemma. □

5.20 Definition. *If $\dot{\lambda}_1(0) \neq 0$ then we say that for the system (5.40) the transversality condition of eigenvalues is fulfilled.*

According to Theorem 3.150, the germ at point 0 represented by the bifurcation equation (5.40) is orbitally C^r-equivalent to that at point 0, which is represented by the parametrized system of differential equations

$$\dot{y}_1 = \alpha \varepsilon y_1 + y_2 + y_1 \left(\sum_{2m+n=2}^{r-1} (b_{1mn} y_1 - b_{2mn} y_2) \varepsilon^n (y_1^2 + y_2^2)^m \right)$$
$$+ \tilde{P}_1(y, \varepsilon) \qquad (5.42)$$
$$\dot{y}_2 = -y_1 + \alpha \varepsilon y_2 + y_2 \left(\sum_{2m+n=2}^{r-1} (b_{1mn} y_1 + b_{2mn} y_2) \varepsilon^n (y_1^2 + y_2^2)^m \right)$$
$$+ \tilde{P}_2(y, \varepsilon),$$

where the numbers $b_{1mn}, b_{2mn} \in R$ are independent of ε, $\tilde{P} = (\tilde{P}_1, \tilde{P}_2)$ is a C^r-mapping, $P(0,0) = 0$, $d_{0,0}^s \tilde{P} = 0$, $s = 1, 2, 3, \ldots, r-1$, $\alpha = \dot{\lambda}_1(0)$. According to Theorem 3.150 this germ is C^r-equivalent to that at point 0, which is represented by the parametrized system of differential equations

$$\dot{\varrho} = R(\varrho, \Theta, \varepsilon) \tag{5.43}$$
$$\dot{\Theta} = 1 + Q_2(\varrho, \Theta, \varepsilon), \tag{5.44}$$

where

$$R(\varrho, \Theta, \varepsilon) = \varrho \left(\alpha \varepsilon + \sum_{2m+n=2}^{r-1} c_{mn} \varepsilon^n \varrho^{2m} + Q_1(\varrho, \Theta, \varepsilon) \right),$$

$c_{mn} \in R$, $Q_j \in C^{r-2}$, $j = 1, 2$, are 2π-periodic functions in the variable Θ, $Q_j(0, \Theta, \varepsilon) \equiv 0, j = 1, 2$ and $Q_1(\varrho, \Theta, \varepsilon) = o(|\varrho|^{r-1})$ for arbitrary Θ, ε.

5.21 Hopf Bifurcation Theorem. *Let $[f]_0$ be a germ of type $m(I)$ represented by the parametrized differential equation (5.27), where $f \in C^r(U \times V, R^N)$, $r \geq 4$, $U \times V \subset R^N \times R$ is an open neighbourhood of the origin and $f(0, \varepsilon) = 0$ for all $\varepsilon \in V$. Let us suppose that $\lambda(\varepsilon) = \lambda_1(\varepsilon) + i\lambda_2(\varepsilon)$ is an eigenvalue of the matrix $A(\varepsilon) = [d_0 f_\varepsilon]$ such that $\lambda_1(0) = 0$ and the transversality condition $\alpha = \dot{\lambda}_1(0) \neq 0$ is satisfied. Then the germ $[f]_0$ is orbitally topologically equivalent to the germ $[G]_0$ which is represented by the parametrized system of differential equations (5.29), (5.30), and by the bifurcation equation (5.30) has the form of the system (5.42). The germ $[H]_0$, represented by the parametrized system (5.42) is orbitally C^{r-1}-equivalent to the germ $[K]_0$ which is represented by the parametrized system (5.43), (5.44). If in the differential equation (5.43) $c_{10} \neq 0$, then there exists a number $a > 0$, an open neighbourhood $\tilde{U}$ of point $y = (0, 0)$, a C^{r-2}-function $\varphi : U \to R, \varepsilon = \varphi(y)$ such that $\varphi(0) = 0$ and the following assertions hold.*

1. If $c_{10}\alpha < 0$ $(c_{10}\alpha > 0)$, then for every $\varepsilon \in (0, a)$ $(\varepsilon \in (-a, 0))$ there exists a unique periodic trajectory $\gamma_\varepsilon \subset \tilde{U}$ of the system (5.42), whereby $\tilde{\gamma}_\varepsilon = \gamma_\varepsilon \times \{\varepsilon\} \subset (y, \varepsilon) : \varepsilon = \varphi(y), y \in \tilde{U}\}$ (see Fig. 39). The periodic trajectory γ_ε is asymptotically stable if $\alpha\varepsilon < 0$ and asymptotically unstable if $\alpha\varepsilon > 0$ (see Fig. 40).

2. For every $\varepsilon \in (-a, a)$ the point $y = 0$ is a singular point of the system (5.42), which is asymptotically stable if $\alpha\varepsilon > 0$ and asymptotically unstable if $\alpha\varepsilon < 0$.

Proof

The orbital topological equivalence of the germs $[f]_0$ and $[G]_0$ follows from the Šošitaišvili Theorem. From Lemma 5.19 and Theorem 3.150 it follows that the germs $[H]_0, [K]_o$ are orbitally C^{r-1}-equivalent. Now it suffices to treat just the existence and stability properties of periodic trajectories of the system (5.42), or the system (5.43), (5.44). Let $(\varrho(t), \Theta(t))$ be a solution of the system (5.43), (5.44) (for simplicity in the notation of this solution we omit ε). Since from equation (5.44) we obtain that

$$\frac{d\Theta(t)}{dt} \neq 0 \quad \text{for all } t \in R,$$

the time t can be transformed by means of the mapping $\vartheta = \Theta(t)$. If $\omega = \omega(\vartheta)$ is the inverse mapping of Θ, then the function $\tilde{\varrho} = \tilde{\varrho}(\vartheta) = \varrho(\omega(\vartheta))$ is the solution of the differential equation

$$\dot{\tilde{\varrho}} = \frac{d\tilde{\varrho}}{d\vartheta} = \tilde{R}(\tilde{\varrho}, \vartheta, \varepsilon) = \frac{R(\tilde{\varrho}, \vartheta, \varepsilon)}{1 + Q_2(\tilde{\varrho}, \vartheta, \varepsilon)}, \qquad (5.45)$$

where R, Q_2 are the functions from the right side of the system (5.43), (5.44). Let us note that the function $\vartheta = \Theta(\omega(\vartheta))$ is the solution of the differential equation

$$\dot{\vartheta} = 1. \qquad (5.46)$$

If $\tilde{\varrho} = \tilde{\varrho}(\vartheta)$ is a solution of differential equation (5.45), then $(\varrho(t), \Theta(t))$, where $\varrho(t) = \tilde{\varrho}(\Theta(t))$, is the solution of system (5.43), (5.44). The systems (5.43), (5.44) and (5.45), (5.46) obviously have the same trajectories which only differ just in the parametrization preserving the orientation of trajectories. From the form of the right side of the system (5.42) it follows that

$$Q_2(0, \vartheta, \varepsilon) \equiv 0, \quad \frac{\partial Q_2(0, \vartheta, \varepsilon)}{\partial \tilde{\varrho}} \equiv 0,$$

and thus, the function $\tilde{Q}_2(\tilde{\varrho}, \vartheta, \varepsilon) \stackrel{\text{def}}{=} (1 + Q_2(\tilde{\varrho}, \vartheta, \varepsilon))^{-1} - 1$ can be written in the form $\tilde{Q}_2(\tilde{\varrho}, \vartheta, \varepsilon) = \tilde{\varrho}^2 A(\tilde{\varrho}, \vartheta, \varepsilon)$, where

$$A(\tilde{\varrho}, \vartheta, \varepsilon) = \int_0^1 \int_0^1 \frac{\partial^2 \tilde{Q}_2(st\tilde{\varrho}, \vartheta, \varepsilon)}{\partial \tilde{\varrho}^2} \, ds \, dt.$$

Thus, we obtain that $(1 + Q_2(\tilde{\varrho}, \vartheta, \varepsilon))^{-1} = 1 + \tilde{\varrho}^2 A(\tilde{\varrho}, \vartheta, \varepsilon)$ for all $(\tilde{\varrho}, \vartheta, \varepsilon)$. Therefore the right side of the differential equation (5.45) can be written in the form

$$\tilde{R}(\tilde{\varrho}, \vartheta, \varepsilon) = \tilde{\varrho}(\alpha\varepsilon + c_{10}\tilde{\varrho}^2 + \tilde{\varrho}^4 P(\tilde{\varrho}, \vartheta, \varepsilon) + \varepsilon\tilde{\varrho}^2 Q(\tilde{\varrho}, \vartheta, \varepsilon)), \qquad (5.47)$$

where P, Q are functions of the class C^{r-2}, 2π-periodic in the variable ϑ. In polar coordinates the system (5.42) is represented by the system (5.43), (5.44) and since this system has the same trajectories as the system (5.45), (5.46), we obtain that the system (5.42) has for a given ε a periodic trajectory γ_ε lying in a neighbourhood $\tilde{U} \subset R^2$ of the origin iff the differential equation (5.45) has a 2π-periodic solution $\tilde{\varrho} = \tilde{\varrho}(\vartheta)$ for which $(\tilde{\varrho}(\vartheta)\cos\vartheta, \tilde{\varrho}(\vartheta)\sin\vartheta) \in \tilde{U}$ for all $\vartheta \in R$ (see Fig. 36). Thus, we can to the number of 2π-periodic solutions of differential equation (5.45). According to Theorem 5.12, every 2π-periodic solution $\tilde{\varrho} = \tilde{\varrho}(\vartheta)$ of differential equation (5.45) has the form $\tilde{\varrho}(\vartheta) = \varrho + u(\varrho, \varepsilon)(\vartheta)$, where ϱ is a real solution of the equation

$$H(\varrho, \varepsilon) = \int_0^{2\pi} \tilde{R}(\varrho + u(\varrho, \varepsilon)(\vartheta), \vartheta, \varepsilon)\,d\vartheta = 0, \qquad (5.48)$$

where $u(\varrho, \varepsilon)$ is the 2π-periodic solution of the differential equation

$$\frac{du}{d\vartheta} = \tilde{R}(\varrho + u, \vartheta, \varepsilon) - \frac{1}{2\pi}\int_0^{2\pi} \tilde{R}(\varrho + u, \vartheta, \varepsilon)\,d\vartheta, \qquad (5.49)$$

where

$$\int_0^{2\pi} u(\varrho, \varepsilon)(\vartheta)\,d\vartheta = 0.$$

Since $\tilde{R}(0, \vartheta, \varepsilon) \equiv 0$, so $u(0, \varepsilon)(\vartheta) \equiv 0$ and thus, $u(\varrho, \varepsilon)(\vartheta) = \varrho\tilde{u}(\varrho, \varepsilon)(\vartheta)$, where $\tilde{u} \in C^{r-2}$, whereby

$$\int_0^{2\pi} \tilde{u}(\varrho, \varepsilon)(\vartheta)\,d\vartheta = 0.$$

Therefore from the equality (5.47) it follows that $H(\varrho, \varepsilon) = \varrho(\alpha\varepsilon + \tilde{c}_{10}\varrho^2 + \varrho^4\overline{P}(\varrho, \varepsilon) + \varepsilon\varrho^2\overline{Q}(\varrho, \varepsilon))$, where $\text{sign}\,\tilde{c}_{10} = \text{sign}\,c_{10}$ and $\overline{P}, \overline{Q} \in C^{r-2}$. From the Implicit Function Theorem it follows that there exists a number $a > 0$ and a unique C^{r-2} function $h : (-a, a) \to R$, $\varepsilon = h(\varrho)$ such that $h(0) = 0$, $H(\varrho, h(\varrho)) = 0$ for all $\varrho \in (-a, a)$. Thus, from equation (5.48) we obtain that

$$\varepsilon = h(\varrho) = -\frac{\tilde{c}_{10}}{\alpha}\varrho^2 + \varrho^4\Psi(\varrho), \qquad (5.50)$$

where $\Psi \in C^{r-2}$. What are we interested in are just the positive solutions $\varrho = \varrho(\varepsilon)$ of the equation (5.48) and these are lying on the graph of the function $\varepsilon = h(\varrho)$ for $\varrho > 0$ (see Figs. 37, 38). For the function $\varphi : \tilde{U} \to R$, $\varepsilon = \varphi(y) = h((y_1^2 + y_2^2)^{1/2}$, where $y = (y_1, y_2)$, assertion 1 holds. The periodic trajectory γ_ε is obviously asymptotically stable (asymptotically unstable) iff its corresponding 2π-periodic solution $\tilde{\varrho}(\vartheta) = \varrho + u(\varrho, \varepsilon)(\vartheta)$

Chapter 5

of differential equation (5.45) is an asymptotically stable (asymptotically unstable). According to Theorem 5.12 this solution is asymptotically stable (asymptotically unstable), iff ϱ is an asymptotically stable (asymptotically unstable) singular point of the differential equation

$$\dot{z} = H(z, \varepsilon), \qquad (5.51)$$

where H is the mapping from the left side of the equation (5.48). Direct verification of stability properties of this singular point will be left to the reader. The same result on stability properties of periodical trajectory γ_ε, however, can also be obtained in the following way. The point $y = 0$ is a singular point of the system (5.42) for all $\varepsilon \in (-a, a)$ and its stability properties are the same as those of the singular point $\varrho = 0$ of the differential equation (5.51) and the latter are identical with those of the singular point $\varrho = 0$ of the linear differential equation $\dot{\varrho} = \alpha \varepsilon \varrho$. It was yet proved that if $c_{10}\alpha < 0$ ($c_{10}\alpha > 0$), then for every $\varepsilon \in (0, a)$ ($\varepsilon \in (-a, 0)$) there exists a unique periodic trajectory $\gamma_\varepsilon \subset \tilde{U}$, and therefore this trajectory must have the stability properties just opposite to those of the singular point $y = 0$ for this ε, and thus assertion 2 holds. □

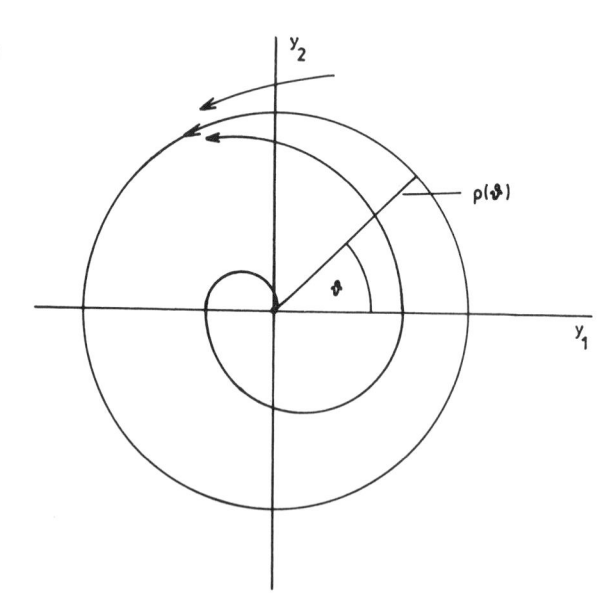

Fig. 36.

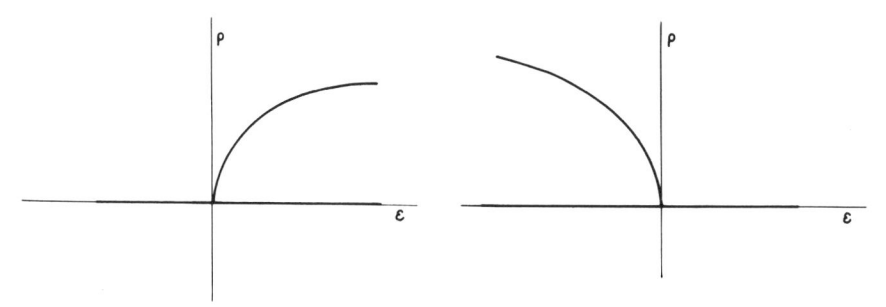

Fig. 37. Supercritical bifurcation ($c_{10}\alpha < 0$).

Fig. 38. Subcritical bifurcation ($c_{10}\alpha > 0$).

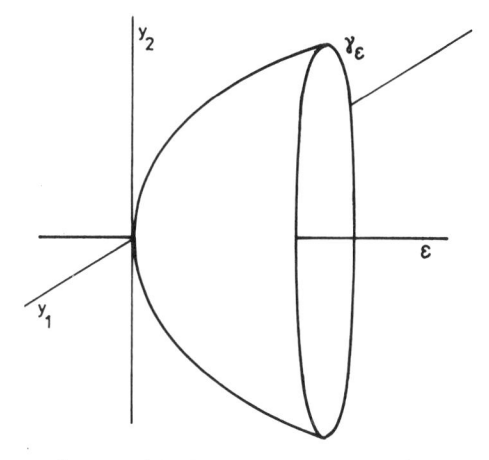

Fig. 39. Supercritical Hopf bifurcation ($c_{10}\alpha < 0$).

The bifurcation of originating of a periodic trajectory from a singular point when changing parameter described in Theorem 5.21 is called the Hopf bifurcation. The Hopf bifurcation corresponding to Fig. 37 (Fig. 38) is called supercritical (subcritical). The theorem on the existence of such a bifurcation for parametrized systems of differential equations in R^2 has already been formulated by H. Poincaré and its exact proof was presented in 1929 by A. A. Andronov (see, e.g., [6]). Therefore in [10] this bifurcation is referred to as the Poincaré–Andronov bifurcation. The multidimensional version of this bifurcation was proved in 1942 by E. Hopf. The translation of his original paper can be found in [98].

For specification of the type of the Hopf bifurcation as well as of the stability properties of the periodic trajectory, the signs of coefficients α and c_{10} of the differential equation (5.43) or the equation (5.48) are of special importance. To find the number α is a trivial task, but the computation of formula for the number c_{10}, i.e. expressing the number c_{10} as a function of

Chapter 5

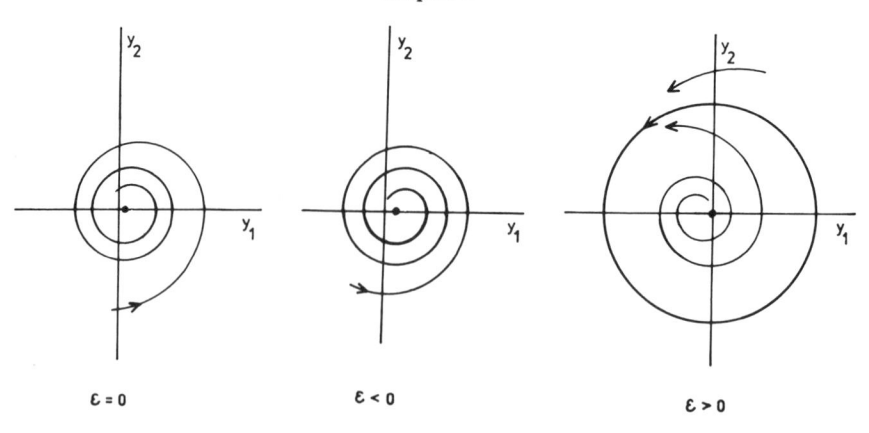

Fig. 40. Hopf bifurcation ($c_{10}\alpha < 0$).

coefficients of the original system (5.41) it is a rather complicated problem. In the papers [6] and [98] the formulae for the coefficients of the Poincaré mapping are calculated, but the computations are far more complicated. In [57], the formula for c_{10} is derived in a relatively simple way (see also [50]). However, the computations are still considerably extensive, that is why we only introduce the resulting formula.

Let the system

$$\dot{y} = -\omega y_2 + f(y_1, y_2),$$
$$\dot{y}_2 = \omega y_1 + g(y_1, y_2) \qquad (5.52)$$

be given, where $f, g \in C^r$, $f = P_2 + P_3 + \tilde{f}$, $g = Q_2 + Q_3 + \tilde{g}$, $\tilde{f}(y) = o(\|y\|^3)$, $\tilde{g}(y) = o(\|y\|^3)$,

$$P_2(y_1, y_2) = a_{20}y_1^2 + a_{11}y_1y_2 + a_{02}y_2^2,$$
$$P_3(y_1, y_2) = a_{30}y_1^3 + a_{21}y_1^2y_2 + a_{12}y_1y_2^2 + a_{03}y_2^3,$$
$$Q_2(y_1, y_2) = b_{20}y_1^2 + b_{11}y_1y_2 + b_{02}y_2^2,$$
$$Q_3(y_1, y_2) = b_{30}y_1^3 + b_{21}y_1^2y_2 + b_{12}y_1y_2^2 + b_{03}y_2^3.$$

According to Theorem 3.150 the germ at point 0 of the system (5.52) is orbitally C^{r-1}-equivalent to the germ at point 0 of a system of the form

$$\dot{\varrho} = A_3\varrho^3 + A_4\varrho^4 + \cdots + A_{r-1}\varrho^{r-1} + g_1(\varrho, \Theta),$$
$$\dot{\Theta} = \omega + g_2(\varrho, \Theta), \qquad (5.53)$$

where g_1, $g_2 \in C^{r-1}$ are 2π-periodic functions in the variable Θ and

$g_1(\varrho, 0) = O(|\varrho|^{r-1})$. According to [57]

$$A_3 = \frac{1}{16}[a_{30} + a_{12} + b_{21} + b_{03}] + \frac{1}{16\omega}[a_{11}(a_{20} + a_{02}) \quad (5.54)$$
$$- b_{11}(b_{20} + b_{02}) - a_{20}b_{20} + a_{02}b_{02}].$$

Let us note that the coefficient c_{10} in system (5.43) or in eq. (5.48) is also a coefficient at ϱ^3.

Now we shall indicate how the normal form of a smooth system of the form (5.40) can be derived with the use of the Takens Theorem. For $\varepsilon = 0$ the matrix of its linearization is of the form $B(0) = A = (a_{ij}) \in M(2)$, where $a_{11} = a_{22} = 0, a_{12} = -a_{21} = -a_{21} = 1$. The mapping $L^m : H^m \to H^m$ from the Takens Theorem is of the form

$$L^m(h)(y) = [d_y h] A y - A h(y) = [d_y h](y_2, -y_1)^* - (h_2(y), -h_1(y))^*, \quad (5.55)$$

where $h = (h_1, h_2)^* \in H^m$, i.e., h_1, h_2 are homogeneous polynomials in the variables y_1, y_2 of degree m. It is necessary to find a complementary space W^m to the space $B^m = \text{Image } L^m$, i.e., such a space that $H^m = B^m + W^m$. Let us note that it is not necessary that this sum is direct. For $m = 2$ the space H^2 has a basis consisting of the following vectors: $v_1 = (y_1^2, 0)^*$, $v_2 = (y_1 y_2, 0)^*$, $v_3 = (y_2^2, 0)^*$, $v_4 = (0, y_1^2)^*$, $v_5 = (0, y_1 y_2)^*$, $v_6 = (0, y_2^2)^*$. Let us denote $w_i = L^2(v_i)$, $i = 1, 2, \ldots, 6$. From the formula (5.55) we obtain that $w_1 = (2y_1 y_2, y_1^2)^*$, $w_2 = (y_2^2 - y_1^2, y_1 y_2)^*$, $w_3 = (-2y_1 y_2, y_2^2)^*$, $w_4 = (y_1^2, 2y_1 y_2)^*$, $w_5 = (-y_1 y_2, y_2^2 - y_1^2)^*$, $w_6 = (-y_2^2, -2y_1 y_2)^*$. These vectors can be expressed in the form of a linear combination of base vectors, i.e. of vectors v_i, $i = 1, 2, \ldots, 6$, and thus, the matrix representing the mapping L^2 with respect to the given basis is obtained. The reader can verify (or he can see from the procedure presented in [50, 69, 151]) that not only this matrix but also the matrices of all the mappings L^m with respect to the analogously chosen basis in H^m, where m is an even number, have non-zero determinants. This means that $B^m = \text{Image } L^m = H^m$ for all the even m. Therefore according to the Takens Theorem in normal form there will be no terms of the form $a y_1^i y_2^j$, where $a \in R$ and $i + j$ is an even number. One can also make sure that for every even number m there is codim $B^m = 2$ and for the complementary subspace W^m the subspace generated by the vectors $(y_1^2 + y_2^2)^n (y_1, y_2)^*, (y_1^2 + y_2^2)^n (y_2, -y_1)^*$ can be chosen, where $m = 2n + 1$. From the Takens Theorem we obtain that the germ at point 0, represented by the differential equation (5.40) for $\varepsilon = 0$ with a smooth right side, is orbitally C^∞-equivalent to the germ at point 0 which is represented by a differential equation of the form

$$\dot{x} = Ax + G(x), \quad (5.56)$$

where G is a smooth mapping whose formal Taylor expansion at point $0 \in R^2$ is of the form

$$T(G)(x) = \sum_{i=1}^{\infty}(x_1^2 + x_2^2)^i \begin{pmatrix} a_i & -b_i \\ b_i & a_i \end{pmatrix} \begin{pmatrix} x_1 \\ x_2 \end{pmatrix}. \quad (5.57)$$

This system is orbitally C^∞-equivalent to a system of the form

$$\dot{\varrho} = \tilde{R}(\varrho, \Theta) \quad (5.58)$$
$$\dot{\Theta} = \omega + Q(\varrho, \Theta), \quad (5.59)$$

where $\tilde{R}, Q$ are smooth functions, 2π-periodic in the variable Θ, whereby the formal Taylor expansion of the function $\tilde{R}$ at the point $(\varrho, \Theta) = (0, 0)$ is entirely independent of Θ and is of the form

$$T(\tilde{R})(\varrho, \Theta) = \sum_{i=3}^{\infty} A_i \varrho^{2i}.$$

From the Hopf bifurcation theorem we know that this bifurcation does not depend upon the terms of the order higher than 3, and thus, it suffices to consider just a one-parameter deformation of the differential equation of the form

$$\dot{\varrho} = R(\varrho), \quad (5.60)$$

where R is a smooth function independent of Θ and its formal Taylor expansion at the point $\varrho = 0$ is of the form

$$T(R)(\varrho) = \sum_{i=3}^{\infty} A_i \varrho^{2i}.$$

Let us then consider a parametrized differential equation of the form

$$\dot{\varrho} = \varrho v(\varrho, \varepsilon), \quad (5.61)$$

where $\varepsilon \in R$, $v \in C^\infty$, $\varrho v(\varrho, 0) = R(\varrho)$. Let us assume that $A_3 \neq 0$ and

$$\frac{\partial v(0,0)}{\partial \varepsilon} = \alpha \neq 0$$

which, in fact are the assumptions of the Hopf bifurcation theorem. From the Malgrange–Weierstrass Theorem it follows that there exists a neighbourhood $U \times V$ of the point $(\varrho, \varepsilon) = (0, 0)$ and smooth functions $a(\varepsilon)$, $b(\varepsilon)$

on V, $B(\varrho,\varepsilon)$ on $U \times V$ such that $v(\varrho,\varepsilon) = (a(\varepsilon) + b(\varepsilon)\varrho + \varrho^2)B(\varrho,\varepsilon)$ for all $(\varrho,\varepsilon) \in U \times V$, where $a(0) = 0$, $b(0) = 0$,

$$\frac{da(0)}{d\varepsilon} = \frac{\alpha}{A_3},$$

$B(0,0) = A_3$. If we define the transformation of coordinates through the mappings $\mu = \sigma[a(\varepsilon) - (1/4)(b(\varepsilon))^2]$, $\tilde{\varrho} = \varrho + (1/2)b(\varepsilon)$, where $\sigma = \text{sign}(\alpha A_3)$, then in these new coordinates the equation (5.61) is of the form $\dot{\tilde{\varrho}} = \tilde{\varrho}(\sigma\mu + \tilde{\varrho}^2)\tilde{B}(\tilde{\varrho},\mu)$, where $\tilde{B}$ is a smooth function such that $\tilde{B}(0,0) = A_3$. From Lemma 5.11 it follows that the germ at the point $(0,0)$ represented by this parametrized differential equation is C^∞-equivalent to the germ at the point $(0,0)$ which is represented by the parametrized equation

$$\dot{\varrho} = \tau\varrho(\sigma\mu + \varrho^2), \tag{5.62}$$

where $\tau = \text{sign}\, A_3$, $\sigma = \text{sign}(\alpha A_3)$. This equation together with the equation $\dot{\Theta} = 1$ is a representation in polar coordinates of the parametrized system of equations

$$\begin{aligned}\dot{x}_1 &= (\tau\sigma)\mu x_1 + x_2 + \tau x_1(x_1^2 + x_2^2) \\ \dot{x}_2 &= -x_1 + (\tau\sigma)\mu x_2 + \tau x_2(x_2^2 + x_2^2).\end{aligned} \tag{5.63}$$

This system has for $\sigma < 0, \tau > 0$ the same bifurcations as shown in Figs. 39, 40, where $\mu = \varepsilon$.

5.22 Theorem. *The germ $[G]_0$ represented by the parametrized system of differential equations*

$$\dot{x} = -x, \dot{y} = y, \tag{5.64}$$

$$\begin{aligned}\dot{z}_1 &= -\mu z_1 + z_2 + z_1(z_1^2 + z_2^2) \\ \dot{z}_2 &= -z_1 - \mu z_2 + z_2(z_1^2 + z_2^2),\end{aligned} \tag{5.65}$$

where $x \in R^m$, $y \in R^n$, is versal.

Proof

According to Theorem 4.40 it is sufficient to prove that the germ $[H]_0$ represented by the bifurcation equation (5.65) is versal. Let a k-parametric deformation $[K]_0$ of a germ represented by system (5.65) for $\mu = 0$ be given and let the germ $[K]_0$ be represented by a parametrized system of differential equations

Chapter 5

$$\dot{z} = v(z, \varepsilon), \tag{5.66}$$

where $v = (v_1, v_2) \in C^\infty$, $z = (z_1, z_2)$, $\varepsilon \in R^k$. Since $v_1(z, 0) = z_2 + z_1(z_1^2 + z_2^2)$ and $v_2(z, 0) = -z_1 + z_2(z_1^2 + z_2^2)$, the germ $[K]_0$ is orbitally C^0-equivalent to the germ $[L]_0$, which is represented by the parametrized system

$$\dot{\varrho} = \varrho w(\varrho, \varepsilon), \quad \dot{\Theta} = 1, \tag{5.67}$$

where $w \in C^\infty, \varepsilon \in R^k, \varrho w(\varrho, 0) = \varrho^3$. Using the Takens Theorem, the system (5.66) can be transformed into the form (5.56), where A and G depend upon ε. The germ at the point 0, represented by this system, is C^∞-equivalent to the germ at point 0 of the system

$$\dot{\varrho} = \varrho \tilde{w}(\varrho, \Theta, \varepsilon), \quad \dot{\Theta} = 1,$$

where the formal Taylor expansion at the point $(0, 0, 0)$ of the function $\tilde{w}$ is entirely independent of Θ. Therefore we obtain the C^0-equivalence of the germs $[K]_0$ and $[L]_0$. From the Malgrange–Weierstrass Theorem it follows, that there exist smooth functions $a(\varepsilon), b(\varepsilon), B(\varrho, \varepsilon)$ such that $a(0) = 0$, $b(0) = 0, B(0, 0) = 1$ and $w(\varrho, \varepsilon) = (a(\varepsilon) + b(\varepsilon)\varrho + \varrho^2)B(\varrho, \varepsilon)$ for all (ϱ, ε) from a small enough neighbourhood of the point $(0, 0)$. From Lemma 5.11 it follows that the germ $[L]_0$ is orbitally C^∞-equivalent to the germ $[M]_0$ which is represented by the parametrized system of differential equations

$$\dot{\varrho} = \varrho(a(\varepsilon) + b(\varepsilon)\varrho + \varrho^2), \quad \dot{\Theta} = 1. \tag{5.68}$$

Since the germ $[H]_0$ is orbitally C^0-equivalent to the germ $[N]_0$ which is represented by the parametrized system $\dot{\varrho} = \varrho(-\mu + \varrho^2), \dot{\Theta} = 1$, it suffices to prove that the germ $[M]_0$ is orbitally C^0-equivalent to a germ which is induced from the germ $[N]_0$. If we choose $\mu = \Psi(\varepsilon) = a(\varepsilon) - (1/4)(b(\varepsilon))^2$ then the germ $[P]_0$ is represented by the parametrized system $\dot{v} = v(-\Psi(\varepsilon) + v^2), \dot{\Theta} = 1$, induced from the germ $[N]_0$ and this is orbitally C^∞-equivalent to the germ $[M]_0$, whereby $\varrho = v + (1/2)b(\varepsilon)$ is the conjugate diffeomorphism of these germs. □

The bifurcation of the saddle-node type and the Hopf bifurcation which were described in Theorems 5.15 and 5.21 are generic bifurcations (see Theorem 5.24). The genericity of these bifurcations was proved in [142, 143 and 102]. In [143, 102] the result on generic bifurcations is formulated in a global way for the set $V^r(P, X)$, i.e., for the set of 1-parameter systems of vector fields on X of the class C^r, where X is an n-dimensional C^r-manifold representing the state space, P is a 1-dimensional C^r-manifold representing the set of parameters, whereby on the set $V^r(P, X)$ the Whitney topology is given. A part of the results contained in these papers is summarized in the following theorem. Before we formulate this theorem, let us introduce the following notations. If $F \in V^r(P, X)$ then $K(F) = \{(x,p) \in X \times P : x$ is the singular point of the vector field $F_p\}$, where $F_p \in V^r(X)$, $F_p(y) \stackrel{\text{def}}{=} F(y,p)$; $K_0(F)$ is the set of such points $(x,p) \in K(F)$ that for arbitrary open neighbourhood $U \times V$ of point (x,p), the point p is a bifurcation point of the parametrized vector field $F/U \times V$.

5.23 Theorem. *Let X, P be smooth manifolds, $\dim P = 1$ and $r \in N$, $r \geq 4$. Then there exists such a massive (open and dense, if X, P are compact) subset W^r in $V^r(P, X)$, such that for $F \in W^r$ it holds that*

1. if the set $K(F) \neq \emptyset$ then it is a one-dimensional C^r-submanifold in $X \times P$;

2. if the set $K_0(F) \neq \emptyset$, then it consists of isolated points and the local bifurcation in a neighbourhood of the point $(x,p) \in K_0(F)$ is either the Hopf's bifurcation or the bifurcation of the saddle-node type.

In Example 3.59 we have defined the so-called second-order differential equation on an n-dimensional C^r-manifold. The study of generic bifurcations for this special class of vector fields is more difficult than that for vector fields of general form, because only such approximations of these vector fields are admissible, which do not break their original form. In [104] a theorem analogous to Theorem 5.23 is proved for the set of 1-parameter system of second order differential equations on an n-dimensional C^r-manifold.

Let us denote by D^r the set of all parametrized differential equations of the form (5.27) whose right sides are of the class C^r.

Though the following theorem is a consequence of Theorem 5.23 its proof does not involve the same methods as Theorem 5.23, and, thus, we can present it along with its proof.

5.24 Theorem. *If $r \geq 4$, then there exists an open and dense subset D_1 in D^r such that if the parametrized differential equation $\dot{u} = f(u, \varepsilon)$ is from the set D_1 then the matrix $A = [d_0 f_0]$ has either no eigenvalue with zero real part or the germ $[f]_0$ represented by this equation is quasi-hyperbolic of degree 1 and the local bifurcation in a neighbourhood of the point $(x, \varepsilon) = (0, 0)$ is either the Hopf bifurcation or the bifurcation of the saddle-node type.*

Proof

Let us denote by D_0 the set of all parametrized differential equations $\dot{u} = f(u, \varepsilon)$ from D^r (further on, we write $f \in D^r$) such that the matrix $A = [d_0 f_0]$ has either no eigenvalue with zero real part or if it has any then the germ $[f]_0$ is quasi-hyperbolic of degree 1, i.e., of the type $m(0)$ or $m(I)$. From Theorem 3.106 it follows that the set D_0 is open and dense in D^r. Let us denote by D_1 the set of all $f \in D_0$, having the following properties.

1. If $[f]_0$ is a germ of the type $m(0)$ and the parametrized differential equation (5.30) is a reduction of f to the centre manifold, then

$$\alpha = \frac{\partial g(0,0)}{\partial \varepsilon} \neq 0 \quad \text{and} \quad \beta = \frac{1}{2} \cdot \frac{\partial^2 g(0,0)}{\partial z^2} \neq 0.$$

2. If $[f]_0$ is a germ of the type $m(I)$ then this germ is orbitally topologically equivalent to the germ $[G]_0$ represented by the system (5.29), (5.30), where the bifurcation equation is of the class C^r. The germ $[\tilde{G}]_0$ represented by the bifurcation equation (5.30) is orbitally C^r-equivalent to the germ at the point 0, which is represented by the parametrized differential equation (5.40) which fulfils the transversality condition $\dot{\lambda}_1(0) \neq 0$, where the coefficient c_{10} at ϱ^3 in its normal form (5.43), (5.44) is non-zero.

From Theorems 5.15, 5.21 it follows that it suffices to prove that the set D_1 is open and dense in the set D_0. Without any loss of generality, it can be assumed that the differential equation f has the form of the system (4.35). Since its centre manifold is defined by C^r-mappings, $x = h_1(z), y = h_2(z)$ which do not contain the terms of order 1 and its reduction to the centre manifold is of the form $\dot{z} = A^0 z + R(h_1(z), h_2(z), z, \varepsilon)$, the mappings h_1, h_2 exert no influence upon the terms of the order lower than the fourth in this reduction. In case 1, just the coefficients up to the second order are significant, while in case 2, only those up to the third order are (see the formula (5.54) for $A_3 = c_{10}$). Thus, it suffices to consider the reduction to the centre manifold, only. Thus, let the sets D^r and D_0 be such as already defined above, but let us assume that in the differential equation (5.27) $\dim u = 1$ or $\dim u = 2$. First, let us assume that $\dim u = 1$. Let us consider the set of 1-jets $J^1 = J^1(U \times V, R)$. The space J^1 can be identified (in the sense of their diffeomorphy; see the Proposition 2.70 and its proof) with the set

$$\left\{ \left(u, \varepsilon, g(u,\varepsilon), \frac{\partial g(u,\varepsilon)}{\partial u}, \frac{\partial g(u,\varepsilon)}{\partial \varepsilon} \right) : (u, \varepsilon) \in U \times V, h \in C^1(U \times V, R) \right\}.$$

Let us define these sets: $K = \{v = (a_0, b_0, c_0, a_{10}, a_{01}) \in J^1 : c_0 = 0\}$, $K_0 = \{v \in K : a_{10} = 0\}$. Obviously, it holds that $j^1(g(u,\varepsilon) \in K$ iff $g(u,\varepsilon) = 0$, $j^1 g(u,\varepsilon) = K_0$. The sets K, K_0 are obviously smooth submanifolds in J^1, where codim $K = 1$, codim $K_0 = 2$, i.e., $\dim K = 1$ and $\dim K_0 = 0$ and $K_0 \subset K$. Let $M \subset R^2$ be an open set containing the origin and it has a compact closure $\overline{M} \subset U \times V$. From the Thom Transversality Theorem it follows that the sets $T = \{g \in C^r(U \times V, R) : j^1 g \overline{\pitchfork}_{\overline{M}} K\}$ and $T_0 = \{g \in C^r(U \times V, R) : j^1 g \overline{\pitchfork}_{\overline{M}} K_0\}$ are open and dense in $C^r(U \times V, R)$, where $j^1 g$ is a 1-jet extension of the mapping g, i.e., $j^1 g : U \times V \to J^1, (u, \varepsilon) \mapsto j^1 g(u,\varepsilon)$. From Proposition 2.76 we obtain that for $g \in T \cap T_0$ the sets $K(g) = (j^1 g)^{-1}(K)$, $K_0(g) = (j^1 g)^{-1}(K_0)$ are C^{r-1} submanifolds in R^2, where $\dim K(g) = 1$, $K_0(g) \subset K(g)$ and $\dim K(g) = 0$ (which is in accordance with the assertions of Theorem 5.29 on sets $K(F), K_0(F)$). From the transversality condition $j^1 g \bar{\pitchfork}_M K$ it follows that for arbitrary $(u, \varepsilon) \in \overline{M}$,

$$\text{rank}\left(\frac{\partial g(u, \varepsilon)}{\partial u}, \frac{\partial g(u, \varepsilon)}{\partial \varepsilon}\right) = 1,$$

i.e. either

$$\frac{\partial g(u, \varepsilon)}{\partial u} \neq 0 \quad \text{or} \quad \frac{\partial g(u, \varepsilon)}{\partial \varepsilon} \neq 0.$$

From this conditions we obtain that if

$$\frac{\partial g(0, 0)}{\partial u} = 0$$

i.e., the germ $[g]_0$ is of the type $m(0)$, then

$$\alpha = \frac{\partial g(0, 0)}{\partial \varepsilon} \neq 0.$$

Now we shall prove that the transversality condition $j^1 g \bar{\pitchfork}_{(u,\varepsilon)} K_0$ is equivalent to the condition

$$\frac{\partial^2 g(u, \varepsilon)}{\partial u^2} \neq 0.$$

Let $z = j^1 g(u, \varepsilon) \in K_0$. Then $g(u, \varepsilon) = 0$,

$$\frac{\partial g(u, \varepsilon)}{\partial u} = 0 \quad \text{and} \quad \frac{\partial g(u, \varepsilon)}{\partial \varepsilon} \neq 0.$$

The transversality condition $j^1 g \bar{\pitchfork}_{(u,\varepsilon)} K_0$ means that $D_{(u,\varepsilon)}(j^1 g)(R^2) + T_z(K_0) = T_z(J^1)$. According to Proposition 2.18 the tangent space $T_z(K_0)$ is isomorphic with $\tilde{K}_0 = \{(a_0, b_0, c_0, a_{10}, a_{01}) \in R^5 : c_0 = 0, a_{10} = 0\}$ and the tangent space $T_z(J^1)$ is isomorphic with R^5. According to Proposition 2.24 the mapping $D_{(u,\varepsilon)}(j^1 g)$ is represented by the Jacobi matrix $[d_{(u,\varepsilon)}(j^1 g)]$, and thus we obtain that the above equality representing the transversality condition $j^1 g \bar{\pitchfork}_{(u,\varepsilon)} K_0$ is fulfilled iff for arbitrary $w = (w_1, w_2, \ldots, w_5) \in R^5$ there exist such $x = (x_1, x_2) \in R^2$ and $y = (y_1, y_2, 0, 0, y_5) \in R^5$ that

$$x_1 + y_1 = w_1,$$
$$x_2 + y_2 = w_2,$$
$$\frac{\partial g(u, \varepsilon)}{\partial \varepsilon} x_2 = w_3,$$
$$\frac{\partial^2 g(u, \varepsilon)}{\partial u^2} x_1 + \frac{\partial^2 g(u, \varepsilon)}{\partial \varepsilon \partial u} x_2 = w_4,$$
$$\frac{\partial^2 g(u, \varepsilon)}{\partial u \partial \varepsilon} x_1 + \frac{\partial^2 g(u, \varepsilon)}{\partial \varepsilon^2} x_2 + y_5 = w_5.$$

This system is solvable iff the system consisting of the third and fourth equations is solvable. This system, however, is solvable iff

$$\frac{\partial^2 g(u,\varepsilon)}{\partial u^2} \neq 0,$$

which was to be proved. The set $D_{10} = D_0 \cap T_0 \cap T$ is open dense in D^r (with $\dim u = 1$) and for all $g \in D_{10}$ we have

$$\alpha = \frac{\partial g(0,0)}{\partial \varepsilon} \neq 0, \qquad \beta = \frac{1}{2}\frac{\partial^2 g(0,0)}{\partial u^2} \neq 0.$$

Let us consider now the set D^r as the set of parametrized differential equations of the form (2.27), where $\dim u = 2$. In this case the bifurcations depend on terms up to order 3, and so, let us consider the set of 3-jets $J^3 = J^3(U \times V, R^2)$, which can be identified with the set $\{(u, \varepsilon, h_1(u,\varepsilon), h_2(u,\varepsilon), a_{10}, a_{01}, a_{20}, a_{02}, a_{11}, a_{30}, a_{21}, a_{12}, a_{03}, b_{10}, b_{01}, \ldots b_{30}, b_{21}, b_{12}, b_{03}) : (u, \varepsilon) \in U \times V, h = (h_1, h_2) \in C^1(U \times V, R^2)\}$, where

$$a_{ij} = a_{ij}(u,\varepsilon) = \frac{\partial^{i+j} h_1(u,\varepsilon)}{\partial u^i \partial \varepsilon^j}, b_{ij} = b_{ij}(u,\varepsilon) = \frac{\partial^{i+j} h_2(u,\varepsilon)}{\partial u^i \partial \varepsilon^j}.$$

Let us define the following sets:

$$H = \{v = (a_0, b_0, c_1, c_2, a_{10}, a_{01}, \ldots, b_{03}) \in J^3 : c_1 = 0, c_2 = 0\},$$
$$H_I = \{v \in H : a_{10} + b_{01} = 0, a_{10}b_{01} - a_{01}b_{10} > 0\},$$
$$H_3 = \{v \in H_I : A_3 = 0\},$$

where A_3 is a function of variables a_{ij}, b_{ij} given by the formula (5.54). Obviously, it holds that $j^3 g(u,\varepsilon) \in H$ iff $g(u,\varepsilon) = 0$, i.e., if $(u,\varepsilon) \in K(g) = \{(v,\mu) \in U \times V : g(v,\mu) = 0\}$; $j^3 g(u,\varepsilon) \in H_I$ iff $u(\varepsilon) \in K(g)$ and $[d_{(u,\varepsilon)}g] = (\alpha_{ij}) \in M(2)$, where $\alpha_{10} + \alpha_{01} = 0, \alpha_{10}\alpha_{01} - \alpha_{01}\alpha_{10} > 0$, i.e., the matrix (α_{ij}) has pure imaginary eigenvalues; $j^3 g(u,\varepsilon) \in H_3$ iff $j^3 g(u,\varepsilon) \in H_I$ and moreover, $A_3 = 0$. Obviously, the sets H and H_I are smooth submanifolds in J^3, whereby $\operatorname{codim} H = 2$ and $\operatorname{codim} H_I = 3$. From the Implicit Function Theorem it also follows that the set H_3 is a smooth submanifold in J^3 (it has several compact components), where $\operatorname{codim} H_3 = 4$. Let $N \subset R^3$ be an open set and it has a compact closure $\overline{N} \subset U \times V$. From the Thom Transversality Theorem it follows that the sets

$$S = \{g \in C^r(U \times V, R^2); j^3 g \, \overline{\pitchfork}_{\overline{N}} H\},$$
$$S_I = \{g \in C^r(U \times V, R^2) : j^3 g \, \overline{\pitchfork}_{\overline{N}} H_I\},$$
$$S_3 = \{g \in C^r(U \times V, R^2) : j^3 g \, \overline{\pitchfork}_{\overline{N}} H_3\},$$

are open and dense in $C^{r-1}(U \times V, R^2)$ where $j^3g : U \times V \mapsto J^3$, $(u, \varepsilon) \to j^3g(u, \varepsilon)$. From Proposition 2.76 it follows that for $g \in S \cap S_I \cap S_3$ the sets

$$K(g) = (j^3g)^{-1}(H),$$
$$K_I(g) = (j^3g)^{-1}(H_I),$$
$$K_3(g) = (j^3g)^{-1}(H_3)$$

are C^{r-3}-submanifolds in R^3, where $\dim K(g) = 1, K_I \subset K(g)$, $\dim K_I(g) = 0$ and $K_3(g) = \emptyset$. Thus if $(0,0) \in K(g)$ is a singular point of type $m(I)$, then $A_3 = c_{10} \neq 0$ (c_{10} is the coefficient at ϱ^3 in the system (5.43), (5.44)). Now it suffices just to prove that for $g \in D_0 \cap S \cap S_I \cap S_3$ the transversality condition for eigenvalues is satisfied. On the basis of Lemma 5.19 it can be supposed without any loss of generality, that g is of the form (5.40). In this, case, $K(g) = \{(u, \varepsilon) \in U \times V : u = 0\}$ and for $(0, \varepsilon) \in K(g)$ we have $[d_{(0,\varepsilon)}g] = B(\varepsilon)$ whereby this matrix is of the form (5.41). It is left to be proved that $\dot{\lambda}_1(0) \neq 0$. In the same way as in the first part of our proof where we expressed the condition $j^1g \overline{\pitchfork}_{(0,0)} H_0$, it is possible to write the transversality condition $j^3g \overline{\pitchfork}_{(0,0)} H_I$, i.e. the condition $D_{(0,0)}(j^3g) + T_Z(H_I) = T_Z(J^3)$, where $Z = j^3g(0,0) \in H_I$ as a certain system of linear equations. We do not intend to present this system as the whole. Let us just note that (a reader can make sure about it very easily) this system is solvable iff its subsystem:

$$\lambda_2(0)y_2 = w_1$$

$$\frac{\partial^2 g_1(0,0)}{\partial u_1^2} y_1 + \frac{\partial^2 g_1(0,0)}{\partial u_1 \partial u_2} y_2 + \dot{\lambda}_1(0)y_3 + z = w_2$$

$$-\lambda_2(0)y_1 = w_3$$

$$\frac{\partial^2 g_2(0,0)}{\partial u_1 \partial u_2} y_1 + \frac{\partial^2 g_2(0,0)}{\partial u_2^2} y_2 + \dot{\lambda}_1(0)y_3 - z = w_4$$

for arbitrary $w = (w_1, w_2, w_3, w_4) \in R^4$, where $g = (g_1, g_2)$. The variables y_1 and y_2 are uniquely defined by the first and third equation and therefore, this system is solvable iff $\dot{\lambda}_1(0) \neq 0$ which was to be proved. Let $D_{20} = D_0 \cap S_1 \cap S_I \cap S_3$ and D_{10} be the set defined in the first part of our proof for the case $\dim u = 1$. Let $\tilde{D}_{10} \subset D^r$ ($\tilde{D}_{20} \subset D^r$) be a set of parametrized differential equations of the form (5.27) with $\dim u = N$ such that if $f \in \tilde{D}_{10}$ ($f \in \tilde{D}_{20}$) then either the matrix $A = [d_0 f_0]$ has no eigenvalue with zero real part or if it has the eigenvalue $\lambda = 0$ (pure imaginary eigenvalue) then the bifurcation equation of the parametrized differential equation belongs to the set D_{10} (to the set D_{20}). The set $D_1 = \tilde{D}_{10} \cap \tilde{D}_{20}$ is open and dense in D^r and for this set the assertion of the above theorem is valid. □

5.25 Remark. The proof of Theorem 5.24 contains, in fact, the procedure of the proof of Theorem 5.23. On the basis of this procedure a reader will be able to understand the proof of Theorem 5.23 published in [102], very easily.

If to the 1-parameter system of differential equations

$$\dot{u} = g(u, \varepsilon) \tag{5.69}$$

the differential equation $\dot{\varepsilon} = 0$ is added, we obtain the differential equation

$$(\dot{u}, \dot{\varepsilon}) = \hat{g}(u, \varepsilon) = (g(u, \varepsilon), 0) \tag{5.70}$$

with the state variable (u, ε). Obviously, the assertion holds that the point (u, ε) is the singular point of the differential equation (5.70) iff $(u, \varepsilon) \in K(g) = \{(v, \mu) : g(v, \mu) = 0\}$. Moreover, the singular point u of the differential equation $\dot{u} = g_\varepsilon(u)$ has the same stability properties as the singular point (u, ε) of the differential equation (5.70).

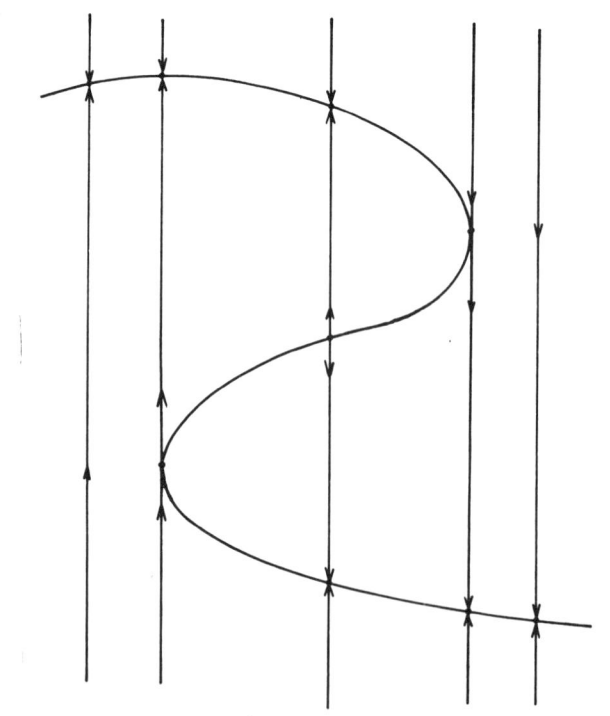

Fig. 41.

Therefore the bifurcation diagram along with the topological structure of trajectories of the 1-parameter system (5.69) can be illustrated in a single figure, which, in fact shows the set of singular points of the differential equation (5.70) and the topological structure of trajectories of this equation in neighbourhoods of these singular points. Generic bifurcations as described in Theorems 5.15 and 5.21 are illustrated in Figs. 41, 42, 43. Fig. 41 shows bifurcation of the saddle-node type. In Fig. 42 the Hopf bifurcation is drawn, whereby in Fig. 43 bifurcation of the saddle-node type is shown for the polar coordinate corresponding to the Hopf bifurcation from Fig. 42.

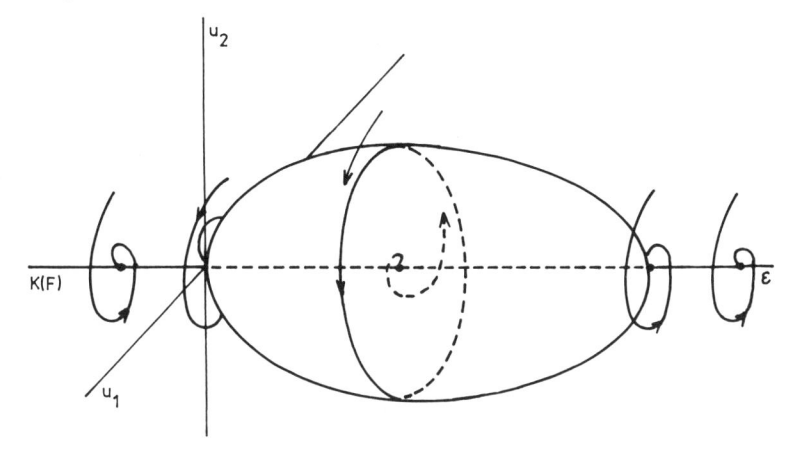

Fig. 42.

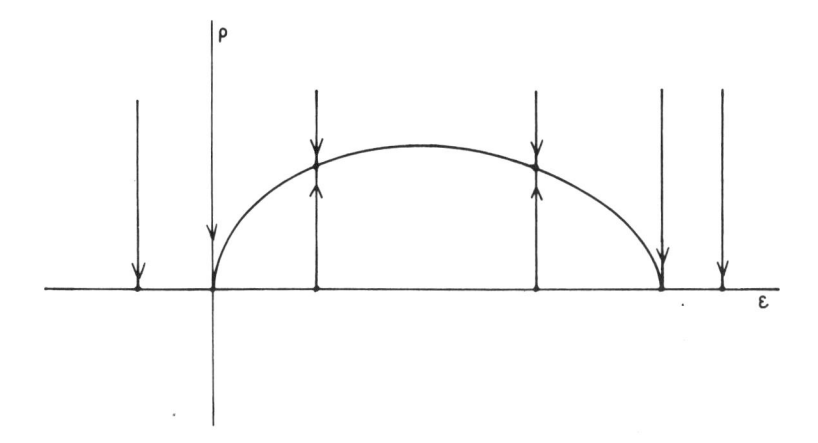

Fig. 43.

Chapter 5

5.3 Generic Bifurcations of 1-parameter Systems of Diffeomorphisms

In this paragraph we shall deal only with such parametrized diffeomorphisms whose linearizations have generic properties in the space of parametrized linear diffeomorphisms.

5.26 Definition. *We say that the fixed point $x = 0$ of the linear diffeomorphism $L : R^n \to R^n$, $L(x) = Ax$, where $A \in M(n)$ is quasi-hyperbolic of degree 1, if one of these cases occurs:*
   1. The matrix A has the eigenvalue $\lambda = 1$ of multiplicity 1 and does not have another eigenvalue on the unit circle S^1.
   2. The matrix A has the eigenvalue $\lambda = -1$ of multiplicity 1 and does not have another eigenvalue on S^1.
   3. The matrix A has an eigenvalue $\lambda \in S^1$, where $\lambda^k \neq \pm 1$ for all $k \in Z$ and beside the complex conjugated eigenvalue to λ it does not have another eigenvalue on S^1.

Let us consider the parametrized diffeomorphism $f \in \text{Diff}^r(V, U)$, where $U \subset R^n$, $V \subset R$ are open sets containing the origin, $f(0, 0) = 0$,

$$f : U \times V \to R^n, (u, \varepsilon) \to f(u, \varepsilon), \qquad (5.71)$$

where u is the state variable and ε is a one-dimensional parameter. Let us also consider the linear diffeomorphism

$$L : R^n \to R^n, L(u) = Au, \qquad (5.72)$$

where $A = [d_0 f_0]$ $(f_0(u) = f(u, 0))$.

5.27 Definition. *Let $[f]_0$ be the germ at the point $0 \in R^n \times R$ represented by the parametrized diffeomorphism (5.71). We say that this germ is of type $m(+1)(m(-1); m(c))$ if the linear diffeomorphism has such a quasi-hyperbolic fixed point of degree 1 that the matrix A has the property 1 (2; 3) from the Definition 5.26.*

5.28 Definition. *The point (u, ε) is called a fixed (m-periodic) point of the parametrized diffeomorphism f if u is a fixed (m-periodic) point of the diffeomorphism f_ε, i.e. $f_\varepsilon(u) = f(u, \varepsilon) = u$. The set of all fixed (m-periodic) points of the parametrized diffeomorphism f will be denoted as $\text{Fix} f$ $(P_m(f))$.*

Let $f \in \text{Diff}^r(V, U)$, where $V \subset R$, $U \subset R^n$ be open sets, $r \geq 2$, $f(0, 0) = 0$ and let the germ $[f]_0$ be quasi-hyperbolic of degree 1. Without loss of generality we can assume that $A = [d_0 f_0] = \text{diag}\{A^s, A^u, A^c\}$, where the matrix A has all eigenvalues with absolute value smaller (greater, equal) than 1. According to Theorem 4.38 there exists a centre C^r-manifold $V_{\text{loc}}^c(f)$ such that $[f]_0$ is orbitally topologically equivalent to the germ at the point 0 which is represented by parametrized diffeomorphism $(g_1, g_2, g_3) \in C^r$,

$$g_1(x, y, z, \varepsilon) = A^s x, \; g_2(x, y, z, \varepsilon) = A^u y, \tag{5.73}$$

$$g_3(x, y, z, \varepsilon) \stackrel{\text{def}}{=} g(z, \varepsilon) = A^c z + h(z, \varepsilon), \tag{5.74}$$

where the parametrized diffeomorphism (5.74) is the reduction of the parametrized diffeomorphism f to the centre manifold $V_{\text{loc}}^c(f)$.

5.29 Theorem on the Saddle-Node Type Bifurcation for Diffeomorphisms. *Let $[f]_0$ be the germ of the type $m(+1)$ represented by the parametrized diffeomorphism (5.71) of the class C^r, where $r \geq 3$. Let us assume that*

$$\alpha = \frac{\partial g(0,0)}{\partial \varepsilon} \neq 0 \quad \text{and} \quad \beta = \frac{1}{2}\frac{\partial^2 g(0,0)}{\partial z^2} \neq 0,$$

where g is the parametrized diffeomorphism (5.74), i.e. g is the reduction of f to the centre manifold $V_{\text{loc}}^c(f)$ from Theorem 4.38. Then the following assertions hold.

1. There exists an open neighbourhood W of the origin in R^2 such that the set $(\text{Fix}\, g) \cap W$ is a one-dimensional C^r-submanifold in R^2.

2. If $\alpha\beta < 0$ $(\alpha\beta > 0)$ then for $\varepsilon > 0 (\varepsilon < 0)$ the diffeomorphism f_ε has two fixed points $K_{1\varepsilon}, K_{2\varepsilon}$, for $\varepsilon = 0$ it has one fixed point $K = (0,0)$ and for $\varepsilon < 0$ $(\varepsilon > 0)$ it does not have a fixed point.

3. Let $L(K_{1\varepsilon})$ $(L(K_{2\varepsilon}))$ be the matrix of linearization of the diffeomorphism f at the point $K_{1\varepsilon}$ $(K_{2\varepsilon})$ and let s_1, s_2 be the number of eigenvalues of the matrix $L(K_{1\varepsilon})$, respectively $L(K_{2\varepsilon})$ which have absolute values greater then 1. Then $|s_1 - s_2| = 1$.

The proof of this theorem is the same as the proof of Theorem 5.15 on bifurcations of the saddle-node type for singular points of vector fields and therefore we do not mention it. The set $\text{Fix}\, g$ is of the same form as the set $K(g)$ for singular points and also the bifurcation of fixed points of singular points shown in Fig. 35 is of the same structure. Therefore this bifurcation is also called the bifurcation of the saddle-node type.

5.30 Period-Doubling Bifurcation Theorem. *Let $[f]_0$ be the germ of the type $m(-1)$ represented by the parametrized diffeomorphism (5.71) of the class C^r, where $r \geq 4$. Let us assume that*

$$\alpha = \frac{\partial^2 g(0,0)}{\partial z \partial \varepsilon} \neq 0,$$

$\beta^2 + \gamma \neq 0$, where

$$\beta = \frac{1}{2}\frac{\partial^2 g(0,0)}{\partial z^2}, \qquad \gamma = \frac{1}{3}\frac{\partial^3 g(0,0)}{\partial z^3},$$

g is the reduction f to the centre manifold $V^c_{loc}(f)$ from Theorem 4.38. Then the following assertions hold.

1. There exists an open neighbourhood W of the origin in R^2 such that the set $(\text{Fix } g) \cap W$ is a one-dimensional C^r-submanifold in R^2, $P_2(g) \cap W$ is a one-dimensional C^{r-1}-submanifold in R^2, where $\overline{(P_2(g)}\setminus P_2(g)) \cap W = \{(0,0)\}$.

2. If $\alpha(\beta^2 + \gamma) < 0$ ($\alpha(\beta^2 + \gamma) > 0$), then for $\varepsilon > 0$ ($\varepsilon < 0$) the diffeomorphism f_ε has just one fixed point $z(\varepsilon)$ and two 2-periodic points $z_1(\varepsilon)$, $z_2(\varepsilon)$ and for $\varepsilon < 0$ ($\varepsilon > 0$) it has just one fixed point.

3. Let

$$a_\varepsilon(z) = \frac{dg_\varepsilon(z)}{dz} \quad \text{and} \quad b_\varepsilon(z) = \frac{dg^2_\varepsilon(z)}{dz},$$

where $g^2_\varepsilon = g_\varepsilon \circ g_\varepsilon$. If $\alpha\varepsilon < 0$ ($\alpha\varepsilon > 0$), then $|a_\varepsilon(z(\varepsilon))| < 1$ for $\varepsilon < 0$, $|a_\varepsilon(z(\varepsilon))| > 1$ for $\varepsilon > 0$ ($|a_\varepsilon(z(\varepsilon))| > 1$ for $\varepsilon < 0$, $|a_\varepsilon(z(\varepsilon))| < 1$ for $\varepsilon > 0$) and $0 < b_\varepsilon(z_i(\varepsilon)) < 1$, $i = 1, 2$, where $|\varepsilon|$ is a sufficiently small number.

Proof

The parametrized diffeomorphism g can be written in a sufficiently small neighbourhood in the form

$$g(z, \varepsilon) = -z + \alpha\varepsilon z + \beta z^2 + \gamma z^3 + q(z, \varepsilon), \qquad (5.75)$$

where $q \in C^{r-1}$, $q(z,0) = o(|z|^3)$ and $q(0,\varepsilon) = O(|\varepsilon|)$. The function $F(z,\varepsilon) = g(z,\varepsilon) - z$ is of the class C^r, $F(0,0) = 0$,

$$\frac{\partial F(0,0)}{\partial z} = -2$$

and therefore according to the Implicit Function Theorem there exists an open neighbourhood $W_1 = U_1 \times V_1$ of the origin in R^2 and a unique C^r-function $\varphi : V_1 \to U_1$, $z = \varphi(\varepsilon)$ such that $\varphi(V_1) \subset U_1$, $\varphi(0) = 0$ and $F(\varphi(\varepsilon), \varepsilon) = 0$ for all $\varepsilon \in V_1$. It means that the set $(\text{Fix } g) \cap W_1$ is a one-dimensional C^r-submanifold in R^2. Without loss of generality we can further assume that $(\text{Fix } g) \cap W_1 = \{(z,\varepsilon) \in W_1 : z = 0\}$. We can introduce new ordinates by means of the mapping $(y,\varepsilon) = h(z,\varepsilon) = (z - \varphi(\varepsilon), \varepsilon)$. In these new ordinates the diffeomorphism g_ε has the representation $\tilde{g}_\varepsilon(y) = h_\varepsilon \circ g_\varepsilon \circ h_\varepsilon^{-1}(y) = g_\varepsilon(y + \varphi(\varepsilon)) - \varphi(\varepsilon)$, where $h_\varepsilon(z) = h(z,\varepsilon)$, $g_\varepsilon(z) = g(z,\varepsilon)$. We obtain that $\tilde{g}_\varepsilon(y) = y$ iff $F(y + \varphi(\varepsilon), \varepsilon) = 0$. From the uniqueness of the function φ it follows that the last equality is fulfilled only if $y = 0$ which means that $(\text{Fix } \tilde{g}) \cap W_1 = \{(y,\varepsilon) \in W_1 : y = 0\}$, where

$\tilde{g}(y,\varepsilon) = \tilde{g}_\varepsilon(y)$. From now on we shall assume that the parametrized diffeomorphism g has this property. Then the function q from (5.75) satisfies the equality $q(0,\varepsilon) = 0$ for all $\varepsilon \in V_1$. Let us look for 2-periodic points of the diffeomorphism g_ε. From (5.75) we obtain that

$$g_\varepsilon^2(z) = z - 2\alpha\varepsilon z - 2(\beta^2 + \gamma)z^3 + q_1(z,\varepsilon), \tag{5.76}$$

where $q_1 \in C^r$, $q_1(0,\varepsilon) = 0$ for all $\varepsilon \in V_1$, $q_1(z,0) = o(|z|^4)$. Let us define the function $q_2(z,\varepsilon) = z^{-1}q_1(z,\varepsilon)$. From the Converse to Taylor Theorem it follows that the function q_2 is of the class C^{r-1}, $q_2(0,\varepsilon) = 0$ for all $\varepsilon \in V$ and $q_2(z,0) = o(|z|^3)$. Therefore $g_\varepsilon^2(z) - z = zG(z,\varepsilon)$, where $G(z,\varepsilon) = -2\alpha\varepsilon - 2(\beta^2 + \gamma)z^2 + q_2(z,\varepsilon)$ is of the class C^{r-1}, $G(0,0) = 0$. From the Implicit Function Theorem it follows that there exists an open neighbourhood $W_2 = U_2 \times V_2$ of the origin in R^2 and a unique C^{r-1}-function $\Psi : U_2 \to V_2, \varepsilon = \Psi(z)$ such that $\Psi(U_2) \subset V_2, \Psi(0) = 0$ and $G(\Psi(z), z) = 0$ for all $z \in U_2$. From the last equality we obtain that

$$\Psi(z) = -\alpha^{-1}(\beta^2 + \gamma)z^2 + o(|z|^4). \tag{5.77}$$

It follows from what we have proved so far that for $W = W_1 \cap W_2$ the assertions 1 and 2 hold. From the equality $\varepsilon = \Psi(z)$ and from (5.77) it follows that if $\alpha(\beta^2 + \gamma)\varepsilon > 0$ then the diffeomorphism g_ε has just two 2-periodic points $z_1(\varepsilon), z_2(\varepsilon)$. The function Ψ_+ (Ψ_-) defined as the restriction of the function Ψ onto the set of positive (negative) real numbers R^+ (R^-) has the inverse function $\Psi_+^{-1}(\Psi_-^{-1})$ and from (5.77) we obtain that for $\varepsilon > 0$ it is

$$\begin{aligned} z_1(\varepsilon) &= \Psi_+^{-1}(\varepsilon) = (-\alpha(\beta^2 + \gamma)^{-1}\varepsilon)^{1/2} + \tilde{q}_1(\varepsilon), \\ z_2(\varepsilon) &= \Psi_-^{-1}(\varepsilon) = -(-\alpha(\beta^2 + \gamma)^{-1}\varepsilon)^{1/2} + \tilde{q}_2(\varepsilon), \end{aligned} \tag{5.78}$$

where $\tilde{q}_i \in C^{r-2}$ and $\lim_{\varepsilon \to 0+} \tilde{q}_i(\varepsilon) = 0, i = 1, 2$. From the formulae (5.78) assertion 3 follows. □

For a 1-parameter system of diffeomorphisms g we define the diffeomorphism $\hat{g}(x,\varepsilon) = (g(x,\varepsilon), \varepsilon)$. Obviously, it holds that $(x,\varepsilon) \in \text{Fix}\,g$ iff $\hat{g}(x,\varepsilon) = (x,\varepsilon)$ i.e. if (x,ε) is a fixed point of the diffeomorphism $\hat{g}$. As $\hat{g}^n(x,\varepsilon) = (g_\varepsilon^n(x), \varepsilon)$ the fixed, respectively periodic point (x,ε) of the diffeomorphism $\hat{g}$ has the same stability properties with respect to $\hat{g}$ as the point x with respect to the diffeomorphism g_ε. Therefore we can draw the bifurcation diagram as well as the topological structure of trajectories of a 1-parametric system of the diffeomorphism g in one figure which shows the set of fixed, respectively periodic points of the diffeomorphism g and the topological structure of trajectories of this diffeomorphisms in their neighbourhoods. For example, if g is a parametrized diffeomorphism from

Theorem 5.30, $\alpha < 0, \beta^2 + \gamma > 0$ then the diffeomorphism g has such a structure of trajectories as is shown in Fig. 44. It can be seen from this figure that for $\varepsilon \leq 0$ the difeomorphism g_ε has one stable fixed point $z(\varepsilon)$, for $\varepsilon > 0$ it has one unstable fixed point $z(\varepsilon)$ and two stable 2-periodic points $z_1(\varepsilon), z_2(\varepsilon)$.

5.31 Lemma. *The germ $[f]_0$ of type $m(c)$ represented by the parametrized diffeomorphism (5.71) is orbitally topologically equivalent to the germ at the point 0 which is represented by the parametrized diffeomorphism (5.73), (5.74), where the mapping (5.74) is of the form*

$$g(z,\varepsilon) = B(\varepsilon)z + R(z,\varepsilon), \qquad (5.79)$$

$B \in C^r(V, M(2)), R \in C^r(U \times V, R^2), U \times V$ is an open neighbourhood of the origin in $R^2 \times R$, $R(0,0) = 0$ for all $\varepsilon \in V$, $R(z,0) = o(\|z\|)$, where

$$B(\varepsilon) = \begin{bmatrix} \lambda_1(\varepsilon) & \lambda_2(\varepsilon) \\ -\lambda_2(\varepsilon) & \lambda_1(\varepsilon) \end{bmatrix}, \qquad (5.80)$$

$\lambda_j \in C^r(V, R), j = 1, 2$.

The proof of this lemma is the same as the proof of Lemma 5.19.

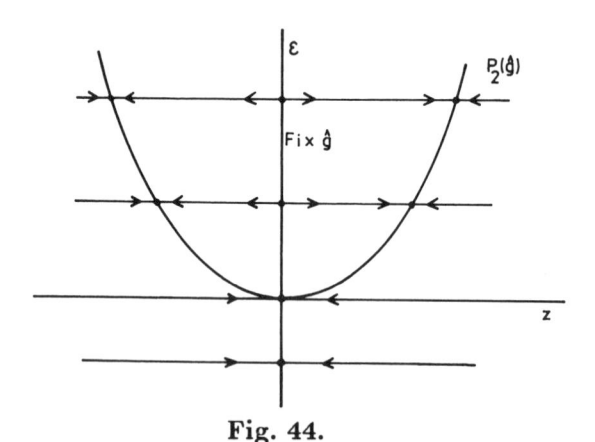

Fig. 44.

5.32 Definition. *Let $\tilde{\lambda} : V \to R^2$, $\tilde{\lambda}(\varepsilon) = (\lambda_1(\varepsilon), \lambda_2(\varepsilon))$, where λ_1, λ_2 are C^r-functions from Lemma 5.31. If $\tilde{\lambda} \bar{\pitchfork}_0 S^1$ then we say that for the parametrized diffeomorphism (5.79) the condition of transversality of eigenvalues is fulfilled.*

5.33 Lemma. *For the parametrized diffeomorphism* (5.79) *the condition of transversality of eigenvalues* $\tilde{\lambda} \pitchfork_0 S^1$ *is fulfilled iff*

$$\frac{d|\lambda(0)|}{d\varepsilon} \neq 0,$$

where $\lambda(\varepsilon) = \lambda_1(\varepsilon) + i\lambda_2(\varepsilon)$, $\tilde{\lambda} = (\lambda_1, \lambda_2)$.

Proof

As $\tilde{\lambda}(0) \in S^1 = c(\langle 0, 2\pi \rangle)$, where $c : \langle 0, 2\pi \rangle \to R^2, c(t) = (\cos t, \sin t)$, there exists $s \in \langle 0, 2\pi \rangle$ such that $c(s) = (\cos s, \sin s) = \tilde{\lambda}(0)$. The vector $\tilde{\lambda}(0)$ is obviously perpendicular to the tangent vector

$$\frac{dc(s)}{dt} = (-\sin s, \cos s)$$

to the circle S^1 at the point $\tilde{\lambda}(0)$. From that it follows that $\tilde{\lambda} \pitchfork_0 S^1$ iff

$$\left(\frac{d\tilde{\lambda}(0)}{d\varepsilon}, \tilde{\lambda}(0) \right) \neq 0,$$

where $(.,.)$ is the scalar product on R^2. As $|\lambda(0)| = \|\tilde{\lambda}(0)\| = 1$ then

$$\frac{d|\lambda(0)|}{d\varepsilon} = \left(\frac{d\tilde{\lambda}(0)}{d\varepsilon}, \tilde{\lambda}(0) \right)$$

and thus we obtain that

$$\tilde{\lambda} \pitchfork_0 S^1 \quad \text{iff} \quad \frac{d|\lambda(0)|}{d\varepsilon} \neq 0. \quad \square$$

Let us give an example which predicts what bifurcations we can expect in the case of germ of the type $m(c)$.

5.34 Example. Let $[g]_0$ be a germ represented by a 1-parameter system of diffeomorphisms g whose complex representation is of the form

$$\tilde{g}(z, \varepsilon) = (\exp(i + \varepsilon + \beta|z|^2))z, \tag{5.81}$$

where ε is a real number and $\beta = \beta_1 + i\beta_2$. Then $g_\varepsilon(x) = B(\varepsilon)x + o(\|x\|)$, where $B(\varepsilon) = (b_{ij}) \in M(2), b_{11} = b_{22} = e^\varepsilon \cos 1, b_{12} = -b_{21} = -e^\varepsilon \sin 1$. The matrix $B(\varepsilon)$ has the eigenvalue $\lambda(\varepsilon) = e^\varepsilon(\cos 1 + i \sin 1)$ and thus it holds that $|\lambda(0)| = 1, |\lambda(\varepsilon)| = e^\varepsilon$,

$$\frac{d|\lambda(0)|}{d\varepsilon} = 1.$$

It means that the condition of transversality of eigenvalues is fulfilled. If $\beta_1 \varepsilon < 0$ then the circle $\tilde{C}(r(\varepsilon)) = \{z \in C : |z| = r(\varepsilon) = (-\beta_1 \varepsilon)^{1/2}\}$. As

$$|\tilde{g}_\varepsilon(z)| = |(\exp(i + \varepsilon + \beta|z|^2))||z| = (\exp(\varepsilon + \beta_2|z|^2))|z|$$

thus for $z \in \tilde{C}(r(\varepsilon))$ we have $|\tilde{g}_\varepsilon(z)| = |z| = (-\beta_1 \varepsilon)^{1/2}\}$. As

$$|\tilde{g}_\varepsilon(z)| = |(\exp(i + \varepsilon + \beta|z|^2))|z| = (\exp(\varepsilon + \beta_2|z|^2))z,$$

thus $|\tilde{g}_\varepsilon(\varepsilon)| = |z| = (-\beta_1 \varepsilon)^{1/2}$ for $z \in \tilde{C}(r(\varepsilon))$. We thus obtain that the circle $C(r(\varepsilon)) = \{x \in R^2 : x_1^2 + x_2^2 = -\beta_1 \varepsilon\}$ is an invariant curve of the diffeomorphism g_ε. If $\beta_1 < 0$ ($\beta_1 > 0$) then the diffeomorphism g_ε has one fixed point $x = 0$ for all ε and one invariant curve (circle) $C(r(\varepsilon))$ for $\varepsilon > 0$ ($\varepsilon < 0$). The bifurcations of the parametrized diffeomorphism g for $\beta_1 < 0$ are shown in Fig. 45.

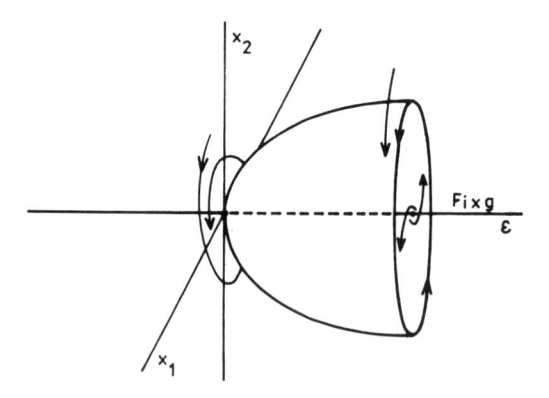

Fig. 45. Nejmark–Sacker bifurcation.

The bifurcation from the previous example is of similar character the Hopf bifurcation for singular points of vector fields and therefore it is sometimes also called the Hopf bifurcation for diffeomorphisms. According to the following theorem this bifurcation is not bound only to some special diffeomorphisms but it is of general character. According to the authors of the first works on this bifurcation we call it the Nejmark–Sacker bifurcation.

5.35 Nejmark–Sacker Bifurcation Theorem. *Let $[f]_0$ be the germ of the type $m(c)$ represented by parametrized diffeomorphism (5.71) of the class C^r, where $r \geq 6$ and $f(0, \varepsilon) = 0$ for all $\varepsilon \in V$. Let us assume that $\lambda(\varepsilon)$ is an eigenvalue of the matrix $A(\varepsilon) = [d_0 f_\varepsilon]$ such that $\lambda(0) \in S^1, \lambda(0))^j \neq 1$ for $j = 1, 2, 3, 4$ and let the condition of transversality of eigenvalues be fulfilled*

Generic Bifurcations of Vector Fields and Diffeomorphisms

$$\frac{d|\lambda(0)|}{d\varepsilon} > 0. \tag{5.82}$$

Then the germ $[f]_0$ is orbitally topological equivalent to the germ $[G]_0$ which is represented by the parametrized diffeomorphism (5.73), (5.74), where the parametrized diffeomorphism (5.74) is of the form (5.80) and the following assertions hold.

1. There exists an open neighbourhood W of the origin in $R^2 \times R$ such that $(\mathrm{Fix}\, g) \cap W$ is a one-dimensional C^r-submanifold in W.

2. Let

$$\tilde{g}_\varepsilon : z_1 = \lambda(\varepsilon)z + U(z, \overline{z}, \varepsilon) \tag{5.83}$$

be the complex representation of the diffeomorphism g_ε. Then there exists a change of coordinates in a neighbourhood of the point $(0, 0) \in C \times R$ defined by the mapping of the form

$$z = cw + P(w, \overline{w}, \varepsilon), 0 < c \leq 1, \tag{5.84}$$

where P is a polynomial of the variables $w, \overline{w}$ whose coefficients are C^r-functions of the variable ε with values in C such that in the coordinates $w, \overline{w}$ the mapping $\tilde{g}_\varepsilon$ is of the form

$$\hat{g}_\varepsilon : w_1 = (\exp(2\pi\alpha(\varepsilon)) + c^2 \beta(\varepsilon)|w|^2))w + R(w, \overline{w}, \varepsilon), \tag{5.85}$$

where α, β are C^r-functions with values in $C, R \in C^r$ and the Taylor series of the function R in a neighbourhood of the point $(w, \overline{w}) = (0, 0)$ does not contain members of degree smaller than 4. The number $\beta(\varepsilon)$ does not depend on the members of the Taylor series of the function U in a neighbourhood of the point $(z, \overline{z}) = (0, 0)$ of the greater order than 3.

3. If $b = \mathrm{Re}\,\beta(0) < 0$ then there exists a positive number γ such that for all $\varepsilon \in \langle 0, \gamma \rangle$ the mapping (5.83) has an asymptotically stable closed invariant curve defined by the function

$$z = a_0 \sqrt{\varepsilon} e^{i\Theta} + \varepsilon f(\Theta, \varepsilon) e^{i\Theta},$$

where $\Theta \in R$, $a_0 = (-b^{-1} a)^{1/2}$,

$$a = 2\pi \mathrm{Re}\, \frac{d\alpha(0)}{d\varepsilon},$$

$\lambda(\varepsilon) = e^{2\pi\alpha(\varepsilon)}$ and $f(\Theta, \varepsilon)$ is a function of the class C^{r-1} defined for all $\varepsilon \in \langle 0, \gamma \rangle$, $\Theta \in R$ and it is 2π-periodic in the variable Θ.

The introductory part of the assertion of the previous theorem is a consequence of Lemma 5.31. The proof of assertion 1 is simple and it follows directly from the Implicit Function Theorem. However, the proof of assertion 2 is technically very difficult and therefore we will not describe it. It can be found in [134].

The first formulation of the Hopf Bifurcation Theorem for diffeomorphisms came from J. I. Nejmark and it can be found in his work [116]. This theorem, in such a form as we have mentioned it, was proved by the American mathematician R. J. Sacker in the work [134]. Other formulations and proofs of this theorem as well as many references regarding this problem can be found in [50, 75, 98].

5.36 Definition. *The bifurcations from Theorems* 5.29, 5.30 *and* 5.35 *we shall also call bifurcations of the type* $m(+1), m(-1),$ *and* $m(c)$ *respectively.*

In the works by P. Brunovský [25, 26] the generic bifurcations of a 1-parameter system of diffeomorphisms are studied firstly. From the results proved in these works it follows that the bifurcations of the type $m(+1), m(-1)$ and $m(c)$ are generic and beside them any other local bifurcations of 1-parameter systems of diffeomorphisms are not generic. The theorems on generic bifurcations are formulated globally in the mentioned works for the set $\text{Diff}^r(P, X)$, where X is an n-dimensional C^r-manifold representing the state space and P is a one-dimensional C^r-manifold representing the set of parameters, where on the set $\text{Diff}^r(P, X)$ the Whitney C^r-topology is given. The part of these results is summarized in the following theorem which, in fact, is of the same character as the Theorem 5.23 for singular points of vector fields. Before we formulate this theorem let us introduce these notations. If $f \in \text{Diff}^r(P, X)$ then

$$Z(f) = \{(x, p) \in X \times P : f(x, p) = x\};$$

$Z_0(f)$ is the set of points $(x, p) \in Z(f)$ such that for an arbitrary open neighbourhood $U \times V$ of the point (x, p) the point p is a bifurcation point of the parametrized diffeomorphism $f/U \times V$.

5.37 Theorem. *Let* X, P *be smooth manifolds,* $\dim P = 1$ *and* $r \in N$, $r \geq 6$. *Then there exists a massive subset* D *in the set* $\text{Diff}^r(P, X)$ *such that for* $f \in D$ *it holds that*

1. if the set $Z(f) \neq \emptyset$ *then it is a one-dimensional C^r-submanifold in* $X \times P$;

2. if the set $Z_0(f) \neq \emptyset$ *then it consists of isolated points and the local bifurcation in a neighbourhood of an arbitrary point* $(x, p) \in Z_0(f)$ *is either of the type* $m(+1)$ *or* $m(-1)$ *or* $m(c)$.

The local version of this theorem can be formulated analogously the way we formulated the Theorem 5.24 for germs of parametrized differential equations. The proof of this local version does not differ from the proof of Theorem 5.24 and the reader can try to prove it himself.

In case of Nejmark–Sacker bifurcations as we have already known, a closed invariant curve from the fixed point originates. On this curve other so-called secondary bifurcations can originate at the change of parameter. In the study of topological structure of these secondary bifurcations in general we can assume, without loss of generality, that the mentioned closed invariant curve is a circle. So we come to the conclusion that in the study of local generic bifurcations, namely in the case of Nejmark–Sacker bifurcation, we are made to solve also the problem of global generic bifurcations of a 1-parameter system of diffeomorphisms on the circle S^1. However, this problem is quite complicated and it requires a deeper analysis. We shall mention some of the most important results necessary for understanding the basic properties of these bifurcations.

Let us consider the diffeomorphism, and the homeomorphism $f : S^1 \to S^1$. The unit circle S^1 is the one-dimensional submanifold in R^2, but now it will be advantageous to represent it as a subset of complex plane, i.e. $S^1 = \{z \in C : |z| = 1\}$. Let us assume that a parametrization on S^1 by means of the mapping $\Theta \mapsto e^{2\pi\Theta i}$ is given. If $z \in S^1$ then $z = e^{2\pi\Theta i}$ for some $\Theta \in R$. Let us denote $\overline{\Theta} = F(\Theta)$, where $f(z) = e^{2\pi\overline{\Theta} i}$. So we obtain the function $F : R \to R$ for which it holds that

$$f(e^{2\pi\Theta i}) = e^{2\pi F(\Theta) i} \tag{5.86}$$

for all $\Theta \in R$ (Fig. 46).

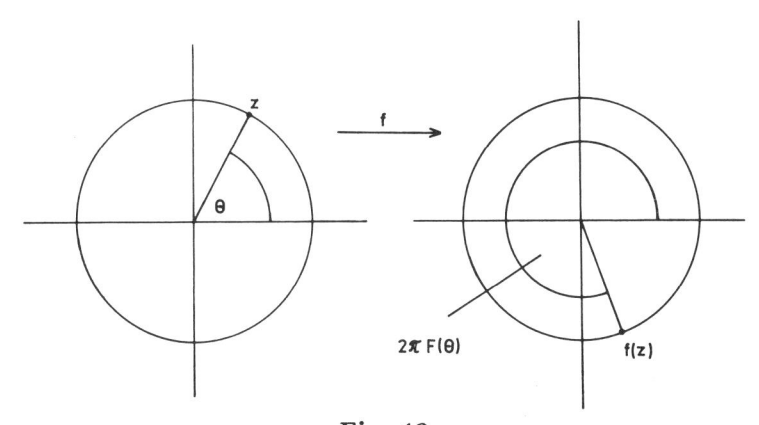

Fig. 46.

5.38 Definition. *The continuous function $F : R \to R$ fulfilling the equality (5.86) is called the lift of the mapping $f : S^1 \to S^1$.*

If F is a lift of the mapping $f : S^1 \to S^1$ and $k \in Z$ then the mapping $F_k : R \to R$, $F_k(\Theta) = F(\Theta) + k$ is obviously a lift of the mapping f and so there exists a countable number of lifts of the mapping f. From (5.86) we obtain that

$$F(\Theta) = \frac{1}{2\pi i} \ln f(e^{2\pi\Theta i})$$

which means that if f is of the class C^r then the lift F is of the class C^r. We shall assume that f preserves orientation, i.e. that its lift F is an increasing function. From the equality (5.86) it follows that

$$F(\Theta + 1) = F(\Theta) + 1 \tag{5.87}$$

for all $\Theta \in R$, i.e. $G : R \to R$, $G(\Theta) = F(\Theta) - \Theta$ is periodic with the period 1.

5.39 Theorem. *Let $f : S^1 \to S^1$ be a homeomorphism preserving the orientation and $F : R \to R$ be its lift. Then it holds that*
1. *for arbitrary $x \in R$ there exists a limit*

$$\varrho(F) = \lim_{|n| \to \infty} \frac{F^n(x)}{n} \tag{5.88}$$

2. *the number $\varrho(F)$ does not depend on the choice of $x \in R$.*

Proof

First we prove assertion 2 under the assumption that assertion 1 holds. From the periodicity of the function $G = F - \text{id}$ and from the fact that f is a homeomorphism we obtain that for arbitrary $x, y \in R$ and $n \in Z$

$$|F^n(x) - F^n(y)| \leq |(F^n(x) - x) - (F^n(y) - y)| + |x - y|$$

$$\leq 1 + |x - y|. \tag{5.89}$$

From this inequality it follows that

$$\lim_{|n| \to \infty} \left(\frac{F^n(x)}{n} - \frac{F^n(y)}{n} \right) = 0$$

and thus the limit (5.88) does not depend on the choice of $x \in R$. On the basis of assertion 2, which we have proved, it is sufficient to prove the existence of the limit (5.88) only for some $x \in R$. Let us assume first that f has the m-periodic point $z = e^{2\pi x i}$. Then there exists $p \in Z$ such that $F^m(x) = x + p$. As for an arbitrary natural number k there exists a natural number q such that $0 \leq q < m$, $k = nm + q$, from the inequality (5.89) the inequality

$$k^{-1}|F^k(x) - F^{nm}(x)| = k^{-1}|F^{nm}(F^q(x)) - F^{nm}(x)|$$

$$\leq k^{-1}(|F^q(x) - x| + 1) \leq k^{-1}(M + 1)$$

follows, where $M = \max_{0 \leq q < m} |F^q(x) - x|$. As the number M does not depend on k, from this inequality we obtain that

$$\varrho(F) = \lim_{|k| \to \infty} \frac{F^k(x)}{k} = \lim_{|n| \to \infty} \frac{F^{mn}(x)}{mn} = \frac{p}{m}. \tag{5.90}$$

Let us assume now that f does not have a periodic point. Then for an arbitrary number m the value $F^m(x) - x$ cannot be an integer for any $x \in R$ and as the function $F^n - \mathrm{id}$ is continuous there exists an integer p_m (independent of x) such that $p_m \leq F^m(x) - x \leq p_m + 1$ for all $x \in R$, i.e.

$$p_m + x \leq F^m(x) \leq p_m + 1 + x \tag{5.91}$$

for all $x \in R$. For $x = 0$ we obtain that

$$p_m \leq F^m(0) \leq p_m + 1, \tag{5.92}$$

$2p_m \leq p_m + F^m(0) \leq F^{2m}(0) \leq p_m + 1 + F^m(0) \leq 2p_m + 1$ and by the induction it can be shown simply that for arbitrary $n \in Z$

$$np_m \leq F^{mn}(0) \leq n(p_m + 1). \tag{5.93}$$

From inequalities (5.92), (5.93) the inequalities below follow

$$\left| \frac{F^{mn}(0)}{mn} - \frac{F^n(0)}{n} \right| \leq \frac{2}{|n|}, \quad \left| \frac{F^{mn}(0)}{mn} - \frac{F^m(0)}{m} \right| \leq \frac{2}{|m|}.$$

From these inequalities we obtain that

$$\left| \frac{F^m(0)}{m} - \frac{F^n(0)}{n} \right| \leq \frac{2}{|m|} + \frac{2}{|n|}.$$

We have proved that the sequence

$$\left\{ \frac{F^n(0)}{n} \right\}_{n=1}^{\infty}$$

is Cauchy and thus the limit (5.88) exists. □

5.40 Example. Let $f : S^1 \to S^1$ be a turn about the angle $2\pi\alpha$, i.e. $f(e^{2\pi x i}) = e^{2\pi(x+\alpha)i}$ for all $x \in R$. Then the function $F : R \to R$, $F(x) = x + \alpha$ is a lift of the diffeomorphism f and

$$\varrho(F) = \lim_{|n| \to \infty} \frac{F^n(x)}{n} = \lim_{|n| \to \infty} \frac{x + n\alpha}{n} = \alpha.$$

5.41 Lemma. *If $\tilde{F}, F$ are two lifts of the homeomorphism $f : S^1 \to S^1$ then there exists $k \in Z$ such that $\varrho(\tilde{F}) = \varrho(F) + k$.*

Proof

There obviously exists $k \in Z$ such that $\tilde{F}(x) + k$ for all $x \in R$, and therefore

$$\varrho(\tilde{F}) = \lim_{|n| \to \infty} \frac{\tilde{F}^n(x)}{n} = \lim_{|n| \to \infty} \frac{F^n(x) + nk}{n} = \varrho(F) + k. \quad \square$$

5.42 Definition. *If F is a lift of the diffeomorphism $f : S^1 \to S^1$ then the number $\varrho(F)(\mod 1)$ is called the rotation number of the homeomorphism f and we denote it as $\varrho(f)$.*

From Lemma 5.41 it follows that the rotation number $\varrho(f)$ of the homeomorphism f does not depend on the choice of its lift.

5.43 Theorem. *If $f : S^1 \to S^1$ is a homeomorphism preserving the orientation, then it holds that:*

1. if f has an m-periodic point then there exists an integer p, such that $\varrho(f) = p/m$;
2. if $\varrho(f) = 0$ then f has a fixed point;
3. if $\varrho(f) = p/m$, where $p \in Z$, $p \neq 0$ and m is a natural number such that the numbers p, m are mutually indivisible then f has an m-periodic point.

Proof

Assertion 1 was proved in the proof of assertion 1 of Theorem 5.39. Let $\varrho(f) = 0$ and let us assume that f does not have a fixed point. Then there exists a lift F of the homeomorphism f such that $\varrho(F) = 0$ and $F(x) > x$ for all $x \in R$. For $x = 0$ we obtain that the sequence $\{F^n(0)\}_{n=1}^{\infty}$ is increasing. We prove that $F^n(0) < 1$ for all $n \in N$. Let there be a natural number r such that $F^r(0) \geq 1$. Then $F^{2r}(0) \geq F^r(1) = F^r(0) + 1 \geq 2$, $F^{3r}(0) \geq F^{2r}(1) = F^{2r}(0) + 1 \geq 3$ and by induction it can be proved that $F^{kr}(0) \geq k$ for an arbitrary natural number k. Thus we obtain that

$$\varrho(F) = \lim_{k \to \infty} \frac{F^{kr}(0)}{kr} \geq \frac{1}{r} > 0$$

and this contradicts the assumption that $\varrho(F) = 0$. Thus we prove that $F^n(0) < 1$ for all $n \in N$ and as the sequence $\{F^n(0)\}_{n=1}^{\infty}$ increases there exists the limit $u = \lim_{n \to \infty} F^n(0)$. As $F(u) = F(\lim_{n \to \infty} F^n(0)) = \lim_{n \to \infty} F^{n+1}(0) = u$ for $z = e^{2\pi u i}$ then $f(z) = e^{2\pi F(u)i} = e^{2\pi u i} = z$

and thus z is the fixed point of the homeomorphism f. Now we prove assertion 3. As $\varrho(f) = p/m$ there exists a lift F of homeomorphism f such that $\varrho(F) = p/m$. As the mapping $G : R \to R$, $G(x) = F^m(x) - p$ is a lift of the homeomorphism f^m and $\varrho(G) = m\varrho(F) - p = 0$ from assertion 2 it follows that f^m has a fixed point. As p, m are mutually indivisible numbers, this fixed point of the homeomorphism f^m is an m-periodic point of the homeomorphism f. $\square$

5.44 Corollary. *If $f : S^1 \to S^1$ is a homeomorphism preserving the orientation then the rotation number of $\varrho(f)$ is irrational iff f does not have either a periodic or a fixed point.*

Let $C(S^1)$ be the set of homeomorphisms of the circle S^1 onto S^1 preserving the orientation. As S^1 is a compact set then

$$d(f, g) = \max_{z \in S^1}(|f(z) - g(z)| + |f^{-1}(z) - g^{-1}(z)|)$$

is a metric on S^1 (let us consider S^1 as the subset in C).

5.45 Theorem. *The function $\varrho : C(S^1) \to R$, $f \mapsto \varrho(f)$, where $\varrho(f)$ is the rotation number of the homeomorphism f, is continuous.*

Proof

Let $f_0 \in C(S^1)$ and $\varepsilon > 0$. As $\varrho(f_0) \in \langle 0, 1 \rangle$ there exists a natural number k and an integer r such that $k > \varepsilon^{-1}$ and

$$\frac{r}{k} < \varrho(f_0) < \frac{r+1}{k}. \tag{5.94}$$

Let F_0 be a lift of the homeomorphism f_0 such that

$$\varrho(f_0) = \lim_{n \to \infty} \frac{F_0^n(x)}{n}.$$

We prove that for all $x \in R$

$$x + r < F_0^k(x) < x + r + 1. \tag{5.95}$$

Let $x_0 \in R$ exists such that $F_0^k(x_0) \leq x_0 + r$. Then

$$\varrho(f_0) = \lim_{p \to \infty} \frac{F_0^{pk}(x_0)}{pk} \leq \lim_{p \to \infty} \frac{x_0 + pr}{pk} = \frac{r}{k}$$

and this is a contaradiction with (5.94) which means that $F_0^k(x) > x + r$ for all $x \in R$. Let us assume that there exists $x_0 \in R$ such that $F_0^k(x_0) \geq x_0 + r + 1$. Then

$$\varrho(f_0) = \lim_{p \to \infty} \frac{F^{pk}(x_0)}{pk} \geq \lim_{p \to \infty} \frac{x_0 + p(r+1)}{pk} = \frac{r+1}{k}$$

and this is again a contradiction with (5.94). By this inequality (5.95) is proved. As the function $F_0^k - \text{id}$ is continuous, from inequality (5.95) it follows that there exists $\eta > 0$ such that

$$r + \eta < F_0^k(x) - x < r + 1 - \eta \tag{5.96}$$

for all $x \in R$. The mapping $\Psi_k : C(S^1) \to C(S^1)$, $\Psi_k(f) = f^k$ is obviously continuous and therefore it holds that there exists $\delta > 0$ such that for each $f \in C(S^1)$ fulfilling the inequality $d(f, f_0) > \eta$ there exists a lift F of the homeomorphism f such that

$$|F^k(x) - F_0^k(x)| < \eta \tag{5.97}$$

for all $x \in R$. From the inequalities (5.96), (5.97) we obtain that $r < F^k(x) - x < r + 1$ for all $x \in R$ and from this inequality the inequality $(pk)^{-1}(x + pr) < (pk)^{-1} F^{pk}(x) < (pk)^{-1}(x + p(r+1))$, for all $x \in R$. Therefore

$$\frac{r}{k} < \varrho(f) < \frac{r+1}{k}. \tag{5.98}$$

From the inequalities (5.94), (5.98) if follows that

$$|\varrho(f) - \varrho(f_0)| < \frac{1}{k} < \varepsilon. \quad \square$$

5.46 Theorem. Let $f : S^1 \to S^1, h : S^1 \to S^1$ be homeomorphisms preserving the orientation and $g = h \circ f \circ h^{-1}$. Then $\varrho(g) = \varrho(f)$, i.e. the rotation number is a topological invariant.

Proof

Let F, G, H be lifts of the homeomorphisms f, g, h, respectively, where

$$\varrho(f) = \lim_{n \to \infty} \frac{F^n(x)}{n}, \quad \varrho(g) = \lim_{n \to \infty} \frac{G^n(x)}{n}.$$

If $\varrho(f) = \varrho(g) = 0$, then there is nothing to prove. Without loss of generality we can suppose that $\varrho(g) \neq 0$. As

$$\lim_{n \to \infty} G^n(x) = \infty, \quad \lim_{z \to \infty} \frac{H(z)}{z} = \lim_{z \to \infty} \frac{H(0) + z}{z} = 1$$

and $F^n = H \circ G^n \circ H^{-1}$ thus

$$\varrho(f)(\varrho(g))^{-1} = \lim_{n\to\infty} \frac{F^n(y)}{n} \left(\lim_{n\to\infty} \frac{G^n(x)}{n}\right)^{-1} = \lim_{n\to\infty} F^n(y)(G^n(x))^{-1}$$
$$= \lim_{z\to\infty} \frac{H(z)}{z} = 1,$$

where $y = H(x)$. □

The assertion of the following lemma belongs to one from the disciplines of number theory, to the so called theory of Diophant approximations. As they are very important for the study of the structure of trajectories of diffeomorphisms on the circle, we present them with proof. Let us note that in the study of the structure of trajectories of diffeomorphisms on the circle and vector fields on the torus assertions stronger than this one are also used (see for example [10]). In these examples connections which very distant mathematical disciplines are shown.

5.47 Lemma. *If α is an irrational number then the following assertions hold.*

1. For an arbitrary natural number q there exists a natural number $m \leq q$ and an integer n such that the numbers m, n are mutually indivisible and

$$|m\alpha - n| < \frac{1}{q}. \tag{5.99}$$

2. There exist infinitely many rational numbers n/m, where m, n are mutually indivisible numbers which satisfy the inequality

$$\left|\alpha - \frac{n}{m}\right| < \frac{1}{m^2}. \tag{5.100}$$

Proof

Without loss of generality we can assume that $\alpha > 0$. Let us denote by (α) the fractional part of the number α. Let q be a natural number. Let us consider the intervals

$$I_j = \left\langle \frac{j-1}{q}, \frac{j}{q} \right\rangle, \quad j = 1, 2, \ldots, q.$$

As $\cup_{j=1}^{q} I_j = \langle 0, 1 \rangle$, each one of the numbers $0, (\alpha), (2\alpha), \ldots, (q\alpha)$ belongs to some of the above-mentioned intervals and as the number of these numbers is $q + 1$ there exists at least one such interval which contains two of these numbers at least. It means that there exists a natural number $s, 0 \leq s \leq q$ and natural numbers k_1, k_2 such that $k_1 \leq q, k_2 \leq q, k_1 > k_2$ and

$$\frac{s}{q} \leqq (k_1\alpha) < \frac{s+1}{q}, \qquad (5.101)$$

$$\frac{s}{q} \leqq (k_2\alpha) < \frac{s+1}{q}. \qquad (5.102)$$

If $m = k_1 - k_2$, then $0 < m \leq q$ and from the inequalities (5.101), (5.102) it follows that

$$0 \leqq (m\alpha) < \frac{1}{q}. \qquad (5.103)$$

Obviously, there exists an integer n such that $m\alpha = (m\alpha) + n$. If m, n are indivisible then from inequality (5.103) we obtain at once assertion 1. If $m = m'k, n = n'k$, where $k \in Z$ and m', n' are mutually indivisible, then from the inequality (5.103) it follows that

$$|m'\alpha - n'| = \frac{(m\alpha)}{|k|} < \frac{1}{q}$$

and thereby assertion 1 is proved. Let us assume now that there exists only a finite number of such fractions n_i/m_i, $i = 1, 2, \ldots, r$ that the numbers n_i, m_i are mutually indivisible and $|\alpha - n_i(m_i)^{-1}| < m_i^{-2}, i = 1, 2, \ldots, r$. As the number α is irrational, there exists a natural number q such that

$$\left|\alpha - \frac{n_i}{m_i}\right| > \frac{1}{q}, \qquad i = 1, 2, \ldots, r. \qquad (5.104)$$

According to assertion 1 there exists a natural number m and an integer n such that the numbers m, n are mutually indivisible $m \leqq q$ and

$$\left|\alpha - \frac{n}{m}\right| < \frac{1}{qm} \leqq \frac{1}{q} \qquad (5.105)$$

and from (5.104) it follows that $n/m \neq n_i/m_i$ for $i = 1, 2, \ldots, r$. As $m \leq q$, from (5.105) it follows that $|\alpha - n/m| < 1/m^2$ and this is a contradiction with the assumption that $n_i/m_i, i = 1, 2, \ldots, r$, are all fractions which satisfy the inequality of this type. $\square$

5.48 Theorem. *If α is an irrational number and $f : S^1 \to S^1$ is a turn about the angle $2\pi\alpha$, i.e. $f(e^{2\pi x i}) = e^{2\pi(x+\alpha)i}$ for all $x \in R$ then each trajectory of the diffeomorphism f is dense in S^1.*

Proof

It must be proved that for arbitrary $z \in S^1$ there is $\overline{\alpha(z)} \cup \overline{\omega(z)} = S^1$, where $\alpha(z), \omega(z)$ are the α-limit and ω-limit sets of the point z for the diffeomorphism f respectively. First, we show that the set $\omega(z)$ does not depend on z, i.e. if y is an arbitrary point from S^1 then $\omega(z) = \omega(y)$. The same assertion can be proved analogously for the set $\alpha(z)$. Let m, n be mutually various natural numbers. Let us denote $I = [f^m(z), f^n(z)]$ an arc on S^1 with initial point $f^m(z)$ and the end point $f^n(z)$ (in sense of orientation on S^1). As α is an irrational number, according to Corollary 5.44 it does not have either a periodic or fixed point, and thus there exist natural numbers $k_1, k_2, \ldots, k_r$ such that

$$\bigcup_{i=1}^{r} F^{-k_i(m-n)}(I) = S^1. \tag{5.106}$$

If $z_0 \in \omega(z)$, then there exists a sequence $\{s_j\}_{j=1}^{\infty}$ of natural numbers such that $\lim_{j \to \infty} s_j = \infty$ and $\lim_{j \to \infty} f^{s_j}(z) = z_0$. From (5.106) it follows that for arbitrary $y \in S^1$ there exists a sequence $\{p_j\}_{j=1}^{\infty}$ of natural numbers such that $\lim_{j \to \infty} p_j = \infty$ and $f^{p_j}(y) \in [f^{s_{j-1}}(z), f^{s_j}(z)]$ for all $j \in N$. As $\lim_{j \to \infty} f^{s_j}(z) = z_0$, we also obtain that $\lim_{j \to \infty} f^{p_j}(y) = z_0$ which means that $z_0 \in \omega(y)$. We have proved that $\omega(z) \subset \omega(y)$. As we can change the tasks z and y, the converse inclusion also holds and thus $\omega(z) = \omega(y)$. As the sets $\alpha(z), \omega(z)$ do not depend on z, it is sufficient to prove that $\overline{\alpha(1) \cup \omega(1)} = S^1$. The mapping $F: R \to R$, $F(x) = x + \alpha$ is a lift of the diffeomorphism f and obviously $F^n(x) = x + n\alpha$. Let $\{\gamma_n\}_{n=-\infty}^{\infty}$, where $\gamma_n = n\alpha - \{n\alpha\}$, $\{n\alpha\}$ is the integer part of the number $n\alpha$. Obviously $\overline{\alpha(1) \cup \omega(1)} = S^1$ iff the sequence $\{\gamma_n\}_{n=-\infty}^{\infty}$ is dense in the interval $\langle 0, 1 \rangle$. This sequence is obviously dense in $\langle 0, 1 \rangle$ iff the set $M = \{y = n\alpha + k : n \in Z, k \in Z\}$ is dense in R. The density of the set M obviously follows from Lemma 5.47. □

5.49 Definition. *Let X be a topological space. The homeomorphism $f: X \to X$ is called topologically transitive if there exists a point $x \in X$ such that the trajectory $\gamma_x = \{f^n(x) : n \in Z\}$ is dense in X. If for an arbitrary point $x \in X$ the trajectory γ_x is dense in X then the homeomorphism f is called minimal.*

The homeomorphism $f: S^1 \to S^1$ from Theorem 5.48 is minimal in the sense of the preceding definition.

5.50 Denjoy Theorem. *If $g: S^1 \to S^1$ is a C^2-diffeomorphism is preserving the orientation and has an irrational rotation number $\alpha = \varrho(g)$ then it is topologically conjugated with the diffeomorphism of turn of the circle S^1 about the angle $2\pi\alpha$.*

This theorem was proved in [31] by the French mathematician A. Denjoy and it was pronounced as a hypothesis even in 1885 by H. Poincaré. The proof of this theorem can be found in works [82, 119, 154]. A. Denjoy also found an example which shows that for validity of Theorem 5.50 it is not sufficient to assume that g is of the class C^1. The analysis of this example is given in work [154].

Now, we mention some results on generic properties of the rotation number of 1-parameter systems of the diffeomorphism of the unit circle S^1. These generic properties are advantageous to be studied in the space of lifts of the diffeomorphism and therefore let us consider the space $\mathscr{F}^r = \mathscr{F}^r(P, R)$ of mappings $F \in C^r(R \times P, R)$, such that for each $\varepsilon \in P$ the mapping $F : R \to R, F_\varepsilon(x) = F(x, \varepsilon)$, is a C^r-diffeomorphism, where $F_\varepsilon(x+1) = F_\varepsilon(x)+1$ for all $x \in R$, i.e. F_ε is the lift of a C^r-diffeomorphism $f_\varepsilon : S^1 \to S^1$. We assume that on $\mathscr{F}^r$ the Whitney C^r-topology is given. If $f \in \operatorname{Diff}^r(P, S)$ and $F \in \mathscr{F}^r$ is a mapping such that for arbitrary $\varepsilon \in P$ the mapping F_ε is the lift of the diffeomorphism f_ε then the mapping F is called a parametrized lift of the parametrized diffeomorphism f.

5.51 Theorem. *Let P be a smooth 1-dimensional manifold. Then there exists a massive subset $\mathscr{F}_1$ in $\mathscr{F}^1(P, R)$ such that it holds that if $F \in \mathscr{F}_1$ and there exists $\varepsilon \in P$ such that the rotation number $\varrho(F_\varepsilon)$ is irrational then there exists an open neighbourhood U of the point ε such that the function $\sigma_F : U \to R, \sigma_F(\mu) = \varrho(F_\mu)$ is monotonous, where for arbitrary $\eta > 0$, such that $J_\eta = (\varepsilon - \eta, \varepsilon + \eta) \subset U$, the function σ_F/J_μ is not constant.*

The proof of this theorem can be found in [110] which is a correction of the proof of the assertion of this theorem from the work [27]. In [110] it is also shown that the method of proof of Theorem 1.51 is not applicable to the proof of the analogous theorem on the generic property of monotonicity of the function σ_F in the set $\mathscr{F}^r(P, R)$ for $r > 1$. The proof of such a theorem still remains an open problem.

5.52 Theorem. *Let P be a smooth one-dimensional manifold. Let us assume that the parametrized lift of the parametrized diffeomorphism $f \in \operatorname{Diff}^1(P, S^1)$ belongs to the set $\mathscr{F}_1$ from Theorem 5.51 and let for $\varepsilon \in P$ the rotation number $\varrho(f_\varepsilon)$ be irrational. Let the positive number M be given. Then there exists a neighbourhood U of the point ε and such a number $N \geq M$ that for an arbitrary number $m > N$ there exists $\mu \in U$ such that the diffeomorphism f_μ has an m-periodic point.*

Proof

Without loss of generality we can assume that $\varrho(f_v) = \varrho(F_v)$ for all $v \in P$. From Theorem 5.51 it follows that there exists an open neighbourhood $\tilde{U}$ of the point ε such that the function σ_F is monotonous and non-constant on $\tilde{U}$. Therefore we can, without loss of generality, assume that $\sigma_F(\tilde{U}) = (\sigma_F(\varepsilon) - \delta, \sigma_F(\varepsilon) + \delta)$, where $\delta > 0$. Let δ be a small number such that $(\sqrt{\delta})^{-1} > M$ and let N be a number such that $N > (\sqrt{\delta})^{-1}$. As the function σ_F is monotonous and it is not constant on $\tilde{U}$, there exists an open neighbourhood U of the point ε such that $U \subset \tilde{U}$ and

$$\sigma_F(U) = (\sigma_F(\varepsilon) - N^{-2}, \sigma_F(\varepsilon) + N^{-2}). \tag{5.107}$$

From Lemma 5.47 it follows that for arbitrary natural number $m > N$ there exists such an integer n such that m, n are mutually indivisible numbers and

$$|\varrho(F_\varepsilon) - \frac{n}{m}| < \frac{1}{m^2} < \frac{1}{N^2}. \tag{5.108}$$

As $\varrho(F_\varepsilon) = \sigma_F(\varepsilon)$ from (5.107), (5.108) it follows that $n/m \in \sigma_F(U)$ and therefore there must exist $\mu \in U$ such that $\sigma_F(\mu) = \varrho(F_\mu) = n/m$. From Theorem 5.43 it follows that the diffeomorphism f_μ has an m-periodic point. $\square$

5.53 Remark. Let g be a parametrized diffeomorphism from the Nejmark–Sacker Theorem and let the assumptions of this theorem be fulfilled. Let us denote by C_ε the closed invariant curve (5.85) of the complexification $\tilde{g}_\varepsilon(z) = \lambda(\varepsilon)z + U(z, \overline{z}, \varepsilon)$ (see (5.83)) of the diffeomorphism g_ε. For the restriction of the diffeomorphism $\tilde{g}_\varepsilon$ on C_ε the rotation number $\varrho(\hat{g}_\varepsilon)$ is defined. From [75, Theorem 3] it follows that

$$\alpha_0 = \lim_{\varepsilon \to 0} \varrho(\tilde{g}_\varepsilon) = \frac{\varphi_0}{2\pi},$$

where $\lambda(0) = \exp(2\pi\varphi_0 i)$, i.e.

$$\alpha_0 = \frac{1}{2\pi i} \ln \lambda(0).$$

As generically there is $\lambda^k(0) \neq \pm 1$ for all $k \in Z$ (that follows from Theorem 3.75), the number α_0 is generically irrational. Let G be a parametrized lift of the parametrized diffeomorphism $\hat{g}$ ($\hat{g}(z, \varepsilon) = \hat{g}_\varepsilon(z)$). There exists $\delta > 0$ such that the function $\sigma_G : J \to R$, $\sigma_G(v) = \varrho(G_v)$ is defined, where $J = (0, \delta)$, which is continuous according to Theorem 5.45. According to Theorem 5.51 this function is generically monotonous, where for arbitrary $\varkappa > 0, 0 < \varkappa < \delta$ the function $\sigma_G/(0, \varkappa)$ is not constant. For the same reasons as mentioned in the proof of Theorem 5.52 we obtain that for arbitrary positive number M there exists a number $N \geq M$ such that for arbitrary natural number $m > N$ there exists $\mu \in (0, \delta)$ such that the diffeomorphism $\hat{g}_\mu$ has an m-periodic point. It means that for an arbitrary neighbourhood V of the point $\varepsilon = 0$ and an arbitrary sufficiently large natural number m there exists $\mu \in V$ such that the diffeomorphism $\tilde{g}_\mu$ has an m-periodic point on its invariant curve C_μ.

Chapter 5

5.4 Generic Bifurcations of 1-parameter Systems of Vector Fields in Neighbourhoods of Periodic Trajectories

As the results on generic bifurcations of vector fields in neighbourhoods of periodic trajectories have, in fact, a local character, in this section we shall consider parametrized vector fields on R^n with the set of parameters R, which are represented by parametrized differential equations. On the basis of results which we have formulated up till now for parametrized vector fields on manifolds we should, however, without any problems, formulate all the results of this part also for parametrized vector fields on manifolds.

Thus, let us consider the set G^r of all parametrized differential equations of the form

$$\dot{x} = f(x, \varepsilon), \qquad (5.109)$$

where $f \in C^r(R^n \times R, R^n)$ and such that for $\varepsilon = 0$ the differential equation (5.109) has a periodic trajectory γ_0 passing through the point $x = 0$. We shall consider the set G^r with the topology induced by the Whitney C^r-topology of the set $C^r(R \times R, R^n)$. As in the previous paragraphs we shall denote by the sign f_ε the differential equation (5.109) for a given fixed ε. If Σ is a transversal to γ_0 at the point $x = 0$ then the Poincaré mapping $H = H[f, \gamma_0, \Sigma, 0, 0, U, V]$ is defined, where U is an open neighbourhood of the point $x = 0$ in Σ and V is an open neighbourhood of the point $\varepsilon = 0$ (see Definition 3.171).

5.54 Definition. *Let $f \in G^r$ and γ_0 be a periodic trajectory of the differential equation $\dot{x} = f(x, 0)$. We say that this periodic trajectory is of the type $\gamma(+1)(\gamma(-1); \gamma(c))$ if the germ $[H]_0$, represented by the Poincaré mapping $H = H[f, \gamma_0, \Sigma, 0, 0, U, V]$, is of the type $m(+1)(m(-1); m(c))$ in the sense of the Definition 5.27.*

5.55 Theorem. *Let $r \in N, r \geq 6$. Then the set G_1 of all such $f \in G^r$ for which the periodic trajectory γ_0 of the differential equation $\dot{x} = f(x, 0)$ passing through the point $x = 0$ is either of the type $\gamma(+1)$ or $(\gamma(-1)$ or $\gamma(c)$, is massive in C^r.*

Proof

Let $f \in G^r$ and γ_0 be a periodic trajectory of the differential equation $\dot{x} = f(x, 0)$ passing through $x = 0$. From Theorem 3.179 it follows that there exists a local transversal Σ to γ_0 at the point $x = 0$ and an open neighbourhood $U \times V$ of the point $(x,\varepsilon) = (0,0)$ in $\Sigma \times R$ such that the parametrized Poincaré mapping is defined and it holds that there exists an open neighbourhood $W^r(H)$ of the mapping H in $C^r(U \times V, \Sigma)$ and a continuous mapping $\varkappa : W^r(H) \to G^r$ such that $f \in \varkappa(W^r(H))$ and for each mapping $\tilde{H} \in W^r(H)$ the parametrized differential equation $\tilde{g} = \varkappa(\tilde{H})$ has the property that the Poincaré mapping $H_{\tilde{g}} = H_{\tilde{g}}[f, \gamma_0, \Sigma, 0, 0, U, V]$ is defined and $H_{\tilde{g}} = \tilde{H}$ (see Definition 3.171). From Theorem 5.38 it follows that in an arbitrary small open neighbourhood $\hat{W}^r(H)$ of the mapping H in $C^r(U \times V, \Sigma)$ there exists a mapping $\hat{H}$ such that $\hat{H}$ is a parametrized diffeomorphism, $\hat{H}(0,0) = 0$, where the germ $[\tilde{H}]_0$ is either of the type $m(+1)$ or $m(-1)$ or $m(c)$. If the neighbourhood $\hat{W}^r(H)$ is so small that $\tilde{W}^r(H) \subset W^r(H)$ then for the parametrized differential equation $g = \varkappa(\hat{H})$ the Poincaré mapping $H_g = H_g[f, \gamma_0, \Sigma, 0, 0, U, V]$ is defined, where $H_g = \hat{H}$. As $\hat{H}(0,0) = 0$, the differential equation $\dot{x} = g(x, 0)$ has a periodic trajectory $\hat{\gamma}_0$ passing through the point $x = 0$ and $H[g, \hat{\gamma}_0, \Sigma, 0, 0, U, V] = H_g = \hat{H}$ which means that the periodic trajectory $\hat{\gamma}_0$ is either of the type $\gamma(+1)$ or $\gamma(-1)$ or $\gamma(c)$. □

5.56 Definition. *Let $f \in G_1$, γ_0 be as in Theorem 5.55. The bifurcation which occurs in the neighbourhood of the periodic trajectory γ_0 of the type $\gamma(+1)(\gamma(-1); \gamma(c))$ is called the bifurcation of saddle-node (type of period-doubling bifurcation; bifurcation of periodic trajectory into an invariant torus).*

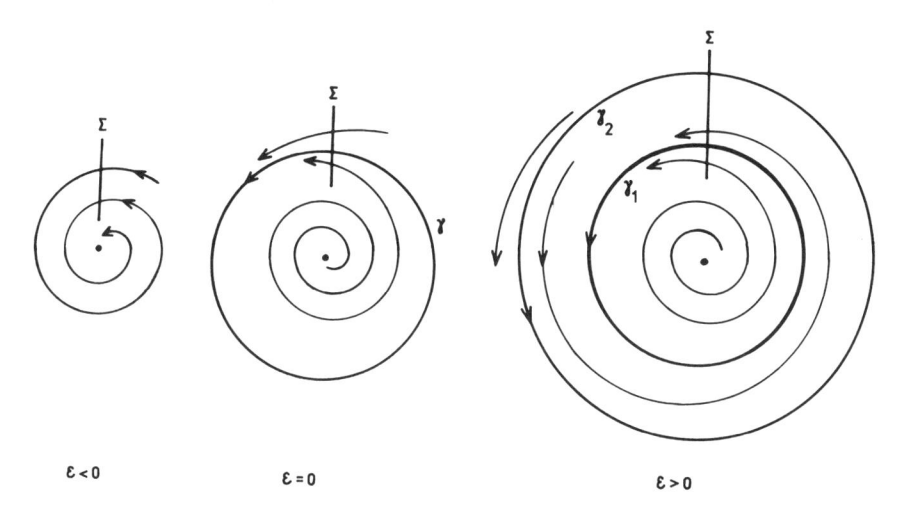

Fig. 47. Saddle-node bifurcation of periodic trajectories.

From the results of Theorems 5.29, 5.30, 5.35 and from Theorem 5.55 it follows that in the case of 1-parameter systems of vector fields in small

Chapter 5

neighbourhoods of periodic trajectories only such bifurcations can occur which correspond to the generic bifurcations of the reduction of Poincaré mappings to their centre manifolds. These are only three types of bifurcations which are shown in Fig. 47, 48, 49. Let us remark that in the case of bifurcation of periodic trajectory into an invariant torus there originate other secondary bifurcations corresponding to the bifurcations of a 1-parameter system of diffeomorphism of the circle which we have described partially in the preceding section. The density of an arbitrary trajectory in the original torus in the case of irrational rotation number as well as the existence of periodic trajectory of arbitrary high period for some values of parameter arbitrarily closed to a bifurcation point (see Theorems 5.48, 5.50, 5.52 and the Remark 5.53), give evidence that the dependence of topological structure on a parameter is in this case considerably irregular. In work [133] the problems of turbulent (irregular, chaotic) flow of liquids are studied by means of this bifurcation.

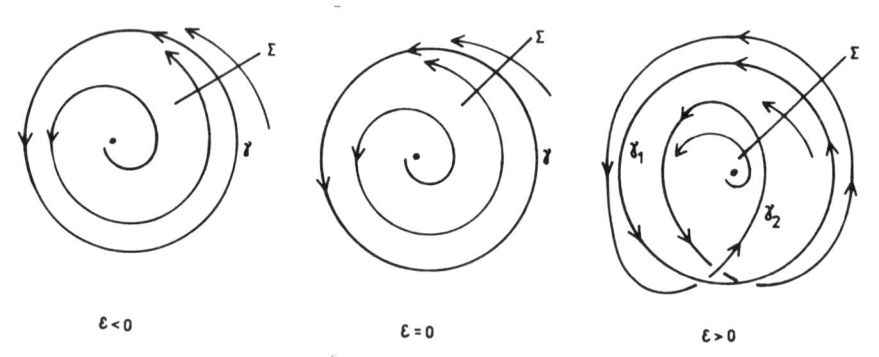

Fig. 48. Period-doubling bifurcation.

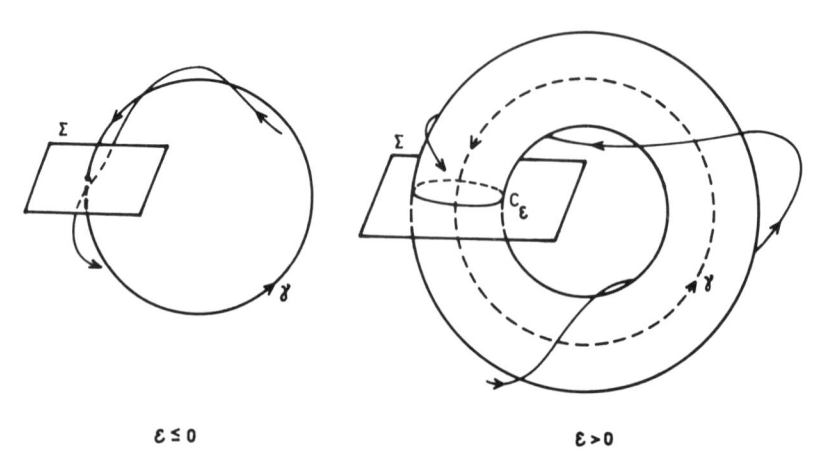

Fig. 49. Bifurcation of periodic trajectory into invariant torus.

5.57 Remark. The theorem on generic bifurcations of 1-parameter systems of vector fields on manifolds in neighbourhoods of periodic trajectories which, in fact, is of the same character as the theorem on 1-parameter systems of diffeomorphisms, was proved in [143 and 103]. We shall not formulate this theorem separately since the reader will be able to formulate it himself. Let us also remark that the analogous theorem for a 1-parameter system of second-order differential equations on manifolds is proved in [105].

6 COMPLEMENTARY NOTES ON THE CONTEMPORARY THEORY OF DYNAMICAL SYSTEMS

This chapter consists of the basic information on local bifurcations of multi-parameter systems of vector fields and some details of the contemporary global theory of dynamical systems. With respect to the limited extent of the book this topic cannot be discussed here in more-detail. We shall confine ourselves here just to the outline of its contents and highlight the information on the basic literature dealing with this topic. The complete list of literature on the present theory of dynamical system is too large to include in this book as it would contain details of dozens of books and hundreds of papers. However, quite a long list of works containing a substantial amount of information on the current theory of dynamical systems is presented at the end of this book.

6.1 Generic Bifurcations of Multi-Parameter Systems of Vector Fields

From Theorem 3.74 it follows that for a two-parameter system of matrices $\{A(\mu)\}_{\mu \in R^2}$, where A belongs to a certain massive subset $N^r(R^2)$ in $C^r(R^2, M(n))$, where $r \geq 1$, there exists a set $X_0 \subset R^2$ having the property as follows. If X_0 is a non-empty set then it consists of isolated points only and for $\mu_0 \in X_0$ the matrix $A(\mu_0)$ has at least two eigenvalues with zero real part (exclusive of their complex conjugate), whereby just one of the following cases occurs (the papers presenting the studies of the corresponding generic bifurcations are given in brackets).

1. $\lambda_1 = 0$ of multiplicity 2 (without any symmetry condition: [10, 18, 19, 151]; with different symmetry conditions: [28, 72, 143, 163]).

2. $\lambda_1 = 0$, $\lambda_2 = i\omega$, $\lambda_3 = -i\omega$, $\omega \neq 0$, all of multiplicity 1: [49, 50, 70, 71, 72, 74].

3. $\lambda_1 = i\omega$, $\lambda_2 = -i\omega_1$, $\lambda_3 = i\omega_2$, $\lambda_4 = -i\omega_2$, $\omega_1 \neq \omega_2$, $\omega_1 \neq 0$, $\omega_2 \neq 0$: [40, 50, 73].

Now we shall introduce the result concerning the case 1, as proved in [18, 19]. Let us consider the set H^r of all 2-parameter systems of differential equations

$$\dot{x} = f(x, \mu), \tag{6.1}$$

where $f \in C^r(R^2 \times R^2, R^2)$, $1 \leq r \leq \infty$, $\mu \in R^2$ is a parameter, $f(0,0) = (0,0)$, where on H^r the Whitney C^r-topology is given.

6.1 Bogdanov Theorem. *Let $r \in N$, $2 \leq r \leq \infty$. Then there exists an open, dense subset H_0 in H^r such that if the parametrized differential equation (6.1) is from the set H_0 and the matrix $[d_0 f_0]$ has a double zero eigenvalue then the germ $[f]_0$, represented by the parametrized differential equation (6.1), is orbitally C^r-equivalent to the germ $[g]_0$ that is represented by a parametrized system of differential equations of the form*

$$\begin{aligned} \dot{y}_1 &= y_2, \\ \dot{y}_2 &= \varepsilon_1 + \varepsilon_2 y_1 + y_1^2 + y_1 y_2 Q(y_1, \varepsilon) + y_2^2 \Phi(y, \varepsilon), \end{aligned} \tag{6.2}$$

where $Q, \Phi \in C^r$, $Q(0,0) = \alpha \neq 0$. The germ $[g]_0$ is orbitally topologically equivalent to the germ $[G]_0$ represented by the parametrized system of differential equations

$$\begin{aligned} \dot{y}_1 &= y_2, \\ \dot{y}_2 &= \varepsilon_1 + \varepsilon_2 y_1 + y_1^2 + b y_1 y_2, \end{aligned} \tag{6.3}$$

where $b = \operatorname{sign} \alpha$, and the system (6.3) is versal.

The system (6.3) is called the Bogdanov normal form.

6.2 Remark. The normal form (6.2) can also be derived with the use of the Takens Theorem (in [18] the Takens Theorem is not used) and the Malgrange–Weierstrass Theorem. In [18, 19] the smoothness of the right side of the differential equation (6.1) is assumed, namely because when deriving the normal form (6.2) the smooth version of the Malgrange–Weierstrass Theorem is used. Reduction of the smooth vector field to the centre manifold, however, need not represent a smooth vector field since the centre manifold need not be smooth (see Chapter 4). Therefore, in general, this result cannot be used even for smooth vector fields with more dimensional state space than 2. In [90], however, the Malgrange–Weierstrass Theorem is also proved for functions of the class C^r, where $1 \leq r \leq \infty$, and thus, the above theorem could be formulated also for the set H^r, where $2 \leq r < \infty$.

6.3 Remark. The Bogdanov normal form (6.3) with $b = -1$ can be transformed into the form (6.3) with $b = +1$ by interchanging the coordinates $(y_1, y_2) \to (y_1, -y_2)$ and transforming time $t \to -t$. This means that it is sufficient to know the bifurcations of the normal form with $b = +1$. The bifurcation diagram and bifurcations corresponding to the above normal form are shown in Fig. 50.

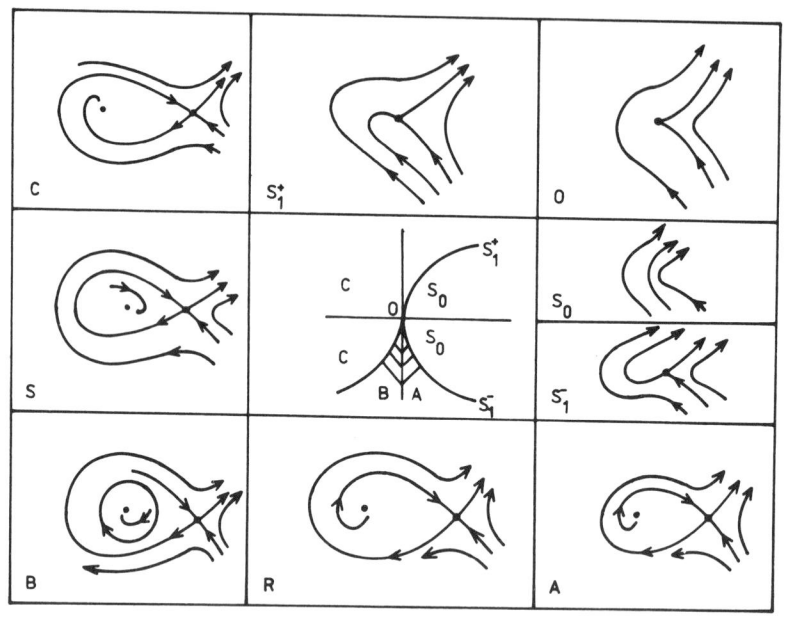

Fig. 50. Bogdanov bifurcation.

In cases 2 and 3 there the reduction to the centre manifold is represented by the vector field on R^3 or R^4, and thus, bifurcations in both cases are rather complicated. Let us consider the case 2, only, i.e. of the 2-parametrized system of differential equations

$$\dot{u} = F(u, \mu), \qquad (6.4)$$

where $F \in C^\infty(R^3 \times R^2, R^3)$, $F(0,0) = 0$, $u = (x, y, z) \in R^3$, $\mu = (\mu_1, \mu_2) \in R^2$ is a parameter and the matrix $A = (a_{ij}) \in M(3)$ of linearization of the differential equation (6.4) for $\mu = 0$ at the point $0 \in R^3$ is such that $a_{12} = -a_{21} = \omega \neq 0$, its other elements being zero. In this case we can proceed as follows. First of all, with the use of the Takens Theorem the best possible normal form of the differential equation (6.4) for $\mu = 0$ is derived. Let us suppose that even the equation (6.4) for $\mu = 0$ is of the normal form like this. Then the cylindrical coordinates $x = r \cos \varphi$,

$y = r\sin\varphi$, $z = z$, $0 \leq \varphi < 2\pi$, $r \geq 0$ are introduced. The convenience of the above normal form rests in the fact that it can be chosen in such a way that the coefficients of its representation in cylindrical coordinates at terms of arbitrarily high chosen order are independent of φ. If in this normal form only these terms dependent on φ are left, thus omitting all other higher order terms and moreover omitting also the differential equation for φ, we obtain a system of two differential equations in the variables r, z whose coefficients are independent of φ. Then the bifurcations of 2-parameter deformations of this system can be studied or an attempt can be made to prove versality of some of them. Now, we shall introduce an example of a deformation like this one which was derived and studied in [49] (see also [41, 50, 74]). The parametrized system of differential equations

$$\dot{r} = \mu_1 r + arz + cr^3 + drz^2,$$
$$\dot{z} = \mu_2 + br^2 - z^2 + er^2 z + fz^3, \qquad (6.5)$$

is concerned, where $a, b, c, d, e, f \in R$ and $\mu_1, \mu_2 \in R$ are parameters. In this case there are several types of bifurcations, of which at least one will be presented here. This type was chosen also because it is illustrative in connection with the considerations that will be used in Section 6.2 on global theory of dynamical systems. If $b = -1, a > 0, B < 0 < A, 3A + B < 0$, where $A = 3c + e, B = d + ef$, then the parametrized system (6.5) has the bifurcation diagram and bifurcations as illustrated in Fig. 51 (see [50]).

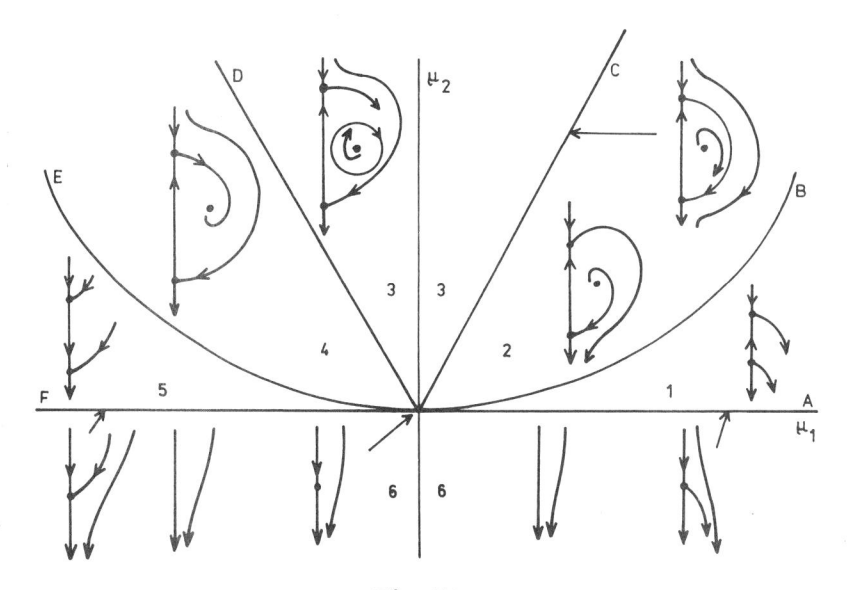

Fig. 51.

The form of the system (6.5) yields that the z-axis is an invariant set of this system for arbitrary μ_1, μ_2. For these reasons the bifurcations of this system cannot be generic. We are able, however, to find bifurcations of perturbations of the above system, allowing us to make conclusions about generic bifurcations of the original system (6.4). Let us note that for $\mu_2 > 0$ the system (6.5) has two saddle type singular points K_1, K_2 (lying on the line $r = 0$) and one focus K having the coordinates (z, r), where $r > 0$. The saddles L_1, L_2 of the system (6.4) (excluding terms of higher order than 3) correspond to the saddles K_1, K_2, while the periodic trajectory γ corresponds to the focus K of the system (6.4). On transition of the parameter through the curve D from region 4 into the region 5 (see Fig. 51) the Hopf bifurcation occurs. From the stable focus K a stable periodic trajectory β is separated, to which the stable invariant torus $T_\mu^2(\gamma)$ of the system (6.4) (without higher order terms than 3) surrounding the trajectory γ corresponds (see Fig. 49 of Chapter 5). On the torus $T_\mu^2(\gamma)$ when the parameter changing the secondary bifurcations depending on the rotation number of the Poincaré mapping P defined on the transversal Σ to γ originate, more precisely, on the rotation number of the mapping $\tilde{P} : S_\mu \to S_\mu$, where $S_\mu = T_\mu^2(\gamma) \cap \Sigma, \tilde{P} = P/S_\mu$ (see Remark 5.53). If then the values of the parameter approach the curve C from Fig. 51, then periods of the periodic trajectories lying on the torus $T_\mu^2(\gamma)$ originating in case of secondary bifurcations converge to infinity. Precisely, on the curve C from the periodic trajectory β the so-called heteroclinic trajectory of the system (6.5) connecting the saddles K_1, K_2 originates. A general definition of this trajectory will be presented below.

6.4 Definition. *Let φ be a dynamic system on the smooth manifold X (either continuous or discrete) of the class C^r, where $1 \leq r \leq \infty$ and $x_1, x_2 \in X$ are its hyperbolic singular points (or hyperbolic fixed or periodic points if φ is discrete). Let $W^u(x_1)$ be stable and $W^s(x_2)$ unstable invariant manifolds of the point x_1 or x_2. The point $p \in W^u(x_1) \cap W^s(x_2)$ different from x_1 and x_2 is called heteroclinic, if $x_1 \neq x_2$ and homoclinic, if $x_1 = x_2$. If $W^t(x_1) \overline{\cap}_p W^s(x_2)$, then the point p is called transversal heteroclinic or transversal homoclinic, respectively (Fig. 52). The trajectory of the dynamical system φ passing through the heteroclinic (homoclinic) point p is called heteroclinic (homoclinic).*

Let us note that from the uniqueness theorem of solutions of differential equations it follows that if x is a singular point of a continuous dynamical system in the plane of the class C^1 and $p \in W^u(x) \cap W^s(x)$ is its homoclinic point then it lies on the homoclinic trajectory for which both the α-limit and ω-limit set is represented by the point x (Fig. 53).

In further considerations the following theorem will be needed.

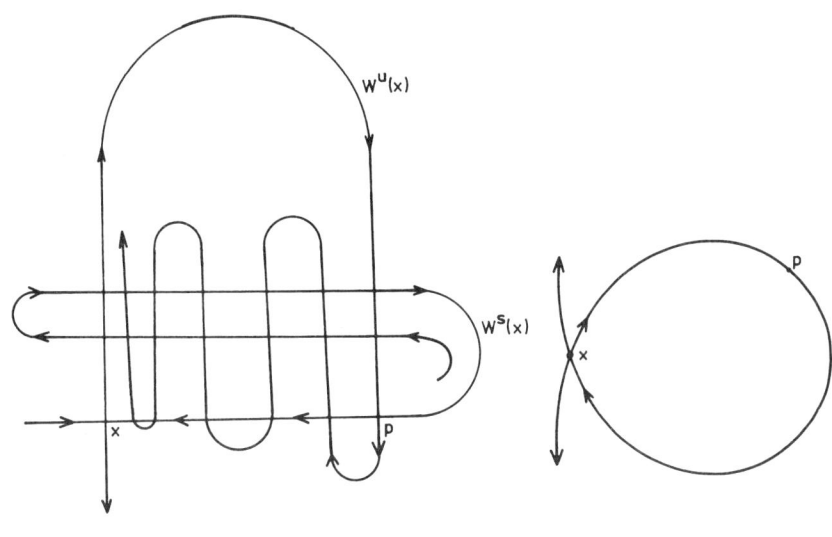

Fig. 52.   Fig. 53.

6.5 Kupka–Smale Theorem. *Let X be a smooth manifold and $r \in N$, $r \geqq 1$. Then there exists a massive subset V_{KS}^r in $V^r(X)$, such that for every vector field $F \in V_{KS}^r$ it holds that:*

1. all critical elements (i.e., the singular points and periodic trajectories) of the vector field F are hyperbolic;

2. if σ_1, σ_2 are critical elements of the vector field F, then $W^s(\sigma_1) \pitchfork W^u(\sigma_2)$, where $W^s(\sigma_1)$ ($W^u(\sigma_2)$) is the stable (unstable) invariant manifold of the critical element $\sigma_1(\sigma_2)$.

Also an analogous theorem for diffeomorphisms, related to the fixed and periodic point of diffeomorphisms holds and it is also called the Kupka–Smale Theorem for diffeomorphisms. The vector field F having the properties 1–3 from Theorem 6.5 is called the Kupka–Smale vector field. The diffeomorphisms having the properties of the diffeomorphisms from the Kupka–Smale Theorem are called the Kupka–Smale diffeomorphisms. Proof of Theorem 6.5 can be found for example in [2]. The part of assertion 1 of this theorem concerning singular points can easily be proved by the reader using the Abraham Transversality Theorem. However, the rest of this assertion concerning periodic trajectories is more complicated. Proof of the Kupka–Smale Theorem for diffeomorphisms can be found, for example in [122].

Let us now come back to bifurcations of the systems (6.4), (6.5). On the curve C in Fig. 51 a heteroclinic trajectory of the system (6.5), connecting the saddle points K_1, K_2 arises. Existence of such a trajectory means that for the system (6.4) (without higher order terms than 3) there exists

Chapter 6

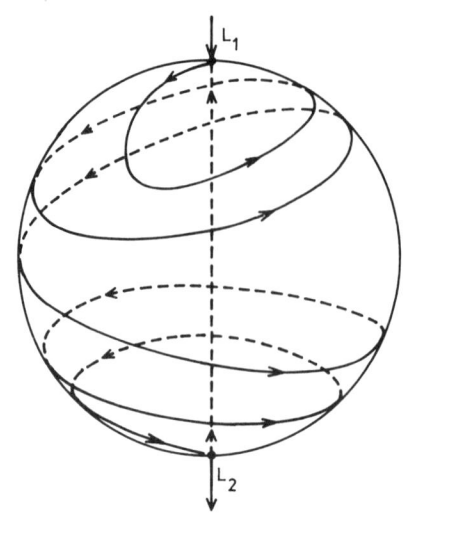

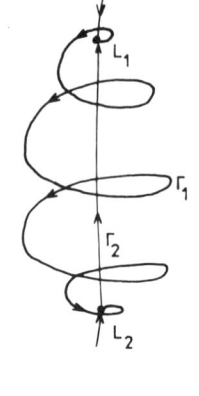

Fig. 54.   Fig. 55.

an invariant sphere S of this system containing the saddle points L_1, L_2 consisting of heteroclinic trajectories connecting the saddles L_1, L_2 with each other, where inside this sphere there exists one more heteroclinical trajectory (see Fig. 54). Since $W^s(L_2) = S \setminus \{L_1\}, W^u(L_1) = S \setminus \{L_2\}$, the invariant manifolds $W^s(L_2), W^u(L_1)$ do not transversally intersect each other at any point and that, with respect to assertion 3 of the Kupka–Smale Theorem is a too non-generic case. Moreover, from the form of the system (6.5) it follows that the z-axis is an invariant set of this system, and thus, for arbitrary values of the parameters μ_1, μ_2 there exists a heteroclinic trajectory connecting the saddles L_1 and L_2. Let us choose one of the heteroclinic trajectories lying on S, denoting it by Γ_1 and let us denote the heteroclinic trajectory lying inside S as Γ_2. The set $\Gamma = \Gamma_1 \cup \Gamma_2 \cup \{L_1\} \cup \{L_2\}$ is a closed curve in R^3 (Fig. 55). Now let us admit arbitrary C^r-perturbations of the vector field we have obtained from the system (6.5) by omitting the terms of order higher than 3, thus also such that will disturb the invariance of the z-axis. From the Kupka–Smale Theorem it follows that a perturbation of this type can be chosen in such a way that for some values of the parameter both the stable and unstable manifolds of the saddles of this perturbation will intersect each other transversally, where there will exist neither heteroclinic nor homoclinic trajectories. In [71] an example of such a perturbation is found. The bifurcations for a perturbed system like this show properties similar to those as introduced later in case of the so-called Šilnikov bifurcation. Let us note that the bifurcations in a neighbourhood of a closed curve formed by two heteroclinic trajectories connecting two saddles (of the type Γ in Fig. 55) are studied in detail in [15]. The bifurcations in a neighbourhood of a homoclinic trajectory of the vector field in R^3 with a certain type of symmetry are studied in [16].

In Section 3.4 we have also introduced generic classification of two-parameter systems of matrices according to their eigenvalues on the unit circle. According to Theorem 3.75 there exist eight types of linear parts of diffeomorphisms for which generic bifurcations of 2-parameter systems of diffeomorphisms can occur. However, all of them are rather complicated, due to the possible existence of homoclinic and heteroclinic points and trajectories, and therefore they have not yet been studied in general.

Generic bifurcations of vector fields with higher order of "degeneracy" have also been studied. For example, in [153] the so-called generalized Hopf bifurcation is studied. This case occurs, when the assumptions of the Hopf Bifurcation Theorem are not fulfilled, namely if in Theorem 5.21 there is $c_{j0} = 0$ for $j = 1, 2, \ldots, m-1$ but $c_{m0} \neq 0$ (see the system (5.43), (5.44)). In this case the parametrized differential equation of the form (5.27) has to be studied, where $\dim \varepsilon = 2m$. For solving the bifurcation equation (5.48) whose number of real solutions is equal to the number of periodic solutions of the differential equation (5.27) the Malgrange–Weierstrass Theorem can be used. After transformation of coordinates in the space of parameters the equation $\varrho^{2m} + \mu_{2m}\varrho^{2m-1} + \cdots + \mu_1 = 0$ is obtained from which the bifurcation diagram for the parametrized differential equation (5.27) can be determined.

In [108] the problem of generic bifurcations of 3-parameter systems of vector fields was studied in a neighbourhood of a value of the parameter for which the matrix of linearization of the vector field has zero eigenvalue of multiplicity 2 having no other eigenvalue with zero real part and the non-linear part of the vector field satisfies a certain "degeneracy" condition. The normal forms of reduction to the centre manifold corresponding to this case are of the form

$$\dot{u}_1 = u_2, \dot{u}_2 = \gamma_1^\sigma(\mu) + \gamma_2^\sigma(\mu)u_1 + \mu_3 u_1^2 + \sigma u_1^3 + u_1 u_2 Q(u_1, \mu) + u_2^2 \Phi(u, \mu),$$

where $\mu = (\mu_1, \mu_2, \mu_3) \in R^3$ is a parameter, $\sigma = \pm 1$, $Q, \Phi \in C^\infty$ (or C^r, $r \geq 3$), $Q(0,0) = \omega > 0$, $\gamma_1^\sigma(\mu) = 2\sigma\mu_1 + \mu_2\mu_3 + (1/27)\mu_3^3$, $\gamma_2^\sigma(\mu) = \sigma\left(3\mu_2 + (1/3)\mu_3^2\right)$ (compare these normal forms with the normal form (6.2), or with the Bogdanov normal form (6.3)).

In paper [107] the generic bifurcations of 3-parameter systems of vector fields in a neighbourhood of a value of the parameter are studied for which the matrix of linearization of the vector field has zero eigenvalue of multiplicity 3 with no other eigenvalue with zero real part and the non-linear part of the vector field satisfies a certain non-degenaracy condition. The normal form of reduction onto the centre manifold corresponding to the above case has the form

$$\dot{y}_1 = y_2, \dot{y}_2 = y_3,$$

$$\dot{y}_3 = \mu_1 + \mu_2 y_1 + y_1^2 + \mu_3 y_2 + y_1 y_2 Q_1(y_1, \mu) + y_1 y_3 Q_2(y_1, \mu)$$

$$+ y_2 Q_3(y_3, \mu) + y_2^2 \Phi_1(y, \mu) + y_3^2 \Phi_2(y, \mu),$$

where $Q_1, Q_2, Q_3, \Phi_1, \Phi_2 \in C^\infty$ (or C^r, $r > 3$), $Q_1(0,0) = \omega_1 \neq 0$, $Q_2(0,0) = \omega_2 \neq 0$ and $\mu = (\mu_1, \mu_2, \mu_3) \in R^3$ is a parameter.

6.2 Global Theory of Dynamical Systems

In Section 6.1 we have formulated the Kupka–Smale Theorem that has a global character. According to this theorem the properties of the Kupka–Smale vector fields (diffeomorphisms) are generic. A question as to arises whether these vector fields (diffeomorphisms) are also structurally stable. For example, a smooth vector field X on a 2-dimensional torus T^2 without singular points defining the diffeomorphisms f of the circle $S = T^2 \cap \Sigma$ (Σ being a cross-sectional transversal to T^2) with an irrational rotation number has neither periodic trajectory, and thus, it is a Kupka–Smale vector field. This vector field is not structurally stable, since by its arbitrarily small C^1-perturbation, a vector field X can be obtained, defining the diffeomorphism $f : S \to S$ with a rational rotation number still having a periodic trajectory (see Chapter 3, part. 3 and also [119]). The definition of the Kupka–Smale vector field does not contain any information on its limit properties, and thus, it is natural that the Kupka–Smale vector fields (it also holds in the case of diffeomorphisms) in general are not structurally stable.

A very important class of vector fields (diffeomorphisms) are the so-called Morse–Smale vector fields (diffeomorphisms).

6.6 Definition. *Let X be a smooth manifold $r \in N$, $r \geq 1$. Then the vector field $F \in V^r(X)$ is called the Morse–Smale vector field if*
  1. *F has a finite number of critical elements and all are hyperbolic;*
  2. *if σ_1, σ_2 are critical elements of the vector field F, then $\overline{W^s(\sigma_1) \cap W^u(\sigma_2)}$, where $W^s(\sigma_1)$ ($W^u(\sigma_2)$) is the stable (unstable) invariant manifold of the critical element $\sigma_1(\sigma_2)$;*
  3. *the set $\Omega(F)$ of all non-wandering points of the vector field F is the union of the set of critical elements of the vector field F.*

Analogously, the Morse–Smale diffeomorphism is defined.

6.7 Theorem. *If X is a smooth, orientable, compact manifold of dimension 2 and $r \geq 1$, then it holds that:*
   1. the vector field $F \in V^r(X)$ is structurally stable iff it is the Morse-Smale vector field;
   2. the set $V_{MS}^r \subset V^r(X)$ of all Morse-Smale C^r-vector fields is open and dense in $V^r(X)$.

Theorem 6.7 was proved by M. M. Peixoto in 1962 (see [123] and also [122]). In [120] the openness of the set V_{MS}^r in $V^r(X)$ is proved as well as the structural stability of arbitrary vector field $F \in V_{MS}^r$ for an orientable compact manifold X of arbitrary dimension. But if $\dim X > 2$, then the set V_{MS}^r need not be dense in $V^r(X)$. In fact on such manifolds structurally stable C^r-vector fields can exist that are not Morse–Smale vector fields. The vector fields like this can be constructed with the use of Theorem 3.24 on a manifold of dimension 3, e.g. with the use of the diffeomorphism $g \in \mathrm{Diff}^r(S^2)$ (S^2 being the 2-dimensional unit sphere) that is constructed with the use of a certain mapping $f : Q \to R^2$, where $Q = \langle 0,1 \rangle \times \langle 0,1 \rangle$, called the Smale horseshoe (see, e.g., [122], Example 4 p. 162). Now we shall outline the definition of the Smale horseshoe, allowing at least intuitive understanding of complicated properties of the set of non-wandering points $\Omega(g)$ of the diffeomorphism g, and thus, also of the set of non-wandering points $\Omega(G)$ of the vector field G constructed from it. A deep analysis of the properties of the Smale horseshoe can be found, for example, in [77, 119, 122].

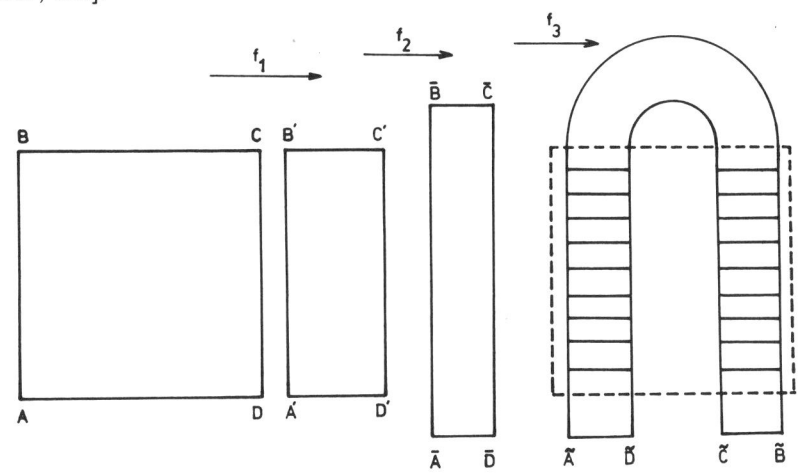

Fig. 56.

Let us consider the square $Q = I \times I$, where $I = \langle 0, 1 \rangle$, whose vertices will be denoted as A, B, C, D and the mapping $f : Q \to R^2, f =$

$f_3 \circ f_2 \circ f_1$, where f_1 is the contraction of the square Q in the horizontal direction, f_2 being the expansion Q in the vertical direction, i.e. the mapping $f_2 \circ f_1$ will map Q into a narrow strip expanded in the vertical direction, while the mapping f_3 will bend this strip into the form of a horseshoe so that $Q \cap f(Q) = A_0 \cup A_1$, where A_0, A_1 are disjoint rectangles situated in Q in the vertical direction (Fig. 56). Then $Q \cap f(Q) \cap f(Q \cap f(Q)) = Q \cap f(Q) \cap f^2(Q) = A_{00} \cup A_{10} \cup A_{01} \cup A_{11}$, where $A_{00}, A_{10} \subset A_0$, $A_{01}, A_{11} \subset A_1$ are the rectangles disjoint with each other situated in Q in the vertical direction (Fig. 57), analogously $Q \cap f(Q) \cap f^2(Q) \cap f^3(Q) = [(A_{000} \cup A_{100}) \cup (A_{010} \cup A_{110})] \cup [(A_{001} \cup A_{101}) \cup (A_{011} \cup A_{111})]$, where $A_{000}, A_{100} \subset A_{00}, A_{010}, A_{110} \subset A_{10}, A_{001}, A_{101} \subset A_{01}, A_{011}, A_{111} \subset A_{11}$ are the rectangles disjoint with each other situated in Q in the horizontal direction. If continuing this process, we get that the set $\cap_{j=0}^n f^j(Q)$ contains 2^n mutually disjoint rectangles situated in Q in horizontal direction that can be denoted analogously as said before. Thus, each of these strips will be denoted by the symbol $A_{i_1 i_2 \ldots i_n}$, where $i_k \in \{0,1\}$ ($k = 1, 2, \ldots, n$). Analogously, it will be shown that the set $\cap_{j=-n}^0 f^j(Q)$ contains 2^n mutually disjoint rectangles (we can denote them as $B_{i_1 i_2 \ldots i_n}$) situated in Q in the vertical direction. The set $\cap_{j=-n}^n f^j(Q)$ consists of the points of intersection of the rectangles $A_{i_1 i_2 \ldots i_n}$ and $B_{i_1 i_2 \ldots i_n}$ ($i_k \in \{0,1\}$, $k = 1, 2, \ldots, n$) and the set $\cap_{j=0}^\infty f^j(Q) = C_1 \times I$, where C_1 has properties identical to those of the Cantor set (see [148]), i.e. it is uncountable, has the Lebesgue measure 0 and is perfect (i.e. closed and every its point is its accumulation point). Analogously, $\cap_{j=-\infty}^0 f^j(Q) = I \times C_2$, where C_2 also has the same properties as the Cantor set. Thus, we get that $\Lambda \stackrel{\text{def}}{=} \cap_{j=-\infty}^\infty f^j(Q) = C_1 \times C_2$. Obviously, the set Λ is an invariant set of the mapping f. It is not difficult to show that there exists a bijective mapping $p : \Lambda \to \Sigma$, where Σ is the set of all the sequences of the form $a = \{a_i\}_{i=-\infty}^\infty$, where $a_i \in \{0,1\}$ for all $i \in Z$. Let us consider Σ as a metric space with the metric

$$d(a,b) = \sum_{i=-\infty}^\infty 2^{-|i|} |a_i - b_i|,$$

where $a = \{a_i\}_{i=-\infty}^\infty, b = \{b_i\}_{i=-\infty}^\infty \in \Sigma$. Let us define the mapping $\sigma : \Sigma \to \Sigma$, $\sigma(a) = \{\sigma(a)_i\}_{i=-\infty}^\infty$, where $\sigma(a)_i = a_{i+1}$ (σ is called *the shift map*). Obviously it holds that the point $a \in \Sigma$ is an n-periodic point of the mapping σ iff $a_{i+n} = a_i$ for all $i \in Z$. Let $a = \{a_i\}_{i=-\infty}^\infty$ be given. For every $k \in N$ let us define the periodic sequence $a_k = \{a_j^{(k)}\}_{i=-\infty}^\infty$, where $a_{i+2kj}^{(k)} = a_j^{(k)} = a_i$ for $-k \leq i \leq k$ ($k = 1, 2, \ldots,$) and for all $j \in N$. The sequence $\{a_k\}_{k=1}^\infty$ converges to a, thus, the set $\text{Per}(\sigma)$ of all periodic points of the mapping σ is dense in Σ. Let us define the mapping $h : \Lambda \to \Sigma, h(x) = \{h(x)_i\}_{i=-\infty}^\infty$, where $h(x)_i = 1$, if $f^i(x) \in A_1$ and $h(x)_i = 0$, if $f^i(x) \in A_0$. It can be shown that h is a homeomorphism and $h \circ f \circ h^{-1}(a) = \sigma(a)$ for all $a \in \Sigma$, i.e., the mapping f/Λ is topologically conjugate with the shift map σ. Therefore the set Per(f) of all periodic points of the mapping f and thus the set of non-wandering points $\Omega(f)$ is dense in Λ which even means that $\Omega(f) = \Lambda$. The example of the Smale horseshoe brings us to the following definition.

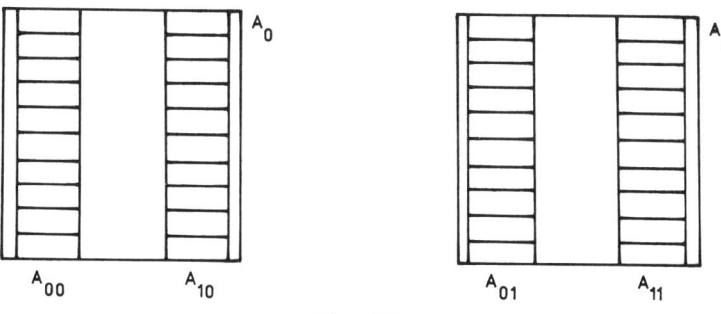

Fig. 57.

6.8 Definition. *Let X be a smooth manifold and $r \in N$, $r \geq 1$. Diffeomorphisms $f, g \in \text{Diff}^r(X)$ are called Ω-conjugated if there exists a homeomorphism $h : \Omega(f) \to \Omega(g)$ ($\Omega(f), \Omega(g)$) are the sets of the non-wandering points of diffeomorphisms f, and g respectively with the topology induced by topology on X such that $\tilde{g} \circ h = h \circ \tilde{f}$, where $\tilde{g} = g/\Omega(g), \tilde{f} = f/\Omega(f)$. We say that the diffeomorphism f is Ω-stable if there exists its neighbourhood $N(f)$ in $\text{Diff}^r(X)$ such that an arbitrary diffeomorphism $g \in N(f)$ is Ω-conjugated with f.*

According to [118, Theorem 4.1] the diffeomorphism f (Smale horseshoe) is Ω-stable. However, in [3] it is proved that even Ω-stability is not a generic property in $\text{Diff}^r(X)$, if dim $X > 2$ and therefore the hope of at least relatively "nice" generic classification of both diffeomorphisms and vector fields fails.

To conclude this section we bring to the reader's attention the fact that beside the books which were quoted here some other important information regarding this question can be found in [8, 50, 56, 117, 118, 139 and 147].

6.3 Šil'nikov Bifurcation

Let us consider the field F of the class C^r, where $r \geqq 1$, defined by system of differential equations of the form

$$\begin{bmatrix} \dot{x} \\ \dot{y} \\ \dot{z} \end{bmatrix} = \begin{bmatrix} \alpha & -\beta & 0 \\ \beta & \alpha & 0 \\ 0 & 0 & \lambda \end{bmatrix} \begin{bmatrix} x \\ y \\ z \end{bmatrix} + \begin{bmatrix} f_1(x,y,z) \\ f_2(x,y,z) \\ f_3(x,y,z) \end{bmatrix}, \qquad (6.6)$$

Chapter 6

where $f_i(0,0,0) = 0, i = 1,2,3$ and let us assume that $\lambda > 0, \alpha < 0$ and $|\alpha| < \lambda$, i.e. the point $K = (0,0,0)$ is the hyperbolic singular point of the vector F. Furthermore, let us assume that there is a homoclinic trajectory Γ of the system (6.6), the α-limit and ω-limit sets of which is the point K. This structure of trajectories may appear generically even for the set of 1-parametric systems of vector fields on R^3 (this follows from [74], Corollary 3.5, p. 383) which can also be understable from Fig. 58.

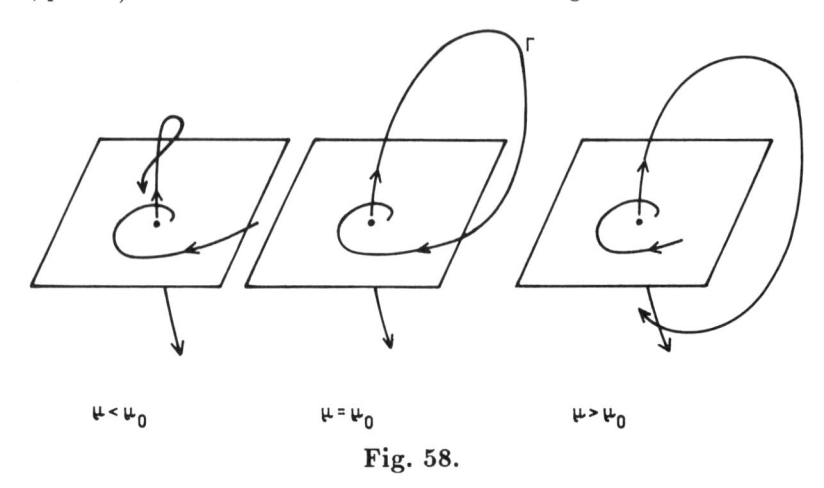

Fig. 58.

We give a result of the structure of trajectories of C^1-perturbations of the system (6.6). The detailed analysis of this case can be found in [149], as well as in [50] from which we choose some important parts of the following considerations and results.

Without loss of generality it can be assumed that in a certain neighbourhood U of the point K the system (6.6) is identical to its linearization at this point. Let r_0, r_1 be small positive numbers such that the region bounded by the surfaces $\Sigma_0 = \{(x,y,z) : x^2 + y^2 = r_0^2, 0 < z < z_1\}, \Sigma_1 = \{(x,y,z) : x^2 + y^2 < r_0^2, z = z_1\}$ belongs to U. Let $p \in \Sigma_0, q \in \Sigma_1$ be cross points of the homoclinic trajectory Γ with Σ_0 and Σ_1 respectively. If the numbers r_0, z_1 are small enough then an arbitrary trajectory of the system (6.6) coming from the point u lying on Σ_0 will cross Σ_1. Let us denote the first of these cross points as $T_1(u)$. Thus, we obtain the mapping $T_1 : \Sigma_0 \to \Sigma_1, u \mapsto T_1(u)$. If $\Sigma_1' \subset \Sigma_1$ is a sufficiently small open neighbourhood of the point q in Σ_1, then an arbitrary trajectory coming from the point $v \in \Sigma_1'$ will cross Σ_0. Let us denote the first such cross point as $T_2(v)$. Thus, we obtain the mapping $T_2 : \Sigma_1' \to \Sigma_0, v \mapsto T_2(v)$. We can introduce new coordinates (Θ, z) on the set Σ_0 by means of the mapping $(x,y) = (r_0 \cos \Theta, r_0 \sin \Theta), 0 \le \Theta < 2\pi$. Let us define the set $V = \{(r_0, \Theta, z) : |\Theta| < \delta, 0 < z < \varepsilon\}$, where δ, ε are positive numbers such that $T_1(V) \subset \Sigma_1'$. Then the mapping $T = T_2 \circ T_1 : V \to \Sigma_0$ (see Fig. 59).

274

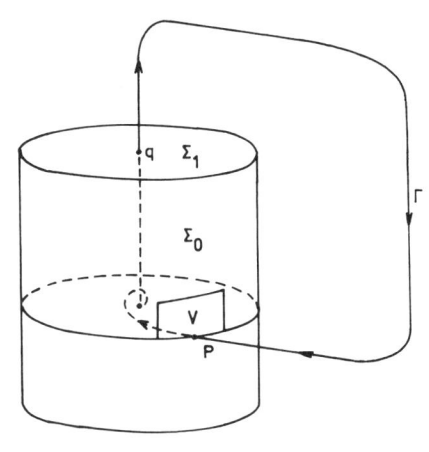

Fig. 59.

6.9 Definition. *The mapping $T = T_2 \circ T_1 : V \to \Sigma_0$ is called Poincaré mapping of the vector field defined by the homoclinic trajectory Γ and it is also denoted as $T[F, \Gamma, V, \Sigma_0]$.*

It can be proved (for example, see [50], pp. 323, 324), that there exists such an arbitrary small C^1-pertubation $\tilde{F}$ of the vector field F such that it holds that

1. $\tilde{F}$ has a homoclinic trajectory $\tilde{\Gamma}$ close to Γ;
2. the Poincaré mapping $\tilde{T} = \tilde{T}[\tilde{F}, \tilde{\Gamma}, \tilde{V}, \Sigma_0] : \tilde{V} \to \Sigma_0$, defined in an analogous way to the way we defined the mapping T, is such that there exist positive numbers $A, B, A < B$ such that the image $\tilde{T}(W)$ of the set $W = \{r_0, \Theta, z\} \in \tilde{V} : z \in (A, B)\}$ is of horseshoe shape shown in Fig. 60 (see [50], p. 323).

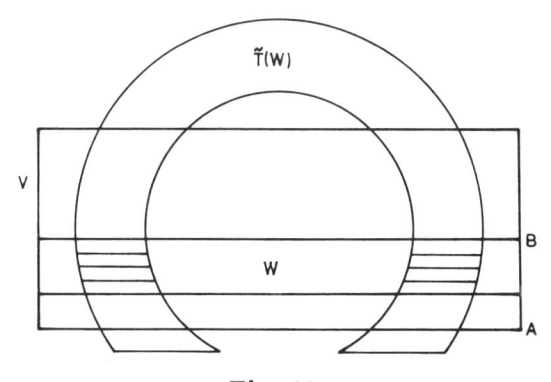

Fig. 60.

The mapping $\tilde{T}$ has the same properties as the Smale horseshoe. This fact gives at least a partial idea of the complexity of the structure of generic bifurcations of parametric deformations of the vector field F. If G is a 1-parametric system of smooth vector fields on R^3 such that $G_0 = F$ and $G_v = \tilde{F}$, then the bifurcation which originates for G when transfering the parameter μ from $\mu = 0$ through $\mu = v$ is called the Šil'nikov bifurcation.

We formulate the theorem which summarized the above considerations.

6.10 Šil'nikov Theorem. *Let F be a C^r-vector field on R^3 defined by the system of differential equations of the form (6.6), where $r \geq 1$, $\alpha < 0$, $\lambda > 0$, $|\alpha| < \lambda$ and let us assume that there exists a homoclinic trajectory Γ of the vector field F the α-limit and ω-limit set of which is the point $K = (0,0,0)$. Then for an arbitrary neighbourhood $U(F)$ of the vector field F in $V^1(F)$ there is a vector field $\tilde{F} \in U(F)$ such that it holds that:*

1. the vector field $\tilde{F}$ has a homoclinic trajectory $\tilde{\Gamma}$ close to Γ;

2. there exist positive numbers $\varepsilon, \delta > 0$ such that on the set $\tilde{V} = \{(r_0, \Theta, z) : |\Theta| < \delta, 0 < z < \varepsilon\}$ the Poincaré mapping $\tilde{T} = \tilde{T}[\tilde{F}, \tilde{\Gamma}, \tilde{V}, \Sigma_0]$ is defined (defined analogously to the mapping T from the definition 6.9) which has the same properties as the Smale horseshoe, i.e. there exists an invariant set Λ of the mapping $\tilde{T}$ which has the properties of a Cantor set, where $\operatorname{Per}(\tilde{T}) = \Lambda = \Omega(\tilde{T})$.

6.4 Global Hopf Bifurcation

In Chapter 5 we studied only local generic bifurcations. However, in the case of local bifurcation, the question arises of what will happen to a periodic trajectory and its period originating at this bifurcation if the parameter value is removed from the given bifurcation value. We present one of the most effective methods of the study of this question.

Let us consider the parametrized differential equation

$$\dot{x} = f(x, \varepsilon), \tag{6.7}$$

where $f \in C^r(R^n \times R, R^n), r \geq 1$. We denote the set of all such equations as D^r and write $f \in D^r$. Let $K(f) = \{(x, \varepsilon) \in R^n \times R : f(x, \varepsilon) = 0\}$ and $K_0(f)$ be the set of all points $(x, \lambda) \in (K(f)$, such that for an arbitrary open neighbourhood $U \times V$ of the point (x, λ) the point λ is a bifurcation point of the parametrized differential equation $\dot{x} = g(x, \varepsilon)$, where $g = f/U \times V$. It follows from Theorems 5.23, 5.55 and from [103], (Theorems 2 and 3) that there exists a massive subset D_1^r in D^r (let us consider the Whitney C^r-topology on D^r) such that the following assertions hold for $f \in D_1^r$.

1. The set $K(f)$ is a one-dimensional C^r-submanifold in $R^n \times R$.

2. The set $K_0(f) \subset K(f)$ consists of isolated points.

3. The parametrized differential equation (6.7) has only generic bifurcations in neighbourhoods of points from $K_0(f)$, i.e. either the generic Hopf bifurcation or generic bifurcation of the saddle-node type.

4. The bifurcations of parametrized differential equation (6.7) which occur in neighbourhoods of their periodic trajectories may be either of the type $\gamma(+1)$ or $\gamma(-1)$, or of the type $\gamma(c)$ (see the Definition 5.56).

6.11 Definition. *We say that the parametrized differential equation (6.7) is generic if it fulfills the conditions 1–4.*

6.12 Definition. *The point $(x, \varepsilon) \in K_0(f)$ is called the centre of the parametrized differential equation (6.7) if the matrix $[d_x f_\varepsilon]$ has at least one pair of mutually complex conjugated pure imaginary eigenvalues and does not have the zero eigenvalue. Let us denote by $C(f)$ the set of all centres of the parametrized differential equation (6.7).*

6.13 Definition. *Let for $(x, \varepsilon) \in K(f)$, $E_f(x, \varepsilon)$ be the number of eigenvalues of the matrix $[d_x f_\varepsilon]$, which have positive real parts. If $(x, \varepsilon) \in C(f)$, then we define the number*

$$\boxplus_f(x, \varepsilon) = (-1)^{E_f(x,\varepsilon)} \chi_f(x, \varepsilon), \tag{6.8}$$

where

$$\chi_f(x, \varepsilon) = \lim_{\delta \to 0+} \frac{1}{2}[E_f(x, \varepsilon + \delta) - E_f(x, \varepsilon - \delta)].$$

The number $\boxplus_f(x, \varepsilon)$ is called the index of the centre (x, ε).

Let $P(f)$ be the set of all points $(x, \varepsilon) \in R^n \times R$ such that a periodic trajectory of the differential equation (6.7) passes through the point x. If the parametrized differential equation (6.7) is generic and $(x_0, \varepsilon_0) \in C(f)$, then it follows from the Hopf Bifurcation Theorem that there exists an open neighbourhood U of the point (x_0, ε_0) such that the set $P(f, U) = P(f) \cap U$ is a two-dimensional C^r-submanifold in $R^n \times R$. Let us denote the maximal connected subset of the set $P(f)$ which contains the set $P(f, U)$ as $Q(f; x_0, \varepsilon_0)$.

6.14 Definition. *We say that there exists the global Hopf bifurcation of the parametrized differential equation (6.7) if there exists its centre $(x_0, \varepsilon_0) \in C(f)$ such that either the set $Q(f; x_0, \varepsilon_0)$ is unbounded or for an arbitrary $K > 0$ there exists $(x, \varepsilon) \in Q(f; x_0, \varepsilon_0)$, such that the periodic trajectory γ_x of the differential equation (6.7) passing through the point x has the period $T_x > K$.*

6.15 Global Hopf Bifurcation Theorem. *Let the parametrized differential equation* (6.7) *be generic. Let us assume that the set of centres $C(f)$ is finite and $\mathbb{D}_f \stackrel{def}{=} \sum_{i=1}^{m} \mathbb{D}_f(x_i, \varepsilon_i) \neq 0$, where $C(f) = \{(x_i, \varepsilon_i)\}_{i=1}^{m}$. Then there is a global Hopf bifurcation of the parametrized equation* (6.7).

Proof of Theorem 6.15 can be found in [96]. An interesting version of the Global Hopf Bifurcation Theorem can be found in [36], where by means of indices of centres the global Hopf bifurcation of parametrized differential equations, on special Banach space, to which bifurcation problems of so-called semi-linear partial differential equations of parabolic type (see [62]) can be reduced, are studied. The methods of the study of global generic bifurcations of the parametrized differential equations by means of properties of their linearization at bifurcation points have found their application in the study of global generic bifurcations of 2-parametric systems of differential equations. In [37] an index of the singular point corresponding to the Bogdan normal form is defined and with its help the problem of global Hopf bifurcation for 2-parametric systems of differential equations is studied.

6.16 Example. Let us consider a 1-parametric system of differential equations (6.7) which is in form of the system

$$\dot{x}_1 = x_2 + [\varepsilon(1-\varepsilon)(2-\varepsilon) - x_1^2 - x_2^2]x_1,$$
$$\dot{x}_2 = -x_1 + [\varepsilon(1-\varepsilon)(2-\varepsilon) - x_1^2 - x_2^2]x_2. \tag{6.9}$$

Obviously, $C(f) = \{(x_i, \varepsilon_i)\}_{i=1}^{3}$, where $(x_1, \varepsilon_1) = (0, 0)$, $(x_2, \varepsilon_2) = (0, 1)$, $(x_3, \varepsilon_3) = (0, 2)$. It follows from (6.8) that $\mathbb{D}_f(x_1, \varepsilon_1) = +1$, $\mathbb{D}_f(x_2, \varepsilon_2) = -1$,

$$\mathbb{D}_f(x_3, \varepsilon_3) = -1$$

and thus $\mathbb{D}_f = \sum_{i=1}^{3} \mathbb{D}_f(x_i, \varepsilon_i) = -1$. It results from Theorem 6.15 that for the system (6.9) there exists a global Hopf bifurcation. If $\varrho = (x_1^2 + x_2^2)^{1/2}$, then from the system (6.9) we obtain a differential equation for ϱ of the form $\dot{\varrho} = \varrho[\varepsilon(1-\varepsilon)(2-\varepsilon) - \varrho^2]$. Bifurcations of singular points of this parametrized differential equation for $\varrho > 0$ are shown in Fig. 61 and the Hopf bifurcations of the system (6.9) in neighbourhoods of centres correspond to them. Connection of singular points $(\varepsilon_1, \varrho_1) = (0, 0), (\varepsilon_2, \varrho_2) = (1, 0)$ by a curve of singular points is possible only because $\mathbb{D}_f(x_1, \varepsilon_1) + \mathbb{D}_f(x_2, \varepsilon_2) = 0$. However, there still remains one centre; $(x_3, \varepsilon_3) = (0, 2)$ the index of which equals -1. Therefore the curve of singular points coming from the point $(\varepsilon_3, \varrho_3) = (2, 0)$ (Fig. 61) cannot end at any singular points. This conclusion also follows from the uniqueness of periodic trajectories which originate at local Hopf bifurcation. However, the Global Hopf Bifurcation Theorem gives more information, especially about the periods of periodic trajectories. We should be able to find an analytical expression of the curve of singular points coming from the point $(\varepsilon_3, \varrho_3)$. According to this theorem there are only two alternatives: this curve is either unbounded or there is a point $(\bar{x}, \bar{\varepsilon})$ lying in its closure such that the periods of periodic trajectories of the system (6.9) corresponding to points laying on this curve converge to ∞ if $\varepsilon \to \bar{\varepsilon}$.

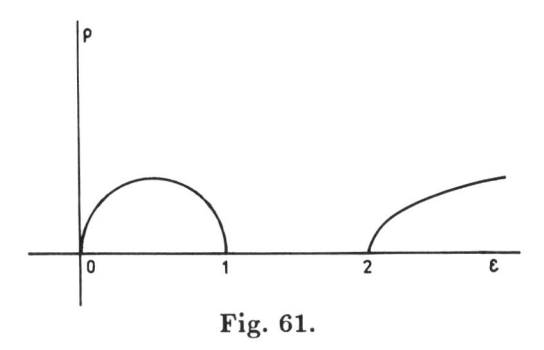

Fig. 61.

6.5 Attractors and Chaotic Sets

The example of a Smale horseshoe which was mentioned in Section 6.2 shows that even dynamical systems on R^3 have an extremely complicated structure of trajectories. However, this example does not sufficiently express the whole complexity of dynamical systems. In this section we show at least what further possible complex structures of dynamical trajectories there are.

In Section 6.2 we mentioned the construction of a vector field on a 3-dimensional manifold such that the mapping of the Smale horseshoe type is its recurrent mapping. The qualitative properties of this vector field are characterized sufficiently by the properties of the Smale horseshoe. Now we mention an example of the vector field the qualitative properties of which are also characterized sufficiently by the properties of a certain recurrent mapping which is not a diffeomorphism. This example represents the whole class of dynamical systems with the so-called chaotic properties. It deals with the so-called Lorenz attractor which belongs to quite a simple dynamical system on R^3 defined by the system of differential equations

$$\dot{x} = -10x + 10y, \quad \dot{y} = 28x - y - xz, \quad \dot{z} = -\frac{8}{3}z + xy. \quad (6.10)$$

With the help of this system E. N. Lorenz modelled certain hydrodynamical processes. He found by numerical calculation with the help of a computer that, as far as their global properties were concerned, this system has a very complicated "irregular" structure of trajectories. The Lorenz work (its translation can be found in the proceedings [138]) was published in 1963 but it attracted mathematicians' attention later. At present they are studing qualitative properties of dynamical systems similar to those possessed by system (6.10). System (6.10) has an invariant set to which all the trajectories of this system coming from its sufficiently small neighbourhood converge for $t \to \infty$, while the global properties of the trajectories lying on it are very "sensitively" dependent on initial conditions and have so-called chaotic global properties. This problem as well as the literature dealing with it can be studied in more detail in [50, 138]. The qualitative properties of a Lorenz attractor can be explained on its topological model, which was suggested by American mathematician R. F. Williams. Before we describe it we shall give a definition of attractor.

6.17 Definition. *Let φ be a C^r-dynamical system on a smooth manifold X. The set $M \subset X$ is called an attractor of dynamical system φ if this set is its invariant set and if there exists its neighbourhood U such that $M = \cap_{t>0} \varphi_t(U), \varphi_t(U) \subset U$ for all $t > 0$ and periodic trajectories of the dynamical system φ lying on M are dense in M.*

This definition of attractor can be found for example in R. F. Williams' article of [138].

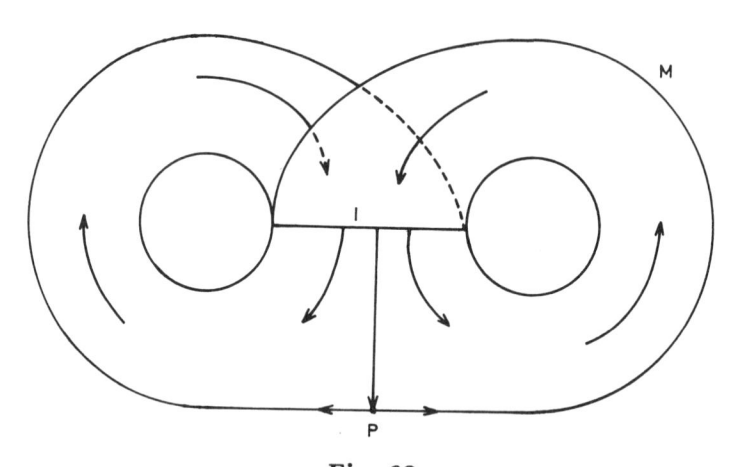

Fig. 62.

Let φ be a dynamical system on R^3 which has an invariant set M as shown in Fig. 62 where the trajectories lying outside M converge to M for $t \to \infty$. The set M is a two-dimensional surface which branches in the segment I, where the dynamical system φ has a hyperbolic singular point P on M. The whole information on the structure of trajectories of the dynamical system φ is contained in the properties of the Poincaré mapping $f : I \to I$, i.e. a recurrent mapping formed by the cross-points of trajectories with the segment I. The segment I can be identified with the interval $\langle -1, 1 \rangle$ for example, and thus it is sufficient to study iterations of the function $f : I \to I$, where $I = \langle -1, 1 \rangle$. Therefore the theory of iterations of functions, which need not be either diffeomorphisms or homeomorphisms, are paid a lot of attention nowadays. The basic information of this problem can be found in [50, 138 and 141]. One can show that under some assumption on the function f, the set M is an attractor in the sense of Definition 6.17. We shall not formulate the obtained results concerning the Lorenz attractor and other similar attractors. It is quite understable that trajectories of the dynamical system φ lying on M can be present for a certain period of time on one branch of the surface M then they transfer to the other branch and after a time they return again etc., where the segment I is the place of transition from one branch to the other. As the sequence of transition moments from one branch of the set M to the other is considerably irregular for many trajectories and iterations of the mapping f have chaotic properties, global properties of trajectories of the dynamical system φ on the set M are sometimes expressed properly by the term deterministic chaos. The Lorenz attractor and similar attractors which are in the form of branching surfaces have been called strange attractors because of their form. However, their inner properties characterized by chaotic properties of the trajectories lying on them are common and substantial features of strange attractors. Their chaotic features can be expressed well by chaotic properties of recurrent mappings defined on the set of branching of these attractors. A definition of the chaotic set of mapping can be found in the article by D. L. Kaplan and J. A. York in [138].

6.18 Definition. *Let (X, d) be a metric-space, $A \subset X$ and $g : A \to X$. The set $B \subset A$ is called an invariant set of the mapping g if $g(B) = B$. We say that the set B is a stable set of the mapping g (in the Ljapunov sense) if for each $\varepsilon > 0$ there is $\delta \in (0, \varepsilon)$ such that $d(x, B) \leq \delta \implies d(g^n(x), B) \leq \varepsilon$ for all $n \in N$, where $d(y, B) = \inf_{z \in B} d(y, z)$ is the distance of the point y from the set B and $g^n(y)$ is the n-iteration of the mapping g in the point y. We say that B is an unstable set of the mapping g (in the Ljapunov sense) if it is not stable. We say that the set B is an inner unstable set of the mapping g, if each trajectory of the mapping g lying in B is an unstable set of the mapping g, i.e. for each $x \in B$ there is $\varepsilon > 0$ and a sequence $\{x_i\}_{i=1}^{\infty}$ of the elements from B converging to x such that it holds that for each $i \in N$ there is $n = n(i) \in N$ such that $d(g^n(x_i), g^n(x)) > \varepsilon$. We say that a compact and invariant set B of the mapping g is a chaotic set of the mapping g if it is inner unstable and there exists a trajectory of the mapping g, dense in B i.e. there exists $x \in B$, such that the closure of the set $\{g^n(x) : n \in N\}$ is equal to B.*

References

[1] Abraham, R. and Marsden, E.J, Foundations of Mechanics, New York, Amsterdam, W.A. Benjamin 1967.
[2] Abraham, R. and Robbin, J., Transversal Mappings and Flows, New York, Amsterdam, W.A. Benjamin 1967.
[3] Abraham, R. and Smale, S., Nongericity of Ω-Stability, Providence R.I., Proceedings of the Symp. in Pure Math. 14, AMS 1970, pp. 5–8.
[4] Adámek, J., Koubek, V. and Reiterman, J., Foundations of General Topology, Prague, SNTL 1977 (In Czech).
[5] Adams, J.F., Lecture on Lie Groups, New York, W.A. Benjamin 1969.
[6] Andronov, A.A., Leontovich, E.A., Gordon, I.I. and Maier, A.G., Theory of Bifurcations of Dynamical Systems on the Plane, Moscow, Nauka 1967 (In Russian).
[7] Andronov, A.A. and Pontryagin, L.C., Rough Systems, Doklady Akad. Nauk 14, 1937, pp. 247–251 (In Russian).
[8] Anosov, D.V., Smooth Dynamical Systems, Proceedings, Moscow, Mir 1977 (In Russian).
[9] Arnol'd, V.I., Ordinary Differential Equations, Moscow, Nauka 1971 (In Russian).
[10] Arnol'd, V.I., Geometrical Methods in the Theory of Ordinary Differential Equations, New York, Heidelberg, Berlin, Springer-Verlag 1983.
[11] Arnol'd, V.I., Catastrophe Theory, Moscow, Moscow Univer. Press 1983 (In Russian).
[12] Arnol'd, V.I., Small Denominators I, On Mappings of the Circle into Itself, Izv. Akad. Nauk SSSR, 25, 1, 1961 pp. 21–26 (In Russian).
[13] Bibikov, J.N., General Course of Differential Equations, Leningrad, Leningrad Univ. Press 1981 (In Russian).
[14] Bibikov, J.N., Local Theory of Nonlinear Analytic Ordinary Differential Equations, Berlin, Heidelberg, New York, LNM 702, Springer-Verlag 1979.
[15] Bykov, V.V., On the Structure of a Neighbourhood of Separatrix Contour with a Saddle-Focus, Proceedings, Methods of Qualitative Theory of Differential Equations, Gorkii, Gorkii Univ. Press 1978, pp. 3–32 (In Russian).
[16] Bykov, V.V., On Bifurcations of Dynamical Systems Close to Systems with Separatrix Contour Containing a Saddle-Focus, Proceedings, Methods of Qualitative Theory of Differential Equations, Gorkii, Gorkii Univ. Press 1980, pp. 44–72 (In Russian).
[17] Birkhoff, G.D., Dynamical Systems, New York, Amer. Math. Soc. Colloquim Publications 1927.
[18] Bogdanov, R.I., Versal Deformations of a Singular Point on the Plane in the Case of Zero Eigenvalues, Trudy Seminara I.G. Petrovskogo, Vol. 2, 1976, pp. 37–65 (In Russian).

References

[19] Bogdanov, R.I., Bifurcations of a Limit Circle of a Family of Vector Fields in the Plane, Trudy Seminara I.G. Petrovskogo, Vol. 2, 1976, pp. 23–36 (In Russian).

[20] Bogoljubov, N.N. and Mitropolskij, J.A., Asymptotic Methods in Theory of Nonlinear Oscillations, New York, Gordon and Breach 1961.

[21] Brin, M.J., On an Embedding of Diffeomorphism into a Flow, Izd. Vuzov, 8, 123, 1973, pp. 19–23 (In Russian).

[22] Bryuno, A.D., Local Methods in Nonlinear Differential Equations, Moscow, Nauka 1979 (In Russian).

[23] Bronštein, I.U., Extensions of Minimal Transformations Groups, Kishinev, Shtiinitsa 1975 (In Russian).

[24] Bröcker, Th. and Lander, L., Differentiable Germs and Catastrophes, Cambridge, Cambridge Univ. Press 1975.

[25] Brunovský, P., On One-Parameter Families of Diffeomorphisms, Comm. Math. Univ. Carolina, 11, 1970, pp. 559–582.

[26] Brunovský, P., On One-Parameter Families of Diffeomorphisms, Comm. Math. Univ. Carolina, 12, 1971, pp. 765–784.

[27] Brunovský, P., Generic Properties of the Rotation Number of One-Parameter Diffeomorphisms of the Circle, Czech. Math. J., 24, 99, 1974, pp. 74–90.

[28] Carr, J., Applications of Center Manifold Theory, Berlin, Heidelberg, New York, Appl. Math. Sciences, 35, Springer-Verlag 1981.

[29] Coddington, E. and Levinson, N., Ordinary Differential Equations, New York, McGraw-Hill 1955.

[30] Demidovich, B.P., Lectures on Mathematical Stability Theory, Moscow, Nauka 1967 (In Russian).

[31] Denjoy, A., Sur les Courbes Definés par les Equations Differentielles de la Surface du Tore, J. Math. Pures et Appl., 11, 1932, pp. 333–375.

[32] Dieudonné, J., Foundations of Modern Analysis, New York, Academic Press 1960.

[33] Diliberto, S., Perturbation Theorems for Periodic Surfaces I, II, Rend. Cir. Math. Palermo, 9, 1960, pp. 265–299, 1961, pp. 111–161.

[34] Dulac, H., Solution d'une Systéme d'Equation Differentielles dans le Voisinage des Valuers Singulliéres, Bull. Soc. Math. France, 40, 1912, pp. 324–383.

[35] Fenichel, N., Persistence and Smoothness of Invariant Manifolds of Flows, Indiana Math., J., 21, 1971, pp. 193–226.

[36] Fiedler, B., An Index for Global Hopf Bifurcation in Parabolic Systems, Journal für die Reine und Angewandte Mathematik, 359, 1985, pp. 1–36.

[37] Fiedler, B., Global Hopf Bifurcation of Two Parameter Flows, Heidelberg, Arch. Rat. Mech. and Anal., 94, 1, 1986, pp. 59–81.

[38] Friedman, A., Generalized Functions and Partial Differential Equations, New Jersey, Prentice Hall, Englewood Cliffs 1963.

[39] Gantmacher, F.R., Theory of Matrices, Moscow, Nauka 1967 (In Russian).

[40] Gavrilov, N.K., On Bifurcations of Equilibria with a Pair of Pure Imaginary Roots, Proceedings: Methods of Qualitative Theory of Differential Equations, Gorkii, Gorkii Univ. Press 1980, pp. 17–30 (In Russian).

[41] Gavrilov, N.K., On Some Bifurcations of Equilibria with one Zero and a Couple of Pure Imaginary Roots, Proceedings: Methods of Qualitative Theory of Differential Equations, Gorkii, Gorkii Univ. Press 1978, pp. 33–40 (In Russian).

[42] Gibson, C.B., Withmüller, K., Du Plessis, A.A. and Liooijenga, E.J.N., Topological Stability of Smooth Mappings, Berlin, Heidelberg, New York, LNM 552, Springer-Verlag 1976.

[43] Godbillon, C., Géometrie Différentielle et Mécanique Analytique, Paris, Collection Méthodes Hermann 1969.

References

[44] Golubitsky, M. and Guillemin, V., Stable Mappings and Their Singularities, New York, Heidelberg, Berlin, Springer-Verlag 1973.

[45] Golubitsky, M. and Langford, W.F., Classification and Unfoldings of Degenerate Hopf Bifurcations, Journal of Differential Equations, 41, 1981, pp. 375–415.

[46] Golubitsky, M. and Schaeffer, D., A Theory for Imperfect Bifurcation via Singularity Theory, Comm. Pure Appl. Math., 32, 1979, pp. 21–98.

[47] Grobman, D.M., On a Homemorphism of Systems of Differential Equations, Dokl. Akad. Nauk 128, 5, 1959, pp. 880–881 (In Russian).

[48] Grobman, D.M., Topological Classification of a Neighbourhood of Singular Point in an n-Dimensional Space, Mat. Sborník 56, 98, 1962, pp. 77–94 (In Russian).

[49] Guckenheimer, J., On a Codimension Two Bifurcation, Dynamical Systems and Turbulence, Berlin, Heidelberg, New York, LNM 898, Springer-Verlag 1981, pp. 99–142.

[50] Guckenheimer, J. and Holmes, P.J., Nonlinear Oscillations, Dynamical Systems and Bifurcations of Vector Fields, New York, Heidelberg, Berlin, Applied Math. Sciences, 42, Springer-Verlag 1983.

[51] Guillemin, V. and Pollack, A., Differential Topology, New Jersey, Prentice-Hall 1974.

[52] Hadamard, J., Sur l'Itération et les Solutions Asymptotiques des Équations Differentielles, Bull. Soc. Math. France, 29, 1901, pp. 224–228.

[53] Hale, J., Introduction to Dynamic Bifurcation, Berlin, Heidelberg, New York, LNM 1057, Springer-Verlag 1984, pp. 106–151.

[54] Hale, J., Integral Manifolds and Perturbed Differential Systems, Ann. Math., 73, 1961, pp. 496–531.

[55] Hale, J., Theory of Functional Differential Equations, New York, Heidelberg, Berlin, Springer-Verlag 1977.

[56] Hale, J., Magalhaes, L.T. and Oliva, M.W., An Introduction to Infinite Dimensional Dynamical Systems – Geometric Theory, New York, Berlin, Heidelberg, Applied Math. Sciences, 47, Springer-Verlag 1984.

[57] Hassard, B.D. and Wan, Y.H., Bifurcation Formulae Derived from Center Manifold Theory, Journal of Math. Anal. Appl., 63, 1, 1978, pp. 297–312.

[58] Hartman, Ph., Ordinary Differential Equations, New York, John Wiley and Sons 1964.

[59] Hartman, Ph., On Local Homeomorphisms of Eulidean Spaces, Proceedings of the Symp. on Ordinary Differential Equations, Mexico City 1961, pp. 220–241.

[60] Hartman, Ph., A Lemma on the Theory of Structural Stability of Differential Equations, Proc. Amer. Math. Soc., 11, 1960, pp. 610–622.

[61] Hejný, M., Kulich, J. and Tvarožek, J., What is the Topology? Bratislava, ALFA 1983 (In Slovak).

[62] Henry, D., Geometric Theory of Semilinear Parabolic Equations, New York, Heidelberg, Berlin, LNM 840, Springer-Verlag 1981.

[63] Herman, M., Measure de Lebesgue et Nombre de Rotation, Berlin, Heidelberg, New York, LNM, Springer-Verlag 1977, pp. 271–293.

[64] Herman, M., Sur la conjugaison differentiable des diffeomorphismes du cercle a des rotations, Publ. I.H.E.S., 49, 1979.

[65] Hirsch, M.W., Differential Topology, New York, Heidelberg, Berlin, Springer-Verlag 1976.

[66] Hirsch, M.W. and Pugh, C.C., Stable Manifolds and Hyperbolic sets, Providence R.I., Proceedings of the Symp. in Pure Math., 14, AMS 1970, pp. 133–164.

[67] Hirsch, M.W., Pugh, C.C. and Shub, M., Invariant Manifolds, Berlin, New York, LNM 583, Springer-Verlag 1977,

References

[68] Hirsch, M.W. and Smale, S., Differential Equations, Dynamical Systems and Linear Algebra, New York, Academic Press 1974.

[69] Holmes, P.J., Center Manifolds, Normal Forms and Bifurcations of Vector Fields, Physica, 2D, 1981, pp. 449–481.

[70] Holmes, P.J., A Strange Family of Three-dimensional Vector Fields Near a Degenerate Singularity, Journal of Diff. Equations, 37, 1980, pp. 382–403.

[71] Holmes, P.J., Unfoldings of a Degenerate Nonlinear Oscillator, a Codimension Two Bifurcation, Annals of the New York Academy of Sciences, 357, 1980, pp. 473–488.

[72] Horozov, E.I., Versal Deformation of Equivariant Vector Fields for the Case of Symmetry of Order 2 and 3, Trudy Seminara I.G. Petrovskogo, 5, 1979, pp. 163–192 (In Russian).

[73] Horozov, E.I., Bifurcations of a Vector Field Near a Singular Point in the Case of Two Pairs of Imaginary Eigenvalues, I and II, Comptes Rendus de l'Academie Bulgare des Sciences, 34, 1981, pp. 1221–1224 and 35, 1982, pp. 142–152 (In Russian).

[74] Chow, S.N. and Hale, J., Methods of Bifurcation Theory, New York, Berlin, Heidelberg, Springer, 1982.

[75] Iooss, G., Bifurcation of Maps and Applications, Amsterdam, New York, Oxford, North-Holland, 1979.

[76] Irwin, M.C., Smooth Dynamical Systems, London, New York, Toronto, Sydney, San Francisco, Academic Press 1980.

[77] Irwin, M.C., On the Stable Manifold Theorem, Bull. London Math. Soc. 2, 1970, pp. 196–198.

[78] Kahn, G.W., Introduction to Global Analysis, New York, London, Toronto, Sydney, San Francisco, Academic Press 1980.

[79] Kelley, A., The Stable, Center-Stable, Center, Center-Unstable and Unstable Manifolds, Journal of Diff. Equations, 3, 1967, pp. 546–570.

[80] Kending, K., Elementary Algebraic Geometry, New York, Heidelberg, Berlin, Springer-Verlag 1977.

[81] Kolmogorov, A.N. and Fomin, S.V., Foundations of the Theory of Functions and Functional Analysis, 5, Moscow, Nauka 1976 (In Russian).

[82] Kornfeld, I.P., Sinai, J.G. and Fomin, S.V., Ergodic Theory, Moscow, Nauka 1980 (In Russian).

[83] Krylov, N.M. and Bogolyubov, N.M., Applications of Methods of Nonlinear Mechanics in the Theory of Stationary Oscillations, Kiev, Ukr. Acad. of Sciences 1934 (In Russian).

[84] Kubíček, M. and Marek, M, Computational Methods in Bifurcation Theory and Dissipative Structure, New York, Berlin, Heidelberg, Springer-Verlag 1983.

[85] Kuroš, A.G., Course of Higher Algebra, Moscow, Gosud. izd. fiziko-matem. literatury, 1962 (In Russian).

[86] Kurzweil, J., Ordinary Differential Equations, Prague, SNTL 1978 (In Czech).

[87] Kurzweil, J., Invariant Manifolds for Flows, Proceedings of the Symp. in Diff. Equations and Dynamical Systems, New York, Mayaguez, Puerto Rico 1965, Academic Press 1967, pp. 431–468.

[88] Kyner, W., Invariant Manifolds, Rend. Cir. Math. Palermo, 10, 1961, pp. 98–100.

[89] Lang, S., Introduction to Differentiable Manifolds, New York, London, Columbia Univ. 1962.

[90] La Salle, M.G., Une Demonstration du Theoréme de Division pour les Functions Differentiables, Topology, 12, 1973, pp. 41–62.

[91] Ljapunov, A.M., Probléme générale de la stabilite du mouvement, Annals of Math. Studies 1, Princeton, Princeton Univ. Press 1949.

References

[92] Lyusternik, L.A. and Sobolev, V.I., Foundations of Functional Analysis, Moscow, Nauka 1965 (In Russian).

[93] Lojasiewicz, S., Ensembles Semi-analytiques, Lecture note et. I.H.E.S., Bires-sur-Yvette, Preprint A, 66, 745, Ecole Polytechnique 1965.

[94] Lu, Y.C., Singularity Theory and an Introduction to Catastrophe Theory, New York, Heidelberg, Berlin, Springer-Verlag 1976.

[95] Malgrange, B., Ideals of Differentiable Functions, Oxford, Oxford Univ. Press 1966.

[96] Mallet-Paret, J. and Yorke, J.A., Snakes: Oriented Families of Periodic Orbits, Their Sources, Sinks and Continuation, Journal of Diff. Equations, 43, 3, 1982, pp. 419–450.

[97] Marsden, J.E., Hamiltonian Mechanics, Infinite Dimensional Lie Groups, Berkeley, Berkeley Notes 1969.

[98] Marsden, J.E. and Mc Cracken, M., The Hopf Bifurcation and Its Application, New York, Springer-Verlag 1976.

[99] Mather, N.J., Stratifications and Mappings, Proc. Dynamical Systems, New York, London, Academic Press 1973, pp. 195–232.

[100] Mather, N.J., Notes on Topological Stability, Harward, Preprint, Harward Univ. 1970.

[101] Mather, N.J., How to Stratify Mappings and Jet Spaces, Proc. Singularités d'applications differentiales, Berlin, LNM 535, Springer-Verlag 1976, pp. 128–176.

[102] Medveď, M., Generic Properties of Parametrized Vector Fields I., Czechosl. Math. J., 25. 1975, pp. 376–388.

[103] Medveď, M., Generic Properties of Parametrized Vector Fields II, Czechosl. Math. J., 26, 1976, pp. 71–83.

[104] Medveď, M., Generic Bifurcations of Second Order Ordinary Differential Equations on Differentiable Manifolds, Math. Slovaca, 1, 1977, pp. 9–24.

[105] Medveď, M., Generic Bifurcations of Second Order Ordinary Differential Equations Near Closed Orbits, Journal of Diff. Equations, 36, 3, 1980, pp. 98–107.

[106] Medveď, M., A Construction of Realizations of Perturbations of Poincaré Maps. Math. Slovaca, 36, 2, 1986, pp. 179–190.

[107] Medveď, M., On a Codimension Three Bifurcation, Časopis pro pěst. matem., 109, 1984, pp. 3–26.

[108] Medveď, M., The Unfoldings of a Germ of Vector Fields in the Plane with a Singularity of Codimension 3, Czechosl. Math. J., 35, 1, 1985, pp. 1–42.

[109] Medveď, M., On Two Parametric Systems of Matrices and Diffeomorphisms, Czechosl. Math. J., 33, 1983, pp. 176–192.

[110] Melo, W. and Pugh, C.C., Generic Arcs of Circle Diffeomorphisms, Rio de Jainero, IMPA, Preprint 198, 1985

[111] Milnor, J., Singular Points of Complex Hypersurfaces, New Jersey, Princeton Univ. Press 1968.

[112] Milnor, J. and Wallace, A.H., Differential Topology, Charlottesvile, The Univer. Press of Virginia 1965.

[113] Narasimhan, R., Analysis on Real and Complex Manifolds, Amsterdam, North-Holland Publ. Company 1968.

[114] Nagy, J., Systems of Ordinary Differential Equations, Prague, SNTL 1983 (In Czech).

[115] Naylor, A.W. and Sell, R.G., Linear Operator Theory in Engineering and Science, New York, Holt, Rinchart and Winston 1971.

[116] Nejmark, J.I., Methods of Points Mapping in the Theory of Nonlinear Oscillations, Proceedings of the International Conference on Nonlinear Oscillations, 2, Kiev, Ukr. Acad. of Sciences 1963 (In Russian).

References

[117] Newhouse, S.E., Palis, J. and Takens, F., Bifurcations and Stability of Families of Diffeomorphisms, Publ. I.H.E.S., 57, 1983, pp. 5–71.

[118] Newhouse, S.E., The Creation of Non-trivial Recurrence in the Dynamics of Diffeomorphisms, Proceedings: Chaotic Behaviour of Deterministic Systems, North-Holland Publ. Company 1983, pp. 384–442.

[119] Nitecki, Z., Differentiable Dynamics, Cambridge, Massachusetts, London, The MIT Press 1971.

[120] Palis, J., On Morse-Smale Dynamical Systems, Topology, 4, 1969, pp. 116–136.

[121] Palis, J., Vector Fields Generate Few Diffeomorphisms, Bull. AMS, 80, 1973, pp. 503–505.

[122] Palis, J. and Mello, W., Geometric Theory of Dynamical Systems, New York, Heidelberg, Berlin, Springer-Verlag 1982.

[123] Peixoto, M.M., Structural Stability on Two-Manifolds, Topology, 1, 1962, pp. 101–120.

[124] Perron, O., Über Stabilität und asymptotische Verhalten der Lösungen eines Systems enslicher Differenzengleichungen, J. Reine Angew. Math., 161, 1929, pp. 41–64.

[125] Perron, O., Über Stabilität und asymptotisches Verhalten der Integrale von Differentialgleichungssystem, Math. Z., 29, 1928, pp. 129–160.

[126] Pliss, V.A., Principal Reduction in the Theory of the Stability of Motion, Izv. Akad. Nauk, 28, 1964, pp. 1297–1324 (In Russian).

[127] Poincaré, H., Sur les courbes définies par les équations Differentielles, C.R. Acad. Sci., 90, Paris 1880, pp. 673–675.

[128] Pugh, C.C., On a Theorem of P. Hartman, Amer. J. Math., 91, 1969, pp. 363–367.

[129] Pugh, C.C., An Improved Closing Lemma and a General Density Theorem, Amer. J. Math., 89, 1967, pp. 1010–1021.

[130] Pugh, C.C. and Shub, M., Linearization of Normally Hyperbolic Diffeomorphisms and Flows, Inv. Math., 10, 1970, pp. 187–198.

[131] Robinson, C., Introduction to the Closing Lemma, Proceedings: The Structure of Attractors in Dynamical Systems, New York, LNM 668, Springer-Verlag 1978, pp. 225–230.

[132] Rudin, W., Functional Analysis, New York, Mc Graw-Hill 1973.

[133] Ruelle, D. and Takens, F., On the Nature of Turbulence, Comm. Math. Phys., 20, 1971, pp. 167–192.

[134] Sacker, R.J., On Invariant Surface and Bifurcation of Periodic Solutions of Ordinary Differential Equations, New York, New York Univ. Report IMN-NYU 333, 1964.

[135] Seidenberg, A., A New Decision Procedure for Elementary Algebra, Ann. of Math., 60, 1954, pp. 365–374.

[136] Schwartz, A.J., A Generalization of a Poincaré-Bendixon Theorem to Closed Two-Dimensional Manifolds, Amer. J. Math., 85, 1963, pp. 453–458.

[137] Siegel, C.L. and Moser, J.K., Lecture on Celestian Mechanics, Berlin, Heidelberg, New York, Springer-Verlag 1971.

[138] Sinai, J.G. and Šil'nikov, L.P., Strange Attractors, Proceedings, Moscow, Mir 1981 (In Russian).

[139] Smale, S., The Mathematics of Time, New York, Heidelberg, Berlin, Springer-Verlag 1980.

[140] Smale, S., Stable Manifolds for Differential Equations and Diffeomorphisms, Ann. Scuola Normale Superiore Pisa, 17, 1963, pp. 97–116.

[141] Smítal, J., On Functions and Fuctional Equations, Adam Hilger, Bristol and Philadelphia 1988.

References

[142] Sotomayor, J., Generic One Parameter Families of Vector Fields in Two-Dimensional, Publ. I.H.E.S., 43, 1973, pp. 1–46.

[143] Sotomayor, J., Generic Bifurcations of Dynamical Systems, Proceedings, Dynamical Systems, New York, London, Academic Press 1973, pp. 561–582.

[144] Sternberg, S., On Behaviour of Invariant Curves Near a Hyperbolic Point of a Surface Transformation, Amer. J. Math., 77, 1955, pp. 526–534.

[145] Sternberg, S., Lecture on Differential Geometry, Englewood Cliffs, Prentice Hall 1964.

[146] Strien, S.J., Center Manifolds are not C^∞, Math. Z., 166, 1979, pp. 143–145.

[147] Szlenk, W., Introduction to the Theory of Smooth Dynamical Systems, Warsaw, Bibl. Mat. 56, PWN 1982 (In Polish).

[148] Šalát, T., Metric Spaces, Bratislava, ALFA 1981 (In Slovak).

[149] Šilnikov, L.P., On a Case of the Existence of a Denumerable Set of Periodic Motions, Dokl. Akad. Nauk 160, 3, 1965, pp. 163–166 (In Russian).

[150] Šošitaišvili, A.N., Bifurcations of Topological Type of a Vector Field Near a Singular Point, Trudy Seminara I.G. Petrovskogo, 1, 1975, pp. 279–308 (In Russian).

[151] Takens, F., Singularities of Vector Fields, Publ. I.H.E.S., Bures-sur-Yvette, Paris, 43, 1974, pp. 47–100.

[152] Takens, F., Forced Oscillations, Ultrecht, Notes 1974.

[153] Takens, F., Unfoldings of Certain Singularities of Vector Fields, Generalized Hopf Bifurcations, Journal of Diff. Equations, 14, 1973, pp. 476–493.

[154] Tamura, J., Topology of Foliations, Moscow, Mir 1979 (In Russian).

[155] Thom, R. and Levin, G., Singularities of Differentiable Mappings, Proceedings of Translated Papers of R. Thom and G. Levin, J.M. Boardman, B. Morin, B. Malgrange, J. Mather, Moscow, Mir 1968, (In Russian).

[156] Trenogin, V.A., Functional Analysis, Moscow, Nauka 1980 (In Russian).

[157] Vanderbauwhede, A., Local Bifurcation and Symmetry, Boston, London, Melbourne, Pitman Advances Publishing Program 1982.

[158] Vanderbauwhede, A. and Van Gils, S.A., Center Manifolds and Contractions on a Scale of Banach Spaces, Journal of Funct. Analysis, 72, 1, 1987, pp. 209–224.

[159] Wall, C.T.C., Regular Stratifications, Proc. Dynamical Systems, Berlin, Heidelberg, New York, LNM 468, Springer-Verlag 1975, pp. 332–244.

[160] Whitney, H., A Function not Constant on a Connected Set of Critical Points, Duke Math, J., 1, 1935, pp. 514–517.

[161] Whitney, H., Elementary Structure of Real Algebraic Varieties, Ann. of Math., 66, 1957, pp. 545–556.

[162] Whitney, H., Tangent to an Analytic Variety, Ann. of Math., 81, 1965, pp. 496–549.

[163] Żoladek, H., On the Versality of a Family of Symmetric Plane Vector Fields, Mat. Sbornik 120, 4, 1983, pp. 473–499 (In Russian).

SUBJECT INDEX

algebraic complement 3
arc of trajectory 162
atlas, C^r, C^∞ 22
– maximal 22
attractor 279, 280
– Lorenz 280
– strange 281

bifurcation 125
– Bogdanov 264
– diagram 125
– generic 125, 203, 214, 238, 258
– Hopf 221, 226
– – generalized 269
– – global 276, 277
– Nejmark–Sacker 244
– of type $m(+1), m(-1), m(c)$ 246
– period doubling 260
– saddle-node type 217, 218, 259
– Šilnikov 273
– subcritical 225
– supercritical 225
boundary of a set 6

centre 115
– of parametrized differential equation 277
characteristic exponent
– of linear periodic differential equation 20
– of periodic trajectory 171
– of singular point 129
characteristic multiplier
– of linear periodic differential equation 20
– of periodic trajectory 169
– of singular point 129
chart, C^r, C^∞ 22
– admissible 23
– compatible 22

closure of set 6
codimension of submanifold 34
complement of set 1
complexification of differential
    equation 145
commutator 153
convex hall 3
critical element 267
curve, C^r 27
– integral 64
– – maximal 65

deformation
– parametric 126
– versal 127
derivative
– Fréchet 12
– Gateaux 13
– r-th 14
– of mapping 12, 31
– of mapping at a point 12
– partial 13
diagonal 7
diffeomorphism, C^r, C^∞ 26
– conjugating 70
– contractive 172, 173
– expansive 172, 173
– Kupka–Smale 267
– local 26
– Morse–Smale 270
– parametrized 74, 75
dimension
– of linear space 3
– of manifold 23
– of submanifold 34
direct sum 3
dynamical system

Subject Index

– continuous 68
– discrete 68
– generated by a diffeomorphism 68
– gradient 13
– linear 103, 104
– – contractive 110
– – expansive 110
– – hyperbolic 106
– – parametrized 123

eigenvalue 11
– resonant 145
eigenvector 11
embedding 32
equation
– bifurcation 206
– characteristic 9
– differential 18
– – autonomous 18
– – generic 277
– – linear 19
– – – periodic 19
– – parametrized 74
– – second-order 85
– in variation 21
equivalence 2
– flow 121
– germ 45, 125
– linear 105
– of curves 27
– orbital 117, 118
– – topological 117, 118
– – – global 117
– – – local 117

flow, C^r 65
– global 65
– gradient 84
– local 65
– parametrized 74
focus 105
– non-degenerate 172
– stable 115
– unstable 116
formula
– Leibnitz 14
– Liouville 19
– variation of constants 19
function
– coordinate 23
– Hamiltonian 85

generic property 40
germ
– conjugating 126
– induced 126
– of diffeomorphism 45, 238
– of vector field 118
– – parametrized 126
– – – of type $m(O), m(I)$ 215
gradient of a function 13
group of diffeomorphisms 68
– discrete 68
– one-parameter 68

hessian of a vector field 129
homeomorphism 8
– conjugating 70
– minimal 255
– topologically transitive 255

immersion 32
index of centrum 277
interior of a set 6
isomorphism 3

jet, C^r, C^∞ 45
jet extension of mapping 45
Jordan canonical form 10

kernel of linear mapping 2

lemma
– Gronwall 21
– Zorn 3
Lie algebra 153
Lie bracket 153
lift of mapping 247
– parametrized 256
limit cycle 83
linearization
– of diffeomorphism 238
– of vector field 140
local representation
– of mapping 26
– of vector field 63

manifold, C^r, C^∞ 22, 23
– centre 183, 185
– – global 185
– – local 185, 194, 201
– differentiable 22, 23
– Grassman 24
– in R^N 35

Subject Index

- invariant 79, 172
- – stable 173, 174
- – – local 173, 174
- – unstable 173, 174
- – – local 173, 174
- of zero dimension 35
- orientable 23
- smooth 23

mapping
- bijective 2
- bilinear 10
- continuous 8
- contractive 16
- coordinate 23
- differentiable, C^r, C^∞ 26
- evaluation 54
- injective 2
- linear 10
- – hyperbolic 131
- Lipschitz 16
- – local 16
- n-linear 10
- Poincaré 159, 165
- – parametrized 165
- proper 32
- recurrent 71
- smooth 14, 26
- surjective 2
- symmetrical 11

matrix
- block diagonal 9
- diagonal 9
- fundamental 19
- inverse 1
- Jacobi 13
- of monodromy 20
- regular 9
- transpose 1
- unit 1
- zero 1

method
- alternative 206
- Ljapunov–Schmidt 203, 206

metric 5
- Riemannian 75

natural projection 2, 30
neighbourhood 6
- of point 6
- tubular 162
node 105
- stable 112

- unstable 112
norm 4
normal form 143, 149
- Bogdanov 263
number
- resonant 145
- rotation 250

order of resonance 146

partition of unity 50
period
- of periodic trajectory 69
point
- accumulation 5
- bifurcation 124
- critical 39
- fixed 16
- heteroclinic 266
- homoclinic 266
- isolated 5
- non-wandering 82, 83
- periodic 69
- – hyperbolic 105, 131
- – quasi-hyperbolic 123, 238
- – – of degree 1, 2 123
- regular 69
- singular 69, 215
- – hyperbolic 130
- wandering 81, 83

polynomial
- characteristic 9
- homogeneous 144
- – vector 144
- resonant 146
- Taylor 15

principle of reduction to centre manifold 195
projector 203
pseudorepresentation 54

reduction to centre manifold 192, 202
region of attractivity 83
relation 2
- of equivalence 2
- of ordering 2
- of partial ordering 2
- on a set 2
representation 54
- local 26, 64
resolvent 19
resonance 146

291

Subject Index

resonant term 146

saddle 106
saddle-node 217
scalar product 3
series
– formal 15
– Taylor 15
set
– α-limit 77
– ω-limit 77
– algebraic 56
– asymptotically 83
– Cantor 272
– chaotic 281
– closed 6
– connected 8
– dense 8
– everywhere dense 8
– invariant 78, 173, 174, 200
– – locally 78
– – orbitally stable 83
– – orbitally unstable 83
– massive 8
– minimal 80
– non-wandering 82
– of zero measure 40
– open 5
– perfect 272
– quotient 2
– semi-algebraic 56
sink 110
Smale horseshoe 271
solution
– of differential equation 18
– – complete 18
– – maximal 18
– – operator 19
source 110
space
– Baire 8
– Banach 5
– complete 5
– Euclidean 3
– Hilbert 5
– linear 3
– – normed 4
– metric 5
– phase 68
– tangent 27, 28
– topological 5
– – Hausdorff 6

– – compact 7
– – fulfilling the second axiom of countability 6
– – connected 8
– vector 3
stability
– Ω-stability 273
– asymptotically orbital 83
– in Lagrange sense 78
– orbital 83
– structural 118
stratification 55
– Whitney 57
– – ordered 57
subspace
– centre 111
– invariant 111
– linear 3
– stable 111
– unstable 111
– vector 3
submanifold, C^r 32, 33
– immersive 34
support of map 50
system
– dynamical 68
– – continuous 68
– – discrete 68
– Hamiltonian 85
– semi-dynamical 68

tangent bundle 30
tangent vector 28
term
– resonant 146
theorem
– Abraham transversality 54
– – to stratification 62
– Banach fixed point 16
– Bogdanov 263
– Borel 15
– centre manifold 191
– – parametrized 194
– closed graph 12
– converse to Taylor 15
– Denjoy 255
– embedding 34
existence and uniqueness 18
– Floquet 20
– global flow box 162
– global Hopf bifurcation 278
– Grobman–Hartman 128, 141

Subject Index

- Hartman 138
- Hopf bifurcation 221
- immersion and submersion 36
- implicit function 16
- inverse mapping 16
- Kupka–Smale 267
- local flow box 160
- Malgrange–Weierstrass 17
- mean value 16
- Nejmark–Sacker bifurcation 244
- on approximation of centre manifold 196
- on closing of trajectories 83
- on invariant manifolds 172
- – of periodic points 174
- – of periodic trajectories 176
- – of singular points 174
- on local structural stability 142
- on realization of C^r-approximation of Poincaré mapping 168
- on the saddle-node type bifurcation 215, 239
- parametric transeversality 53
- period doubling bifurcation 239
- Poincaré–Bendixon 81
- Sard 40
- Schwarz 81
- Šilnikov 276
- Šošitaišvili 194
- Takens 157
- Tarski–Seidenberg 60
- Taylor 14
- Thom transversality 55
- weak transversality 49
- Whitney stratification 59
topology 6
- induced by a metric 7
- of uniform convergence 43
- product 6
- quotient 7
- relative 6
- strong 44

- weak 44
- Whitney, C^r, C^∞ 44
torus 26
- n-dimensional 26
trace of a matrix 1
trajectory 69
- closed 69
- heteroclinic 266
- homoclinic 118, 266
- negative 69
- oriented 69
- periodic 69
- – hyperbolic 169
- – of type $\gamma(+1)$, $\gamma(-1)$, $\gamma(c)$ 258
- positive 69
T–periodic 69
transversal 71
transversality 46
- of manifolds 46
- of mapping 46
- – to a manifold 46
- to stratification 61

unfolding 126
- parametric 126
- versal 127

value
- critical 39
- regular 39
vector field, C^r, C^∞ 63
- complete 65
- gradient 84
- Hamiltonian 84
- Kupka–Smale 267
- linear 104
- – hyperbolic 106
- – – parametrized 123
- Morse–Smale 270
- parametrized 73, 74